Methods in Molecular Genetics

Volume 1

Gene and Chromosome Analysis

Part A

Methods in Molecular Genetics

Edited by

Kenneth W. Adolph

Department of Biochemistry
University of Minnesota Medical School
Minneapolis, Minnesota

Volume 1

Gene and Chromosome Analysis Part A

ACADEMIC PRESS, INC.
Harcourt Brace & Company
San Diego New York Boston London Sydney Tokyo Toronto

Front cover photograph: Fluorescence *in situ* hybridization with chromosome 4 library showing a complex chromosome rearrangement. Courtesy of Denise A. S. Tatista, The Johns Hopkins University School of Medicine, Department of Gynecology and Obstetrics, and the University of São Paulo State, Biosciences Institute, Department of Genetics, Botucatu, São Paulo, Brazil.

This book is printed on acid-free paper. ∞

Academic Press, Inc.
1250 Sixth Avenue, San Diego, California 92101

United Kingdom Edition published by
Academic Press Limited
24–28 Oval Road, London NW1 7DX

International Standard Serial Number: 1067-2389

International Standard Book Number: 0-12-044301-5

PRINTED IN THE UNITED STATES OF AMERICA

93 94 95 96 97 98 EB 9 8 7 6 5 4 3 2 1

Table of Contents

Contributors

Article numbers are in parentheses following the names of contributors. Affiliations listed are current.

STUART A. AARONSON (1), Laboratory of Cellular and Molecular Biology, National Cancer Institute, Bethesda, Maryland 20892

FAREED ABOUL-ELA (14), CRC Nucleic Acid Structure Group, Department of Biochemistry, The University, Dundee DD1 4HN, Scotland

IRIS ALROY (16), Cell Biology and Genetics Program, Sloan-Kettering Institute and Cornell Medical College, New York, New York 10021

STEVEN ARTANDI (15), Department of Microbiology and Department of Biochemistry and Molecular Biophysics, Columbia University College of Physicians and Surgeons, New York, New York 10021

KAREN ARTZT (19), Department of Zoology, The University of Texas at Austin, Austin, Texas 78712

NOELLE BECUWE (12), Généthon, 91000 Evry, France

RICHARD BOWATER (14), CRC Nucleic Acid Structure Group, Department of Biochemistry, The University, Dundee DD1 4HN, Scotland

PETER M. BUTTRICK (20), Department of Medicine, Albert Einstein College of Medicine, Bronx, New York 10461

KATHRYN CALAME (15), Department of Microbiology and Department of Biochemistry and Molecular Biophysics, Columbia University College of Physicians and Surgeons, New York, New York 10021

NATHALIE CHERON (12), Généthon, 91000 Evry, France

STEVEN F. DOWDY (8), The Whitehead Institute, Cambridge, Massachusetts 02142

CHANTAL DUPRAZ (12), Généthon, 91000 Evry, France

SARAH C. R. ELGIN (18), Department of Biology, Washington University, St. Louis, Missouri 63130

PAUL B. FISHER (4), Departments of Neurosurgery, Pathology, and Urology, Comprehensive Cancer Center/Institute of Cancer Research, College of Physicians and Surgeons, Columbia University, New York, New York 10032

R. E. K. FOURNIER (6), Department of Molecular Medicine, Fred Hutchinson Cancer Research Center, Seattle, Washington 98104

LEONARD P. FREEDMAN (16), Cell Biology and Genetics Program, Sloan-Kettering Institute and Cornell Medical College, New York, New York 10021

ERROL C. FRIEDBERG (7), Laboratory of Molecular Pathology, Department of Pathology, The University of Texas Southwestern Medical School, Dallas, Texas 75235

JOHN D. GEARHART (3), Department of Gynecology and Obstetrics, The Johns Hopkins University School of Medicine, Baltimore, Maryland 21205

PETER M. GLAZER (5), Department of Therapeutic Radiology, Yale University School of Medicine, New Haven, Connecticut 06510

ERIC D. GREEN (11), Departments of Pathology, Genetics, and Medicine, Washington University School of Medicine, St. Louis, Missouri 63110

GABOR GYAPAY (12), Généthon, 91000 Evry, France

JAMILÉ HAZAN (12), Unité de Génétique Moléculaire Humaine, Institut Pasteur, 75724 Paris Cedex, France

KARLA A. HENNING (7), Laboratory of Molecular Pathology, Department of Pathology, The University of Texas Southwestern Medical School, Dallas, Texas 75235

ALASDAIR C. IVENS (10), Department of Biochemistry, Imperial College of Science, Technology, and Medicine, London SW7 2AZ, United Kingdom

MATTHEW L. KAPLAN (20), Department of Medicine, Albert Einstein College of Medicine, Bronx, New York 10461

ALYSON A. KASS (20), Department of Microbiology and Immunology, Albert Einstein College of Medicine, Bronx, New York 10461

LARRY KEDES (17), Institute for Genetic Medicine, Department of Biochemistry and Molecular Biology and Department of Medicine, University of Southern California School of Medicine, Los Angeles, California 90033

TIMOTHY F. KING (10), Department of Biochemistry, Imperial College of Science, Technology, and Medicine, London SW7 2AZ, United Kingdom

RICHARD N. KITSIS (20), Department of Medicine, Albert Einstein College of Medicine, Bronx, New York 10461

ERIC LADER (19), Department of Zoology, The University of Texas at Austin, Austin, Texas 78712

AMELIA A. LANGSTON (6), Department of Molecular Medicine, Fred Hutchinson Cancer Research Center, Seattle, Washington 98104

ANN M. LAWLER (3), Department of Gynecology and Obstetrics, The Johns Hopkins University School of Medicine, Baltimore, Maryland 21205

LESLIE A. LEINWAND (20), Department of Microbiology and Immunology, Albert Einstein College of Medicine, Bronx, New York 10461

DAVID M. J. LILLEY (14), CRC Nucleic Acid Structure Group, Department of Biochemistry, The University, Dundee DD1 4HN, Scotland

PETER F. R. LITTLE (10), Department of Biochemistry, Imperial College of Science, Technology, and Medicine, London SW7 2AZ, United Kingdom

QIN LU (18), Department of Biology, Washington University, St. Louis, Missouri 63130

JAMES MCCLELLAN (14), CRC Nucleic Acid Structure Group, Department of Biochemistry, The University, Dundee DD1 4HN, Scotland

CARL R. MERRIL (9), Laboratory of Biochemical Genetics, National Institute of Mental Health Neuroscience Center at Saint Elizabeth's, Washington, D.C. 20032

TORU MIKI (1), Laboratory of Cellular and Molecular Biology, National Cancer Institute, Bethesda, Maryland 20892

COREY H. MJAATVEDT (3), Department of Gynecology and Obstetrics, The Johns Hopkins University School of Medicine, Baltimore, Maryland 21205

ALASTAIR I. H. MURCHIE (14), CRC Nucleic Acid Structure Group, Department of Biochemistry, The University, Dundee DD1 4HN, Scotland

SIMON NGUYEN (12), Généthon, 91000 Evry, France

SATISH PARIMOO (2), Boyer Center for Molecular Medicine, Yale University School of Medicine, New Haven, Connecticut 06536

SANKHAVARAM R. PATANJALI (2), Boyer Center for Molecular Medicine, Yale University School of Medicine, New Haven, Connecticut 06536

MIHAEL H. POLYMEROPOULOS (9), Laboratory of Biochemical Genetics, National Institute of Mental Health Neuroscience Center at Saint Elizabeth's, Washington, D.C. 20032

HOWARD PRENTICE (17), Institute for Genetic Medicine, Department of Biochemistry and Molecular Biology and Department of Medicine, University of Southern California School of Medicine, Los Angeles, California 90033

P. Gopal Reddy (4), Department of Urology, Comprehensive Cancer Center/Institute of Cancer Research, College of Physicians and Surgeons, Columbia University, New York, New York 10032

Vittorio Sartorelli (17), Institute of Chemistry, University of Brescia School of Medicine, 25100 Brescia, Italy, and Institute for Genetic Medicine, Department of Biochemistry and Molecular Biology and Department of Medicine, University of Southern California School of Medicine, Los Angeles, California 90033.

Eric J. Stanbridge (8), Department of Microbiology and Molecular Genetics, College of Medicine, University of California, Irvine, Irvine, California 92717

Gail Stetten (13), Cytogenetics Laboratory, Department of Gynecology and Obstetrics, The Johns Hopkins School of Medicine, Baltimore, Maryland 21205

Zao-Zhong Su (4), Department of Urology, Comprehensive Cancer Center/Institute of Cancer Research, College of Physicians and Surgeons, Columbia University, New York, New York 10032

William C. Summers (5), Department of Therapeutic Radiology, Yale University School of Medicine, New Haven, Connecticut 06510

Martine Tranchant (12), Généthon, 91000 Evry, France

Alain Vignal (12), Généthon, 91000 Evry, France

Lori L. Wallrath (18), Department of Biology, Washington University, St. Louis, Missouri 63130

Jean Weissenbach (12), Unité de Génétique Moléculaire Humaine, Institut Pasteur, 75724 Paris Cedex, France

Sherman M. Weissman (2), Boyer Center for Molecular Medicine, Yale University School of Medicine, New Haven, Connecticut 06536

Preface

The new series "Methods in Molecular Genetics" will provide practical experimental procedures for use in the laboratory. Because the introduction of molecular genetic techniques and related methodology has revolutionized biological research, a wide range of methods will be covered. The power and applicability of these techniques have led to detailed molecular answers to important biological questions and have changed the emphasis of biological research, including medical research, from the isolation and characterization of cellular material to studies of genes and their protein products.

Molecular genetics and related fields are concerned with genes: DNA sequences of genes, regulation of gene expression, and the proteins encoded by genes. The consequences of gene activity at the cellular and developmental levels are also investigated. In medical research, knowledge of the causes of human disease is reaching an increasingly sophisticated level now that disease genes and their products can be studied. The techniques of molecular genetics are also being widely applied to other biological systems, including viruses, bacteria, and plants, and the utilization of gene cloning methodology for the commercial production of proteins for medicine and industry is the foundation of biotechnology. The revolution in biology that began with the introduction of DNA sequencing and cloning techniques will continue as new procedures of increasing usefulness and convenience are developed.

In addition to the basic DNA methods, instrumentation and cell biology innovations are contributing to the advances in molecular genetics. Important examples include gel electrophoresis and DNA sequencing instrumentation, *in situ* hybridization, and transgenic animal technology. Such related methodology must be considered along with the DNA procedures.

The focus of this first volume, "Gene and Chromosome Analysis" Part A, is on eukaryotic cells, and, in particular, on human cells. The volume is divided into five sections—Gene Cloning; Cell Hybrids; Chromosome Mapping; DNA and DNA–Protein Interactions; Gene Expression—each of which presents selected new methods relevant to important topics such as genome organization, development, and cancer.

"Methods in Molecular Genetics" will be of value to researchers, as well as to students and technicians, in a number of biological disciplines because of the wide applicability of the procedures and the range of topics that will be covered.

KENNETH W. ADOLPH

Section I

Gene Cloning

[1] cDNA Cloning Method Using Stable Expression in Mammalian Cells

Toru Miki and Stuart A. Aaronson

Introduction

In bacterial and yeast systems, genes can be cloned using complementation of mutant alleles. This type of approach requires neither nucleic acid nor antibody probes. In mammalian systems, the introduction of cDNAs placed under a suitable promoter has been utilized to complement mutations in recipient cells or to alter cell phenotype in specific ways. Although eukaryotic expression cloning procedures have been widely used to isolate cDNAs for cell surface antigens, growth factors, and their receptors, these systems have generally been based on transient expression in COS cells. In the COS cell system, expression is efficient due to replication of introduced vectors from the simian virus 40 (SV40) origin of replication. However, selection of cells showing a specific stable phenotype is not possible because transfected cells subsequently die, presumably due to the uncontrolled replication of the introduced plasmid. The development of efficient cDNA cloning systems using stable expression would be required to isolate cDNAs that cause phenotypic changes in the host cells. In a stable expression cloning system, a single copy of a particular cDNA molecule integrated in the chromosome of the host cell expresses the phenotype. Because only a fraction of the introduced cDNA molecules might be stably integrated into the host chromosome, the possibility of cloning desired molecules would be much lower than in transient expression systems. Thus, efficient synthesis of full-length cDNAs, unidirectional insertion of cDNA fragments into vectors, and high-level expression of cDNA inserts would be required to establish an optimal stable cloning system.

We have developed an efficient expression cDNA cloning system that fulfills such requirements. The system includes the automatic directional cloning (ADC) method, an efficient procedure for constructing unidirectional cDNA libraries (1), and an expression vector with the capability of efficient plasmid rescue (2). We have successfully utilized this procedure for the isolation of a number of cDNAs with transforming potential for fibroblasts (2–5). This article describes technical details for this expression cloning system.

Methods in Molecular Genetics, Volume 1

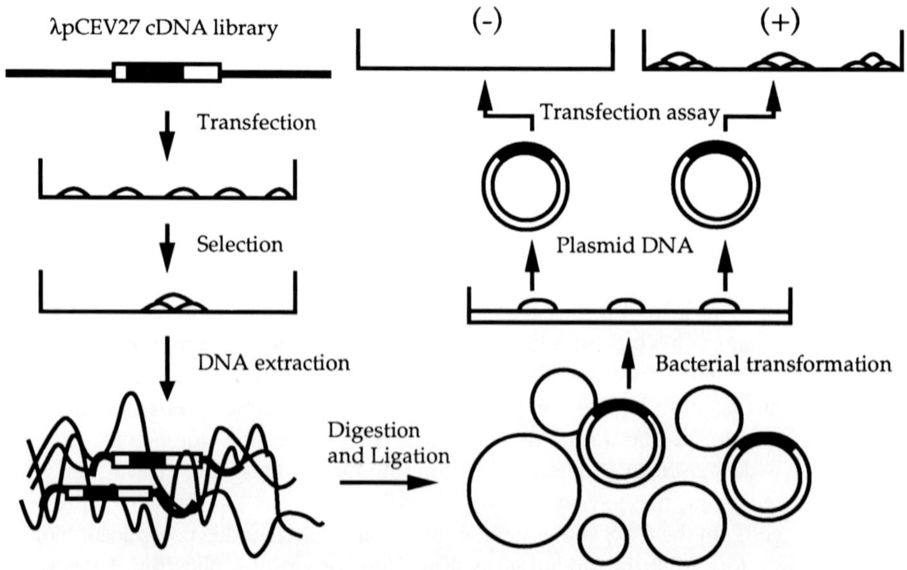

FIG. 1 Expression cloning using the λpCEV27 system. Host cells are transfected with a λpCEV27 cDNA library and selected for a phenotype of interest. Alternatively, cells can be first selected for G418 resistance and then screened for a desired phenotype. Following expansion of selected cells to mass culture, genomic DNA is isolated and subjected to plasmid rescue by digestion with either *Not*I, *Xho*I, or *Mlu*I, followed by ligation at low DNA concentration and transformation of a suitable bacterial strain. Colonies resistant to both the ampicillin and kanamycin are isolated. Plasmid DNA extracted from each colony is tested by transfection analysis to identify the cDNA clone responsible for the phenotype.

cDNA Cloning Using Stable Expression

Figure 1 shows a scheme for applications of this cloning vector. A cDNA library is constructed in λpCEV27 (Fig. 2) and transfected into host cells. Cells showing the new phenotype are directly selected or screened from marker-selected colonies. After cells are expanded to mass cultures, the genomic DNA is extracted. The DNA is cleaved by one of the restriction enzymes that can cut the λ-plasmid junction of λpCEV27 and ligated at a low concentration to circularize the excised plasmids. The ligated DNA is used to transform *Escherichia coli* competent cells, and plasmid DNAs are extracted from the colonies. Then, individual plasmids are reintroduced into the host cells to examine whether the plasmid can induce the same phenotype.

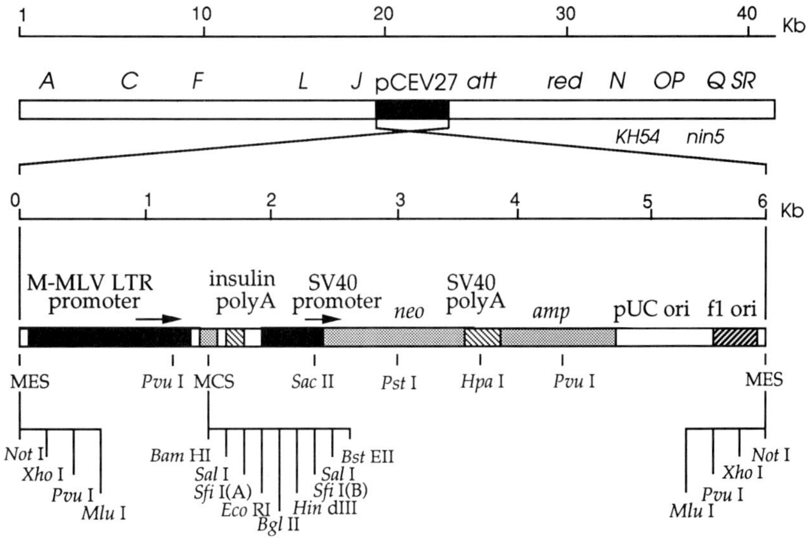

FIG. 2 Structure of the λpCEV27 cDNA expression cloning vector. Structure of the λpCEV27 vector is shown above, with the location of λ genes. *KH54* and *nin5* are the deletions in the λ genome. The plasmid portion is enlarged and shown below. The vector contains several features, including two *Sfi*I sites in multiple cloning sites (MCSs) for construction of cDNA libraries using the automatic directional cloning (ADC) method, a Moloney murine leukemia virus long terminal repeat (MMLV LTR) promoter suitable for cDNA expression in mammalian cells, the rat preproinsulin poly(A) signal downstream from the cDNA cloning site, the SV40 promoter-driven *neo* gene as a selectable marker, and multiple excision sites (MESs) for plasmid rescue from genomic DNA. A stuffer fragment encoding a part of the *lacZ* gene (*lacZ'*) is present in the *Bgl*II site in the MCS. The f1 replication origin as well as SP6 and T7 phage promoters are also included to facilitate construction of subtraction libraries as well as analysis of cDNA inserts by production of single-stranded DNA and RNA transcripts, respectively.

Principles of Automatic Directional Cloning Method

λpCEV27 has been designed to use the ADC method for cDNA library construction. This method achieves efficient synthesis of cDNA, insertion of cDNA into vectors unidirectionally, and construction of libraries with high complexity. When typical vectors and insert DNA fragments are mixed and ligated, several molecules are produced in addition to those desired. These include self-ligation products of vectors or inserts, head-to-tail or head-to-head dimers of vector or insert, and vectors containing multiple inserts. The formation of such molecules reduces the clon-

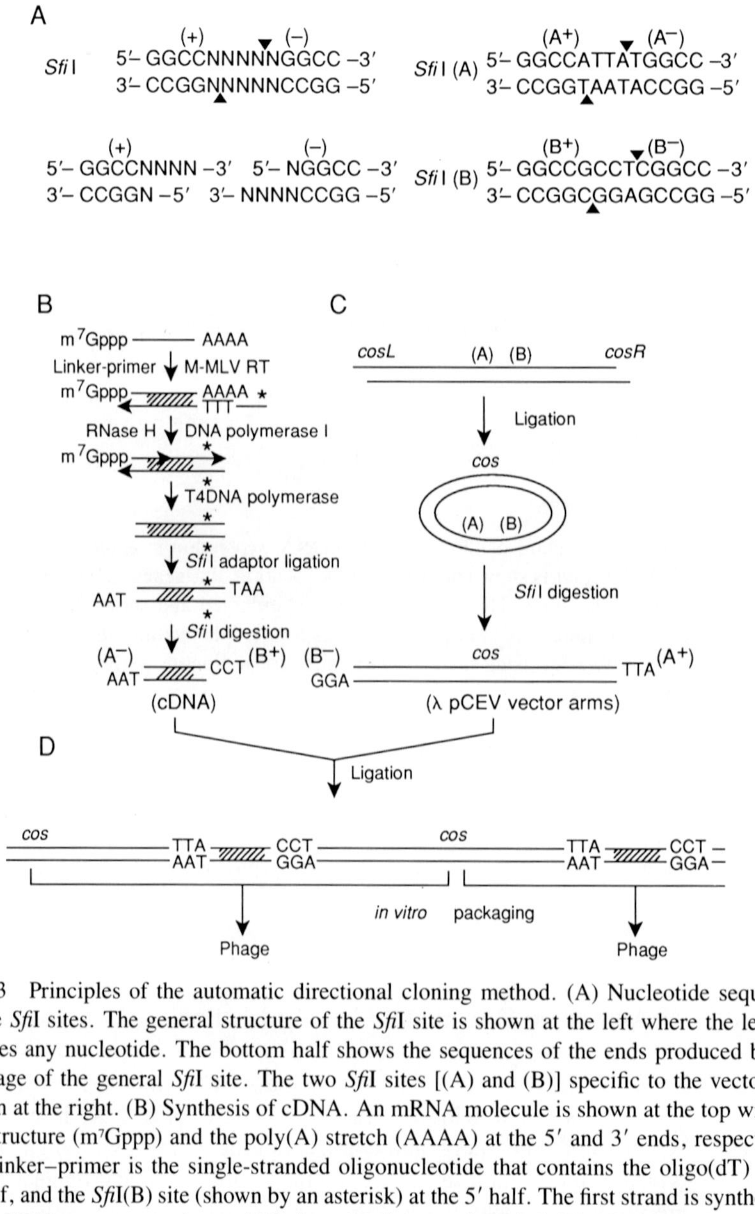

FIG. 3 Principles of the automatic directional cloning method. (A) Nucleotide sequences of the *Sfi*I sites. The general structure of the *Sfi*I site is shown at the left where the letter N denotes any nucleotide. The bottom half shows the sequences of the ends produced by *Sfi*I cleavage of the general *Sfi*I site. The two *Sfi*I sites [(A) and (B)] specific to the vectors are shown at the right. (B) Synthesis of cDNA. An mRNA molecule is shown at the top with the cap structure (m^{7}Gppp) and the poly(A) stretch (AAAA) at the 5′ and 3′ ends, respectively. The linker–primer is the single-stranded oligonucleotide that contains the oligo(dT) at the 3′ half, and the *Sfi*I(B) site (shown by an asterisk) at the 5′ half. The first strand is synthesized by M-MLV reverse transciptase (RT) from the linker–primer hybridized with the RNA molecule. The second strand is synthesized by DNA polymerase I from the nicks on the RNA moiety introduced by RNase H. The linker–primer is converted to the double-stranded form

ing efficiency. Even when vector and insert DNAs are prepared by digestion with two different enzymes, it is not possible to eliminate formation of ligation products such as head-to-head dimers, although the population of desired molecules is significantly higher and insertion occurs in a defined orientation. The reason why these self-litigation products and dimers are made is that the majority of restriction enzymes recognize sequences of dyad symmetry. In contrast, the ADC method utilizes *Sfi*I, an infrequent cutter, which lacks symmetry in the 3′ extension of three nucleotides (Fig. 3A). Because the central 5 bases in the recognition site can be any sequence, two *Sfi*I sites [(A) and (B)] were designed (Fig. 3A) and introduced into the vector (Figs. 2 and 3C). The cDNA fragments to be inserted into the vector are oriented by use of oligo(dT) primers having the sequence of the *Sfi*I(B) site.

The steps for cDNA synthesis are schematically shown in Fig. 3B. During the first-strand synthesis reactions, the single-stranded linker–primer is repaired so as to be double-stranded. After cDNA molecules are blunt-ended, an adaptor, designated the *Sfi*I adaptor and having the 3′ extension that fits to the *Sfi*I(A^+) end, is ligated. After cleavage by *Sfi*I, the resulting cDNA molecules should have different 3′ extensions that can fit on the ends of the vector to achieve directional cloning (Fig. 3D).

In a model experiment utilizing the ADC method, neither the vector nor insert fragments were self-ligated. Only when both were present were they efficiently ligated (Fig. 4A). Excess fragments were left unligated (Fig. 4B). Thus it is not necessary to dephosphorylate the vector fragments.

---

in the first- or second-strand synthesis step. T4 DNA polymerase treatment removes any single-stranded extensions from the synthesized cDNA molecule. The *Sfi*I adaptor ligation adds one 3′ single-stranded extension, the *Sfi*I(A^-) sticky end, to the cDNA molecule. Another 3′ extension, the *Sfi*I(B^+) sticky end, is exposed by *Sfi*I cleavage of the *Sfi*I(B) site on the repaired linker–primer portion of the cDNA molecule. (C) Preparation of the λpCEV vector arms. λpCEV vector DNA is shown at the top, where *cosL* and *cosR* represent the left and right cohesive ends of λ, respectively. The locations of the *Sfi*I(A) and *Sfi*I(B) sites are shown as (A) and (B), respectively. Following ligation to seal the cohesive ends (in the middle), the DNA is cleaved by *Sfi*I to expose the *Sfi*I(A^+) and *Sfi*I(B^-) sticky ends of the vector molecules. The small stuffer fragment is removed to prepare the vector arms shown at the bottom. The sequences of the single-stranded extensions are shown at both the 3′ ends of the cDNA and vector arms. (D) Formation of the λ concatemers containing cDNA inserts. When cDNA fragments are ligated with the prepared vector arms, alternating concatemers consisting of the cDNA inserts and vector arms in the defined orientation are produced automatically due to the base-pairing specificity. *In vitro* packaging extracts cut out the DNA segments hemmed by the two *cos* sites from the concatemer to form the active λ phage particles.

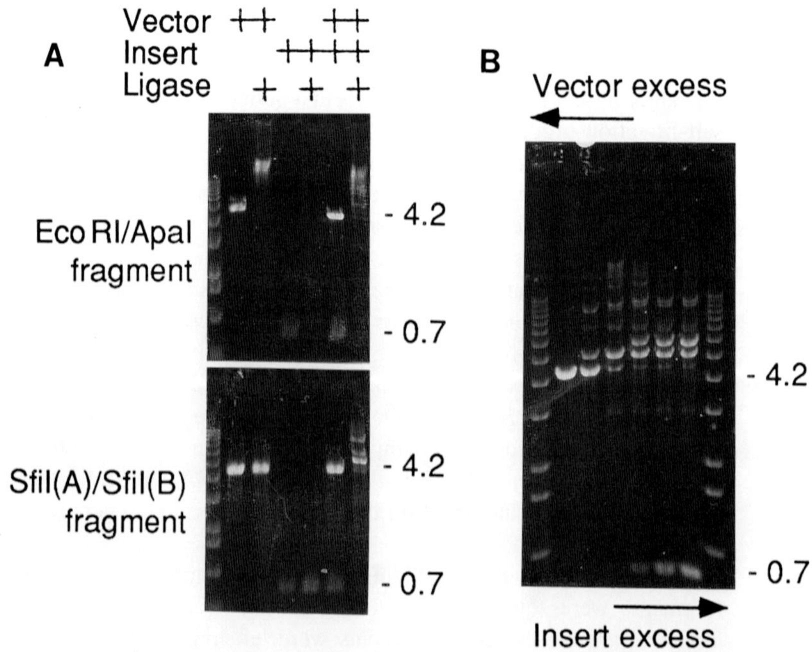

FIG. 4 Comparison of ADC and forced cloning methods. (A) Analysis of ligation products. The insert and vector fragments used in this experiment were prepared from a plasmid pCEV15-*ras* (1). The plasmid DNA was digested with *Sfi*I, and vector (4.2 kb) as well as insert (0.7 kb) fragments were purified from the gel. Similarly, *Eco*RI/*Apa*I vector and insert fragments, which were nearly identical to the *Sfi*I fragments, were also prepared as controls. The vector (plasmid part) and/or insert (*H-ras*) *Sfi*I fragments (bottom half) or *Eco*RI/*Apa*I fragments (upper half) were incubated in a ligation mixture with or without T4 DNA ligase as indicated. The products were analyzed by agarose gel electrophoresis. Sizes of the fragments are shown (kb). (B) The *Sfi*I fragments were mixed in different ratios, ligated, and ligation products analyzed by agarose gel electrophoresis.

Preparation of λpCEV27 Arms and *Sfi*I Adaptor

Because λpCEV27 was designed as a rapidly growing phage without amber mutations, many λ-sensitive host strains are available. However, strains harboring multicopy plasmids may not be used to avoid homologous recombination between λpCEV27 and other plasmids. As far as we have examined, LE392 has shown the highest plating efficiency. A single plaque of λpCEV27 is used to prepare a high-titer lysate. Phage particles are purified from 1 to 2 liters of a high-titered lysate prepared in a liquid culture by standard procedures. λpCEV27 contains a small stuffer fragment at the *Bgl*II site in the multiple cloning sites. Following DNA preparation, the stuffer fragment is removed by *Sfi*I digestion and potassium acetate gradient centrifugation. These techniques can be found in standard protocol books (e.g., Ref. 6). A protocol for preparation of phage DNA is also found in a later section. To prepare λ arms, the phage DNA is ligated to seal cohesive ends and then cleaved sequentially by *Sfi*I and *Eco*RI. After phenol–chloroform extraction, λpCEV27 arms were purified by centrifugation through a 5–20% potassium acetate gradient (6).

*Sfi*I adaptors should contain a 3′ AAT 5′ single-stranded extension at one end, and the other end should be blunt. The two oligonucleotides are synthesized and purified by high-performance liquid chromatography (HPLC). An example of the sequences of the oligos is shown below. This adaptor contains a restriction site for *Nru*I. Sequences containing other restriction sites, phage promoters, initiation and/or termination codons, or epitope tags for monoclonal antibodies, can be included in the adaptor.

*Sfi*I adaptor (type 4): 5′ CCAATCGCGAAACC 3′
3′ AATGGTTAGCGCTTTGG 5′

To prepare the adaptor, about 1 nmol of each oligonucleotide is separately phosphorylated by T4 polynucleotide kinase in a 20-μl reaction mixture, mixed, heated at 80°C for 5 min, and then slowly cooled to 4°C.

cDNA Synthesis from Linker–Primer

We usually perform cDNA synthesis reactions without incorporation of radiolabeled precursors. When labeling is required, reactions are carried out in separate tubes. The cDNA products should show a size range of 0.5–12 kilobases (kb), with an average size of approximately 2 kb (Fig. 5). The products of the first- and second-strand synthesis should be similar. When a fraction of the second-strand synthesis product is much longer than the first-strand product, hairpin–loop formation should be suspected. *Sfi*I adaptors cannot be ligated to such molecules. Poly(A)$^+$ RNA is purified by passage twice through an oligo(dT)-cellulose column. The RNA preparations should be of high quality, because the quality of cDNA libraries is largely dependent on the purity of the RNA. Thus contamination of RNA by enzyme inhibitors, such as sodium dodecyl sulfate (SDS), should be avoided. Contamination of RNA by oligo(dT) leaked from oligo(dT)-cellulose must also be avoided, because such molecules can act as primers, and cDNA synthesized from oligo(dT) cannot be inserted into λ arms.

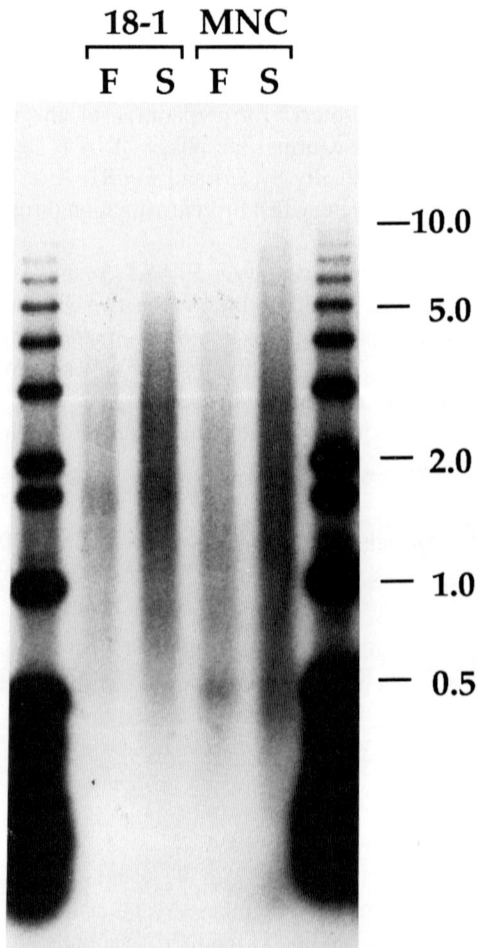

FIG. 5 Analysis of ^{32}P-labeled first and second cDNA products. Poly(A)$^{+}$RNAs were prepared from an NIH 3T3 transformant (18-1), or human peripheral blood mononuclear cells (MNC). The cDNA was labeled with [^{32}P]dCTP in the first- (F) or second (S)-strand synthesis reactions, and analyzed by alkaline agarose gel electrophoresis and autoradiography. As a molecular size standard, a 1-kb ladder (Bethesda Research Laboratories) was labeled by fill-in reaction of Klenow fragment and run in the leftmost and rightmost lanes. Locations of the bands are shown (kb).

Reagents

Linker–primer: 5′ GGACAGGCCGAGGCGGCC$(T)_{40}$ 3′ (0.5 μg/μl)
FS buffer (5×): 250 m*M* Tris-HCl (pH 8.3), 375 m*M* KCl, 15 m*M* $MgCl_2$
Dithiothreitol (DTT): 100 m*M*
SS buffer (10×): 188 m*M* Tris-HCl (pH 8.3), 906 m*M* KCl, 46 m*M* $MgCl_2$, 50 m*M* $(NH_4)_2SO_4$, 38 m*M* DTT, 1.5 m*M* β-NAD
dNTP mix (10 m*M*): 10 m*M* each of dGTP, dATP, dTTP, and dCTP
Ligation buffer A (10×): 660 m*M* Tris-HCl (pH 7.5), 100 m*M* $MgCl_2$, 50 m*M* DTT, 500 μ*M* ATP
Ligation buffer B (5×): 500 m*M* Tris-HCl (pH 7.6), 100 m*M* $MgCl_2$, 10 m*M* ATP, 10 m*M* DTT, 50% (w/v) polyethylene glycol 8000 (PEG 8000)
*Sfi*I adaptor (0.5 μg/μl)
*Sfi*I buffer (10×): 100 m*M* Tris-HCl (pH 7.9), 500 m*M* NaCl, 100 m*M* $MgCl_2$, 60 m*M* 2-mercaptoethanol, 1 μg/μl bovine serum albumin
TE: 10 m*M* Tris-HCl (pH 8.0), 1 m*M* ethylenediaminetetraacetic acid (EDTA)
STE: 100 m*M* NaCl, 10 m*M* Tris-HCl (pH 8.0), 1 m*M* EDTA

Protocol

Template Denaturation

Combine the following in a siliconized microcentrifuge tube:

Poly$(A)^+$ RNA (~0.5–1 μg/μl, ~1–5 μg)	*x* μl
H_2O	10 − *x* μl
Methylmercury (II) hydroxide (100 m*M*)	0.5 μl
	10.5 μl

Incubate at room temperature for 10 min and add 0.5 μl of 1.4 *M* 2-mercaptoethanol.

First-Strand Synthesis

Assemble the following:

Denatured poly$(A)^+$ RNA	11.0 μl
RNasin (Promega, Madison, WI) (40 units/μl)	0.6 μl
H_2O	0.8 μl
FS buffer (5×)	5.0 μl

DTT (0.1 M)	2.5 μl
dNTP mix (10 mM)	1.3 μl
Linker–primer (1 μg/μl)	2.5 μl
Moloney murine leukemia virus (MMLV) RNase H^- reverse transcriptase (RT) [Superscript (Bethesda Research Laboratories (BRL), Gaithersburg, MD), 200 units/μl]	1.3 μl
	25.0 μl

Incubate at 37°C for 1 hr and chill on ice.

Second-Strand Synthesis

Assemble the following:

FS synthesis reaction mix	25.0 μl
H_2O	$146.2 - x - y$ μl
dNTP mix (10 mM)	3.8 μl
SS buffer (10×)	20.0 μl
DNA polymerase I (BRL) (10 units/μl)	5.0 μl
RNase H (BRL)	x μl (1 unit)
E. coli DNA ligase (BRL)	y μl (5 units)
	200.0 μl

Incubate at 16°C for 2 hr.

T4 DNA Polymerase Treatment

Add 5 units of T4 DNA polymerase (Pharmacia, Piscataway, NJ) and incubate at 37°C for 10 min. Add 200 μl of phenol–chloroform (1:1), mix well for 3 min, and centrifuge for 10 min. Transfer the upper layer to a fresh tube and extract the aqueous phase with 200 μl of phenol–chloroform once more and then with 200 μl of chloroform once.

Add 100 μl of 7.5 M NH_4Ac, mix, and add 0.7 ml of cold ethanol. Invert the tube several times and incubate at room temperature for 5 min. Centrifuge at maximum speed for 30 min. Wash the pellet with 70% (v/v) ethanol and dry the pellet by incubating the tubes at 37°C for 5 min. Dissolve the cDNA into 5 μl of water.

Modification of Double-Stranded cDNA

The double-stranded cDNA is modified to have different 3′ extensions at each end. First the *Sfi*I adaptor is attached to both the ends and then a part of the repaired linker–primer is trimmed by *Sfi*I digestion. cDNA molecules are separated by gel filtration from small oligonucleotides.

Protocol

*Attachment of Sfi*I *Adaptor*

Combine the following:

cDNA	5 μl
*Sfi*I adaptor (0.5 μg/μl)	2 μl
Ligation buffer B (5×; see page 11)	2 μl
T4 DNA ligase (1 unit/μl)	1 μl
	10 μl

Incubate at 14°C overnight.

Sfi*I Cleavage*

Inactive DNA ligase by heating at 68°C for 15 min. Centrifuge briefly and combine the following:

cDNA with the *Sfi*I adaptor	10 μl
*Sfi*I buffer (10×)	2 μl
*Sfi*I (New England Biolabs, Beverly, MA) (10 units/μl)	2 μl
H_2O	6 μl
	20 μl

Incubate at 50°C for 1 hr.

Purification of cDNA Fragments by Gel Filtration

Extract with phenol–chloroform and back-extract the organic phase with 20 μl of TE. Combine both the aqueous phases.

Prepare a Sepharose CL-4B spin column. In a plastic column (e.g., Select-D column; 3 Prime →5 Prime, Inc., Paoli, PA), add Sepharose CL-4B suspended in STE, and drain excess STE. Adjust the column height to the white line near the top.

Spin the column at room temperature twice for 3 min at 1000 rpm in a tabletop centrifuge and discard the flowthrough each time. Load the cDNA sample. Spin at 1000 rpm for 4 min and collect the flowthrough. In advance, perform a calibration of the column using the mixture of DNA size markers and the *Sfi*I adaptor as a reference.

Analysis of cDNA Fragments

Analyze 8 μl of the sample by agarose gel electrophoresis. Usually a smear ranging from 0.7 to 15 kb is visible (Fig. 6). The average size of the cDNA molecules should be around 2 kb. Use 2 μl of the sample to estimate the concentration with a DNA fluorometer (TKO 100; Hoefer, San Francisco, CA). Typically, a concentration of 10–80 ng/μl is expected.

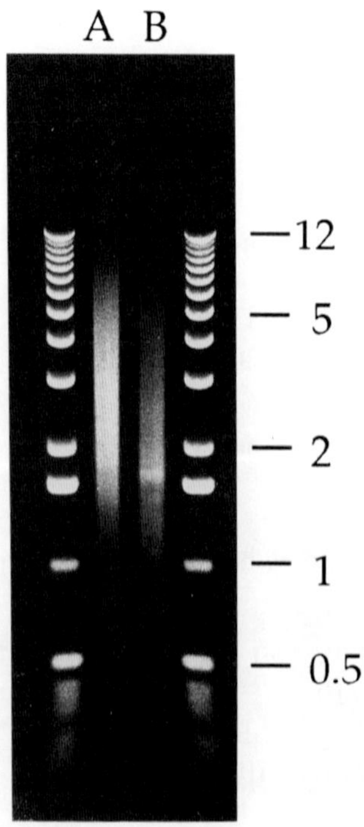

FIG. 6 Analysis of cDNA products after gel filtration. cDNA was synthesized from (A) mouse testis or (B) a derivative of HL60 poly(A)$^+$ RNAs according to the protocol described. After gel filtration, 8 μ of each sample was analyzed by agarose gel electrophoresis. As a molecular size standard, a 1-kb ladder was run in the leftmost and rightmost lanes. Locations of the bands are shown (kb).

Ligation with λ Arms and *in Vitro* Packaging

cDNA is EtOH precipitated with λ arms as a carrier. The DNA is ligated and used to package *in vitro*. Because the ratio of cDNA and vector is dependent on the quality of λ arms and cDNA, it should be determined empirically.

Protocol

Ligation with λpCEV Arms

Mix 0.5 μg of cDNA and 2.5 μg of λ arms. Add a half sample volume of 7.5 *M* NH_4Ac and a four-sample volume of cold ethanol. Incubate at 4°C for 5 min, and centrifuge at room temperature for 30 min. Remove the supernatant carefully. (The pellets are easily dislodged.) Wash the pellet in cold 70% ethanol, and drain excess ethanol. Briefly dry the pellet in a 37°C heat block.

Add ligation reaction mixture (1 × ligation buffer A with 0.05 unit/μl T4 DNA ligase) to the tube. Mix and incubate at 14°C overnight.

In Vitro Packaging

Carry out the packaging reaction as directed by the manufacturer. Titer the library on LE392 cells. Usually a titer between 1×10^7 and 2×10^8 plaque-forming units (pfu)/ml is expected. The presence of cDNA inserts should increase the titer by 10- to 100-fold.

Quality Check of Library

Add 1 μl of the library to 0.3 ml of the LE392 cell suspension in a tube (e.g., Falcon 2070; Becton Dickinson, Oxnard, CA), and incubate at 37°C for 15 min. Add 10 ml of prewarmed NZY medium containing 2 m*M* $MgCl_2$ and shake at 37°C until clear lysis occurs (~6 hr). Follow the small-scale λ DNA preparation protocol (1), and resuspend DNA in 50 μl of TE. Digest 8 μl of the library DNA with *Not*I. Usually a smear ranging from 6.5 to 15 kb is visible under the bands of λ arms (Fig. 7, lane A). When the background is higher, a 6-kb band of the plasmid vector without cDNA insert is visible (Fig. 7, lane B).

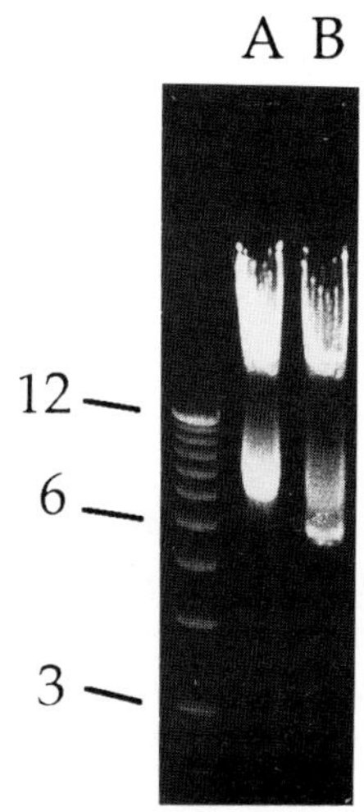

FIG. 7 Examination of cDNA inserts. One microliter each of two libraries (A and B) was amplified in a 10-ml culture. The DNA was extracted, digested by *Not*I, and analyzed by agarose gel electrophoresis. As a molecular size standard, a 1-kb ladder was run in the leftmost lane. Locations of the bands are shown (kb).

Library Amplification and DNA Preparation

cDNA libraries should be amplified on agar plates to prevent possible homologous recombination between cDNA clones. Preparation of high-quality DNA is required for high-efficiency transfection. Use soft agarose instead of agar.

Protocol

1. Prepare NZY agar plates in 150-mm dishes on a flat surface to amplify each clone equally. Leave the plates at 37°C overnight or at room temperature for 1–2 days to dry them and warm at 42°C before plating phages.
2. Inoculate 150 ml of NZY broth with a fresh single colony of a host strain (LE392) and shake at 37°C overnight. Centrifuge at 3000 rpm at room temperature for 15 min. Resuspend the pellet in 150 ml of SM buffer [50m*M* Tris-HCl (pH 7.5) 8m*M* $MgSO_4$, 100m*M* NaCl, 0.01% (w/v) gelatin].
3. Centrifuge the phage (packaging reaction mix) to remove cell debris. Add 1×10^4 pfu to 1 ml of the cell suspension in a tube. Incubate at 37°C for 15 min.
4. To each tube, add 8 ml of liquid 0.7% (wv) NZY soft *agarose* containing 10 m*M* $MgCl_2$, which has been kept at 48°C. Spread evenly onto a NZY hard agar plate (150 mm) on a flat surface.
5. Incubate plates at 37°C for approximately 6 hr in an inverted position. Do not allow the plaques to fuse, to avoid the possibility of rearrangements due to homologous recombination.
6. Pipette 15 ml of SM buffer onto each plate. Incubate the plates at 4°C overnight on a flat surface. Care should be taken to extract all the phages evenly. (Shake the plates periodically and/or change the orientation of the plates several times.)
7. Recover the overlay and rinse the plate with an additional 5 ml of SM buffer. Pool the phage suspension and add several drops of chloroform. Incubate for 15 min on ice.
8. Remove chloroform with a Pasteur pipette. Centrifuge at 8000 rpm for 15 min at 4°C. Transfer the supernatants into a fresh flask.
9. Store a portion of the phage lysate in aliquots as the amplified library at 4°C with several drops of chloroform. For longer storage, add dimethyl sulfoxide (DMSO) at a final concentration of 7% (v/v), mix, and store at −70°C.
10. Add 1 μg/ml each of DNase I and RNase A to the remainder of the phage lysate, and incubate at 37°C for 1 hr.
11. Per liter of lysate, add 40 g of NaCl and 140 g of PEG 8000 (use the powder form, not crystals), and stir gently until all PEG powder goes into solution. Store in a cold room overnight.
12. Centrifuge at 8000 rpm for 30 min at 4°C. Discard the supernatants. Keep the tubes in an inverted position on paper towels for 5 min.
13. Resuspend the pellets into 20 ml of SM buffer per 500-ml tube. Pool the suspension and dispense 20 ml into plastic tubes (e.g., Falcon 2070). Add 20 ml of

chloroform, mix well, and centrifuge the suspension at 16,000 rpm for 20 min at room temperature.

14. Transfer the upper phase into plastic tubes. Remove chloroform by passing air bubbles through the lysate. Transfer the phage solution to 30-ml centrifuge tubes and spin at 25,000 rpm for 2 hr at 4°C.

15. Immediately after centrifugation, discard the supernatant and keep the tubes in an inverted position for 5 min. Add 0.5 ml of SM buffer per tube, keeping the tube at an angle so that the buffer covers the entire surface of the pellet. Incubate at 4°C overnight.

16. Transfer the suspension into microcentrifuge tubes and centrifuge at 14,000 rpm for 5 min. at room temperature. Collect the supernatant and back-extract the pellets in a small amount of SM buffer. Centrifuge and pool the supernatants.

17. Measure the volume of the phage solution and add 1.31 vol of CsCl solution saturated at 0°C in SM buffer. Transfer to nitrocellulose or ultraclear tubes for the SW50.1 rotor (Beckman, Palo Alto, CA) and overlay with paraffin oil.

18. Centrifuge in the SW50.1 centrifuge at 23,000 rpm for 22 hr at 4°C. Recover the phage bands by using a needle attached to a syringe.

19. Dialyze the phage solution in SM buffer. This can be conveniently done in a Centricon 30 (Amicon, Beverly, MA). Transfer the phage solution into a Centricon 30. Cover the upper end of the Centricon with Parafilm and dip the bottom into SM buffer in a plastic tube. Remove air bubbles at the bottom of the Centricon and incubate at room temperature for 1–2 hr. Change the buffer and incubate at 4°C overnight.

20. Transfer the phage solution into microcentrifuge tubes. Add proteinase K solution to a final concentration of 50 μg/ml, mix well, and incubate at 37°C for 15 min. Add 1/10 vol of 100 m*M* Tris-HCl (pH 8.0), 100 m*M* EDTA, 1% SDS, and incubate at 65°C for 1 hr.

21. Transfer the phage solution into a plastic tube and fill to 20 ml with TE [10 m*M* Tris-HCl (pH 8.0), 1 m*M* EDTA]. Extract with same volumes of phenol–chloroform twice and chloroform once. In each extraction, separate the phases by centrifugation at 16,000 rpm for 15 min at room temperature.

22. Add $\frac{1}{2}$ vol of 7.5 *M* NH_4Ac, mix, and then add the same volume of cold 2-propanol. Mix gently several times to form a clear, viscous precipitate. Recover the precipitate with a heat-sealed Pasteur pipette. Drain the 2-propanol and wash the DNA precipitate with cold 70% ethanol twice.

23. Air dry the DNA pellet partially, and dissolve in 1–2 ml of TE. Incubate at 4°C overnight. If it is hard to dissolve DNA, heat the solution to 55°C. Estimate the concentration by optical density (OD) at 260nm (6), and store at 4°C.

Introduction of Expression Libraries into Host Cells

Calcium phosphate transfection (7) is used to stably introduce library DNA into cells. NIH 3T3 cells are maintained in Dulbecco's modified Eagle's medium (DMEM) containing 5% (v/v) calf serum for focus assay or 10% calf serum plus 750 μg/ml G418 for antibiotic selection. When other cells are used, optimal conditions should be determined empirically.

Reagents

HN buffer: 50 m*M* *N*-2-hydroxyethylpiperazine-*N*′-2-ethanesulfonic acid (HEPES), 280 m*M* NaCl (pH 7.1)
$NaPO_4$ (70 m*M*)
$CaCl_2$ (2 *M*)
High molecular weight calf thymus DNA (1 mg/ml)

Protocol

1. Plate cells 1 day prior to transfection at a density of 1×10^5 cells/100-mm culture dish.
2. Add 5 μg/plate of a library DNA and 40 μl/plate of high molecular weight calf thymus carrier DNA (1 mg/ml) to plastic tubes (e.g., Falcon 2070). Add double-distilled (dd) H_2O to tubes such that the total volume of carrier plus library DNA plus ddH_2O is 0.5 ml/plate.
3. Dilute 2 *M* $CaCl_2$ to a concentration of 0.5 *M*. Add 0.5 ml of 0.5 *M* $CaCl_2$ to a tube containing DNA. Mix and leave for 15–20 min.
4. Mix HN buffer with 70 m*M* $NaPO_4$ at a ratio of 1.95 ml:45 μl to prepare HNP buffer. Add 1.0 ml of HNP buffer to a plastic tube.
5. While vigorously bubbling air into the HNP buffer, slowly add (drop by drop) the DNA–$CaCl_2$ solution to the tube. Allow the tubes to stand for 15–20 min.
6. Slowly add 2 ml of the transfection mixture to the medium overlying the cells in a plate and swirl gently.
7. Return all plates to a 37°C incubator.
8. After incubation in a 37°C incubator for 18 hr, aspirate off the medium and precipitates, wash once with DMEM containing 10% (v/v) calf serum, and change the medium twice weekly.

Plasmid Rescue

Efficient rescue of cDNA clones from transfectants is critical to stable expression cloning. When plasmid cDNA libraries are used to transfect mammalian cells, single plasmids integrated in genomic DNA are difficult to release (Fig. 8A). Because λpCEV27 is a λ–plasmid composite vector, cDNA clones are integrated by recombination between λ DNA and host genomic DNA. For plasmid rescue, geno-

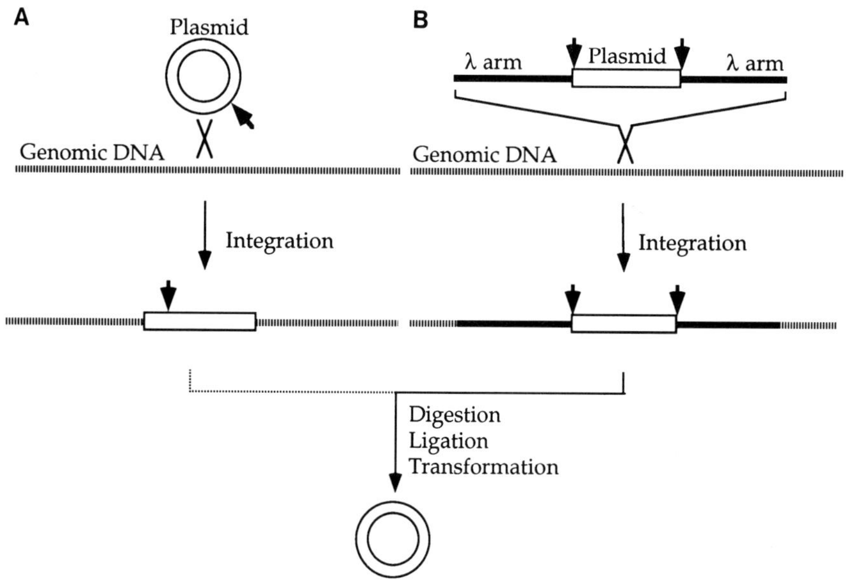

FIG. 8 Comparison of the use of (A) plasmid and (B) λ–plasmid composite vector for plasmid rescue. Short arrows represent sites for an infrequent cutter for plasmid rescue. Plasmid parts, λ arms, and host genomic DNA are shown by open boxes, thick lines, and thick dotted lines, respectively.

mic DNA extracted from the transformant can be subjected to digestion with the infrequent cutters *Not*I, *Xho*I, or *Mlu*I, which can cleave the λ–plasmid junctions. The digested DNA can then be circularized and used for bacterial transformation (Fig. 8B). When genomic DNA extracted from a transformant induced by a λp-CEV27 library was cleaved with one of these enzymes (Fig. 9A, three left lanes), plasmid DNAs with cDNA inserts were efficiently released (Fig. 9B, three left lanes). From the transformant used in this experiment, five plasmids with distinct cDNA inserts were rescued (Fig. 9C). It is important to select *E. coli* cells resistant to both ampicillin and kanamycin to avoid isolation of truncated plasmid clones.

Protocol

1. Digestion

Digest 50 μg of genomic DNA with one of the enzymes (30 units) in 300 μl and incubate at 37°C for 2 hr. Extract with phenol–chloroform and then with 300 μl of chloroform. Add 150 μl of 7.5 *M* NH_4Ac and 1 ml of cold EtOH. Leave on ice for 10 min and centrifuge for 30 min. Wash the pellet with 70% EtOH. Drain the

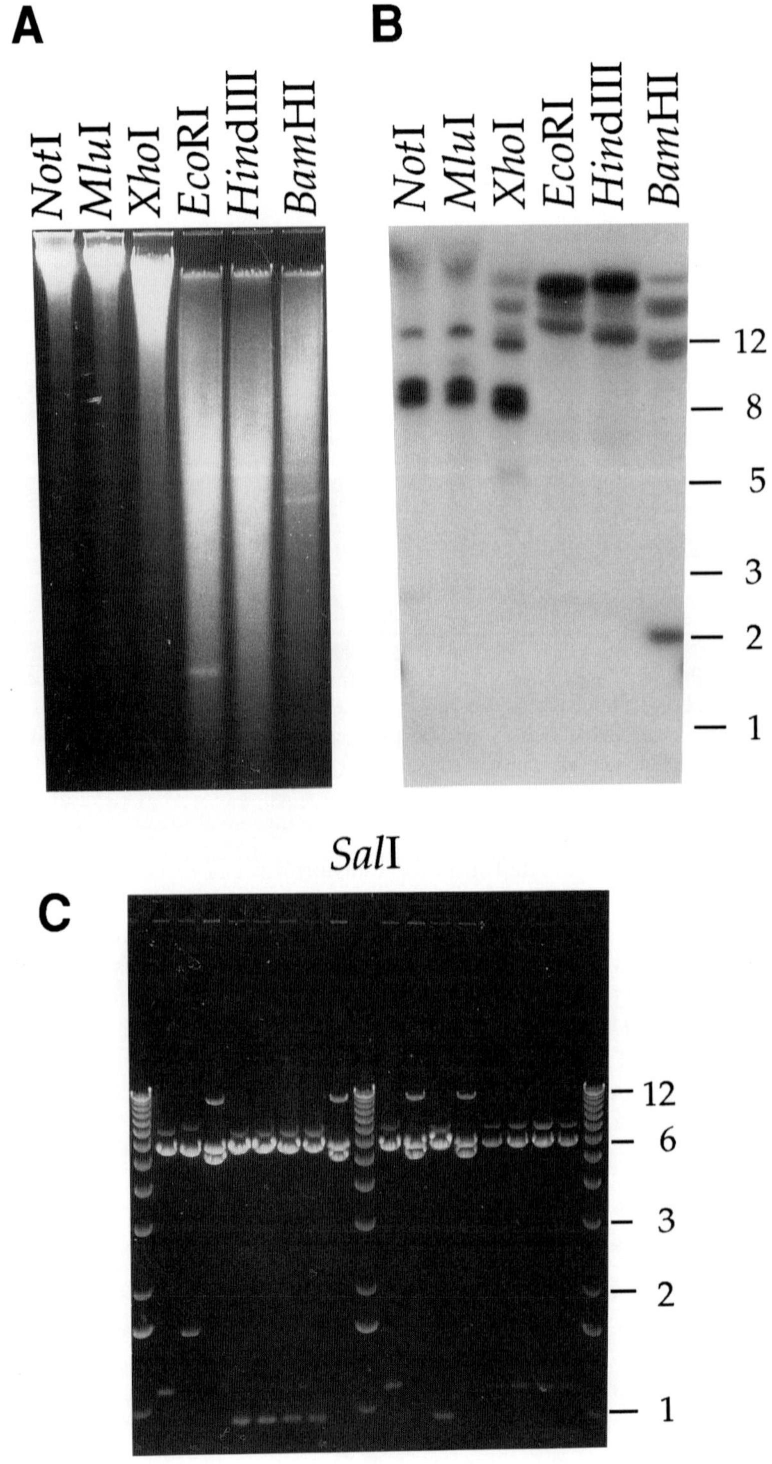

A
B
C
*Not*I
*Mlu*I
*Xho*I
*Eco*RI
*Hind*III
*Bam*HI
*Not*I
*Mlu*I
*Xho*I
*Eco*RI
*Hind*III
*Bam*HI
12
8
5
3
2
1
*Sal*I
12
6
3
2
1

supernatant and briefly dry the pellet at 37°C in a heat block. Resuspend the DNA in 100 μl of TE. Estimate the concentration using a DNA fluorometer.

2. Ligation

Combine the following:

Digested DNA (1.2 μg)	*x* μl
Ligation buffer A (10×)	40 μl
T4 DNA ligase (1 unit/μl)	5 μl
H_2O	355 − *x* μl
	400 μl

Incubate at 15°C overnight. Extract with 400 μl of phenol–chloroform followed by 400 μl of chloroform. Add 40 μl of 3 *M* NaAc and 1 ml of cold EtOH. Following incubation on ice for 10 min, centrifuge at 14,000 rpm for 30 min at room temperature. Wash the pellet with 70% EtOH. Drain the supernatant and briefly dry the pellet at 37°C in a heat block. Resuspend the DNA in 10 μl of TE.

3. Transformation

Add 2 μl of DNA to 100 μl of competent cells of an *mcrA*$^-$ *mcrB*$^-$ strain [DH5αMCR (Bethesda Research Laboratories), DH10B (Bethesda Research Laboratories), PLK-F′ (Stratagene, La Jolla, CA), or SURE (Stratagene)]. Following incubation at 0°C for 30 min and heat shock for 45 sec, chill the cells on ice for 2 min and add SOC medium (Bethesda Research Laboratories) containing 1 mM isopropylthio-D-galactopyranoside (IPTG) (to induce expression of the *tac-neo* gene on pCEV27). Rotate at 225 rpm for 1 hr at 37°C in a rotary shaker (New Brunswick Scientific, Edison, NJ).

Spread the cells on 150-mm plates of NZY agar containing 100 μg/ml ampicillin, 25 μg/ml kanamycin, and 1 m*M* IPTG. Incubate the plates at 37°C. PLK-F′ and SURE cells sometimes need longer incubation (for more than 1 day), because the sizes of the colonies are smaller and heterologous.

Rescued plasmids are subjected to restriction analysis (Fig. 9C) and then analyzed for their ability to induce the phenotype expected.

FIG. 9 Plasmid rescue from the cells transfected with a λpCEV27 cDNA library. (A) DNA was digested by infrequent cutters (*Not*I, *Mlu*I, and *Xho*I) or frequent cutters (*Eco*RI, *Hind*III, and *Bam*HI) and separated by agarose gel electrophoresis. (B) The DNA bands in the gel were transferred to a nitrocellulose filter, and probed with the plasmid pCEV27. (C) The digested DNA was ligated and used for plasmid rescue. Sixteen plasmids isolated were cleaved by *Sal*I to release the inserts, and separated by agarose gel electrophoresis. Locations of the bands are shown (kb).

Applications

This stable expression cloning system has been successfully utilized to isolate and identify oncogenes of tumor cells (2). It has also been effective in the isolation of a growth factor receptor, whose introduction into a recipient cell expressing the growth factor resulted in autocrine transformation (3, 4). Thus this system should be applicable to the isolation of both growth factors or receptors using an appropriate strategy. It is also conceivable that introduction of genes encoding downstream targets of growth factor signaling might induce transformation in a recipient cell if such a gene played a critical regulatory role (5). Other possible applications of this system include the isolation of genes for drug resistance and viral cell surface receptors as well as genes that complement specific mutations.

References

1. T. Miki, T. Matsui, M. A. Heidaran, and S. A. Aaronson, *Gene* **83,** 137 (1989).
2. T. Miki, T. P. Fleming, M. Crescenzi, C. J. Molloy, S. B. Blam, S. H. Reynolds, and S. A. Aaronson, *Proc. Natl. Acad. Sci. U.S.A.* **88,** 5167 (1991).
3. T. Miki, T. P. Fleming, D. P. Bottaro, J. S. Rubin, D. Ron, and S. A. Aaronson, *Science* **251,** 72 (1991).
4. T. Miki, D. P. Bottaro, T. P. Fleming, C. L. Smith, W. H. Burgess, A. M.-L. Chan, and S. A. Aaronson, *Proc. Natl. Acad. Sci. U.S.A.* **89,** 246 (1992).
5. T. Miki, C. L. Smith, J. E. Long, and T. P. Fleming, *Nature (London)* in press (1993).
6. T. Maniatis, E. F. Fritsch, and J. Sambrook, "Molecular Cloning: A Laboratory Manual." Cold Spring Harbor Lab. Press, Cold Spring Harbor, New York, 1982.
7. M. Wigler, S. Silverstein, L.-S. Lee, A. Pellicer, Y. C. Cheng, and R. Axel, *Cell* **11,** 223 (1977).

[2] Normalization and Selection with Short-Fragment cDNAs

Satish Parimoo,* Sankhavaram R. Patanjali,* and
Sherman M. Weissman

Introduction

A mammalian genome encodes between approximately 40,000 and 150,000 different mRNAs, some of which may be found in multiple forms because of alternative splicing or the use of alternative 3′ or 5′ ends. Only a few percent of the total DNA of the genome encodes mRNA sequences, the remainder consisting of intron sequences, intergenic sequences, pericentromeric and telomeric repetitive sequences, and so forth (1). Therefore the analysis of the sequence of the genome is at least one to two orders of magnitude more difficult than the analysis of sequences of representatives of each type of mRNA. The highly repetitive nature of some noncoding DNA imposes additional difficulties on genomic analysis. The analysis of randomly selected cDNA clones could initially give a great deal of information (2). However, the levels of expression of individual mRNAs can range over several orders of magnitude, so that an extensive analysis of scarce cDNAs could be performed only by complete characterization of large cDNA libraries (3). Such libraries would be too large for many types of functional studies. An approximately equimolar set of intron-free clones of coding sequences would be useful for mapping of gene positions within large DNA fragments, for identification of the bulk of coding sequences, for cataloging cDNAs expressed selectively in some cell types or tissues, and for various applications that require functional expression of cDNA fragments.

Our laboratory, as well as others (4, 5), has been working on procedures to cope with the problems mentioned above. In the following pages we describe protocols for preparing normalized libraries of cDNA fragments, that is, libraries in which fragments of different sequence are represented in a similar abundance, regardless of their level of expression in the source tissues, and for identifying cDNA fragments encoded by specific segments of genomic DNA.

A situation frequently arises in positional cloning studies in which a substantial region of genomic DNA that spans a disease locus has been cloned. The task is then to identify transcription units in this DNA and determine which carry mutations responsible for the disease. Numerous approaches for identifying expressed sequences have been employed or proposed. These include examination of individual segments of genomic DNA for cross-species homology to detect conserved se-

* Both authors contributed equally. Their names appear alphabetically.

Methods in Molecular Genetics, Volume 1

quences that are presumably under evolutionary constraint because of their coding capacity (6); identification of regions with a high content of CpG dinucleotides (HTF islands) as these frequently occur at the 5′ end of genes (7); direct use of large DNA fragments such as yeast artificial chromosomes (YACs) for screening cDNA libraries (8); and exon trapping methods (9, 10).

Each of the methods for ifentifying expressed sequences has limitations. Detection of genes by screening DNA "zoo blots" to find conserved sequences is a time-tested method that has resulted in success a number of times. However, not all genes are sufficiently conserved to show cross-species hybridization. In addition, such hybridization requires isolation and preparation of a substantial number of probes from genomic DNA fragments, and still requires isolation of cDNA clones for proof. Screening for HTF islands has also been successful in several cases. However, not all genes have HTF islands and not all HTF islands are located immediately adjacent to and overlapping the 5′ ends of genes (7). Using large DNA fragments as hybridization probes for screening cDNA libraries has worked well in some cases, but experience varies with regard both to the amount of false-positive material obtained and to the efficiency of detecting true positives. In addition, any method that involves screening unnormalized libraries is cumbersome when applied to scarce cDNA species and unlikely to detect cDNA species present at much less than one copy per several million clones. The exon trapping methods have the theoretical advantage of detecting genes not transcribed in available tissues. However, in addition to being somewhat cumbersome, they may generate false-positive results so that putative exons must be checked against mRNA populations to determine if they are really expressed. In addition, these procedures require multiple transfections of animal cells. We have reported a method for using large fragments of genomic DNA to select cDNA clones directly by the hybrid selection method (11).

Normalized cDNA Libraries

Two general approaches have been proposed for creating equimolar or normalized libraries of cDNA fragments (12). In one approach, genomic DNA is hybridized to a large excess of cDNA fragments, so that all genomic coding sequences represented in the library would be saturated. After removal of the excess unhybridized cDNA, the hybridized material is eluted from the genomic DNA and cloned. In general, this approach would fail to the extent that saturation was not achieved for scarce cDNAs. It would also pick up cDNAs in proportion to the number of genes plus pseudogenes within a closely homologous family. As a result, unique genes would be underrepresented relative to those that are part of a family.

The second approach to creating normalized libraries is based on the kinetics of renaturation of a mixture of fragments of various abundances. In theory, the rate of annealing of a denatured double-stranded molecule is proportional to the square of its concentration prior to denaturation (13). After extensive but incomplete renaturation of cDNA fragments, amplified from the library using vector primers (Fig. 1), the relative abundance of the different species in the single-stranded fraction would differ much less than in the initial mixture. For example, after the least abundant species was 50% annealed, the amount of the most abundant species left in single-stranded form would be only twice that of the least abundant form. In practice, certain complications occur. Fragments of cDNA from the same mRNA do not necessarily share identical 5′ and 3′ ends and complex concatenated structures may be formed between three or more strands of cDNA. As a consequence, the most abundant species may actually become less concentrated in the single-stranded fraction than are the less abundant species. Also, for uncertain reasons, even after relative short C_0t values [C_0t = initial concentration of DNA × time, mol·sec/liter] of reassociation, a significant fraction of the rare species may appear in the double-stranded fraction, as judged by hydroxyapatite chromatography.

A representative tissue culture cell line may contain 200,000–300,000 mRNA molecules per cell and these may correspond to some 10,000 different mRNA species occurring at abundances between 1 and tens of thousands per cell (3). Considering that each mRNA may be represented by several different nonoverlapping se-

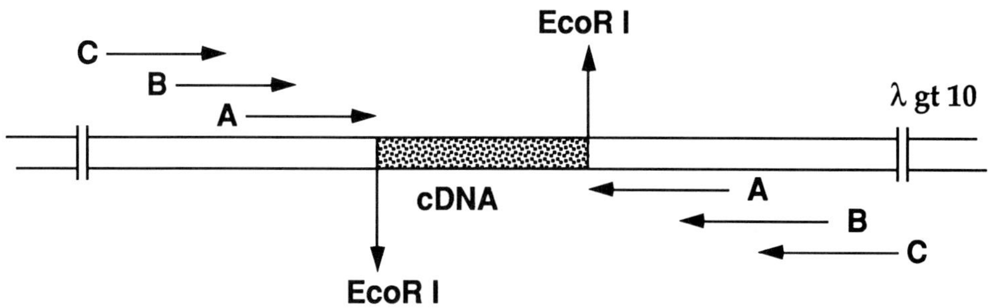

PRIMER SETS :

SET A : AGCCTGGTTAAGTCCAAGCTG
CTTCCAGGGTAAAAAGCAAAAG

SET B : AGCAAGTTCAGCCTGGTTAAG
CTTATGAGTATTTCTTCCAGGGTA

SET C : CCACCTTTTGAGCAAGTTCAG
GAGGTGGCTTATGAGTATTTC

FIG. 1 Orientation of PCR primers used to amplify cDNA inserts from λgt10.

quence fragments, a fully normalized library would include perhaps 45,000 or more different fragments at equal abundance. The least abundant species could have been increased only from 1/300,000 to 1/10,000–1/30,000, depending on its length. Therefore, enrichment of a scarce cDNA obtained by normalization of a library (Fig. 2) for a single type is somewhat limited. However, when RNA is prepared from a complex tissue or organ, some cell types may be present as only a small fraction of the total cell content, and a scarce cDNA peculiar to such a cell type may be orders of magnitude lower in relative abundance and therefore be correspondingly more enriched by the normalization procedure.

If the normalization procedure is carried too far, chance inclusion of a few molecules of genomic DNA in the original cDNA library will eventually lead to a relative increase of those fragments in the normalized library. More seriously, leakage from

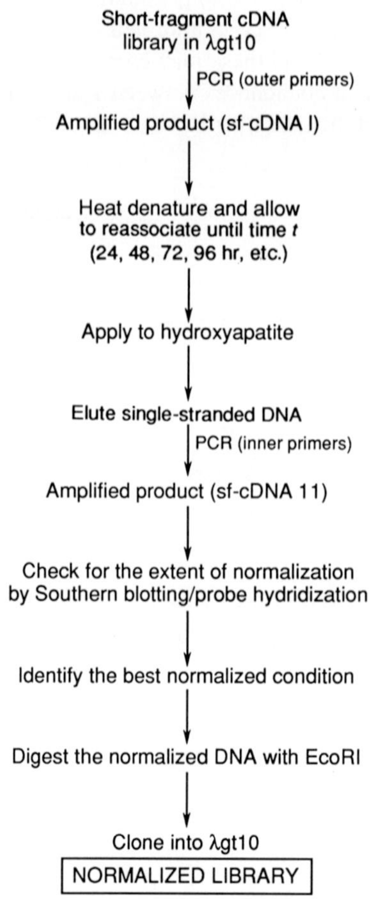

FIG. 2 Schematic representation of normalization.

nuclei of incompletely spliced or unspliced mRNAs will result in the normalization of some intronic and exonic fragments. This is more likely to occur for the more abundant transcripts, or for those that are inefficiently exported from the nucleus or rapidly degraded in the cytoplasm.

After any approach to normalization of a cDNA library, the normalized mixture of fragments must be converted into a library within minimum distortion of their abundances. Methods that require cloning of fragments of widely differing lengths can defeat the purpose of normalization because of the bias against cloning large inserts in some vectors. For this reason we have chosen to use procedures that depend on normalization of cDNA fragments of a defined size range, and to use random priming in an effort to ensure that all portions of long mRNAs are represented. Sampling of any normalized library is subject to statistical laws, so that any randomly chosen sample will no longer be entirely equimolar. Regardless of these difficulties, amplified normalized libraries prepared by the methods described below have been found to contain representatives of each moderately to very low abundance cDNA for which we have searched.

The method described below has been used in our laboratory to prepare a number of normalized libraries. The process of preparation of short-fragment cDNA libraries and their normalization is described in detail.

cDNA Selection Techniques

The general approach toward identification of expressed sequences is to immobilize the target genomic DNA fragment, whether derived from cosmid or YAC cloning, pulsed-field gel electrophoresis, or other approaches. cDNA fragments are amplified by polymerase chain reaction (PCR) (14) from cDNA libraries and annealed to the immobilized target in the presence of an excess of quenching reagents. The immobilized DNA is washed to remove nonspecific material, and the cDNAs are recovered by PCR and cloned (11) (Fig. 3). Several difficulties are dealt with in order to develop an efficient protocol that permits screening of multiple large DNA fragments in parallel.

A significant fraction of cDNAs include medium- and low-level repetitive sequences, particularly in their 3′ untranslated regions. Because these sequences are more abundant than any gene sequences in the target DNA such cDNAs would be overrepresented and also would be misassigned to multiple targets if these sequences were selected. If full-length cDNAs were used and efforts were made to initially remove repetitive sequences, there would be a risk of depleting the libraries of those cDNAs that contain repeats. Also, PCR could introduce an unacceptable bias in the representation of different cDNAs because of their length or content of GC-rich regions. To avoid these problems as well as the potential underrepresentation of the 5′ ends of long cDNAs, we have worked with randomly primed, relatively short cDNA sequences cloned in phage vectors and recovered by PCR. Be-

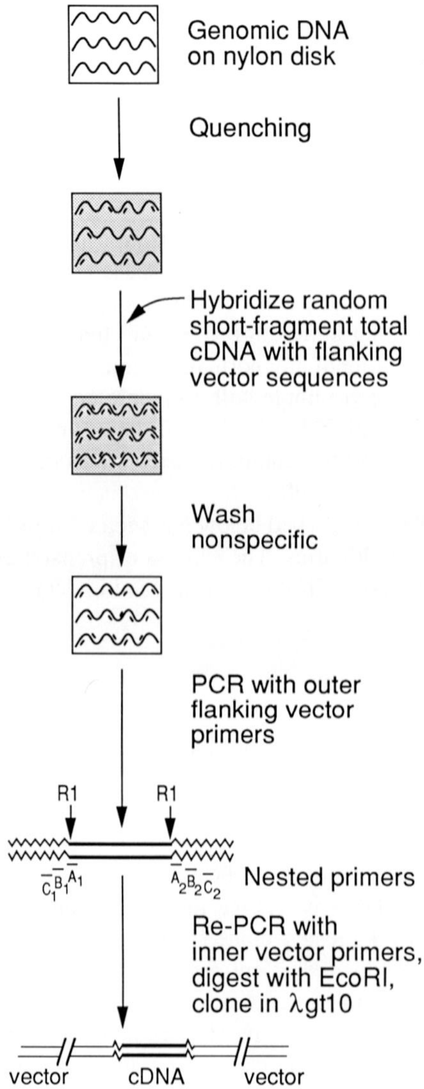

FIG. 3 Schematic representation of cDNA selection.

cause of the high sensitivity of PCR, it is possible to carry out the selection with small amounts of target DNA and relatively small amounts of cDNA. As a consequence, the selection can be carried out in the presence of a much greater excess of quenching sequences than could be used if other methods of recovery were employed.

To avoid bias in the amplification of genomic DNA fragments, we first optimized the procedure for use with small amounts of target DNA. This makes it possible to use YAC DNA isolated from a single lane of pulsed-field gel electrophoresis directly. The target DNA is directly immobilized on small pieces of nylon filter and hybridized to cDNA libraries. The sensitivity of the procedures evolved is sufficient, so that it is possible to blot YAC DNA from electrophoretic runs in agarose onto nylon and use a segment of the appropriate transferred DNA band for selection. Also, the sensitivity in reconstruction experiments approaches levels that make it possible to use whole genomic DNA from somatic cell hybrids for selection.

Normalization

Materials Required

Short-Fragment cDNA Libraries (Average Size, 400 to 2000 bp)

Isolation of Total RNA

Total RNA is isolated from the tissue or cell line of interest by a combination of guanidine thiocyanate–phenol extraction (15) and selective precipitation of RNA (16). The steps involved are as follows.

1. Homogenization is performed for 60–90 sec with a Polytron tissue homogenizer (Brinkmann, Westbury, NY) and 20–25 ml of the homogenization buffer [4.0 *M* guanidine thiocyanate, 25 m*M* sodium citrate (pH 7.0), 0.5% sarcosine (w/v), 0.1 *M* 2-mercaptoethanol] per gram tissue. Alternatively, 10^7–10^8 cells are homogenized using a sterile Dounce homogenizer (Wheaton, Millville, NJ).
2. Add sequentially $\frac{1}{10}$ vol of 2.0 *M* sodium acetate (pH 4.0), an equal volume of water-saturated phenol, and $\frac{1}{5}$ vol of chloroform and isoamyl alcohol (24:1), with mixing by inversion after each addition. The final suspension is shaken vigorously and chilled on ice for 10 min.
3. The aqueous phase is recovered by centrifugation at 10,000 *g* for 10 min at 4°C and precipitated with an equal volume of 2-propanol at −20°C for 1 hr. The DNA and total RNA in the mixture are pelleted down by centrifugation at 10,000 *g* for 10 min at 4°C.
4. The nucleic acid pellet is resuspended in 0.5 ml guanidine thiocyanate [5.0 *M* guanidine thiocyanate, 50 m*M* Tris-HCl (pH 7.5), 10 m*M* ethylenediaminetetraacetic acid (EDTA), 0.1 *M* 2-mercaptoethanol]. Seven volumes of 4.0 *M* lithium chloride is added and the lysate is mixed thoroughly and stored at 4°C overnight. RNA alone is selectively precipitated by this treatment.
5. The pelleted RNA is recovered by centrifugation, washed with 1.0 ml of 8.0 *M* urea containing 4.0 *M* lithium chloride, and pelleted at 10,000 *g* for 5 min at 4°C.
6. The pellet is redissolved in TE buffer [10 m*M* Tris-HCl (pH 8), 1 m*M* EDTA] at pH 7.5 and the total RNA concentration is estimated by absorbance at 260 nm.

Isolation of Poly(A)$^+$ RNA

The total RNA isolated as described is treated with 1.0 unit of DNase RQ 1 (Promega, Madison, WI) per milligram of RNA in 40 m*M* Tris-HCl (pH 7.9), 10 m*M* NaCl, 6.0 m*M* $MgCl_2$, 0.1 m*M* $CaCl_2$ at 37°C for 15 min. The RNA is then applied to oligo(dT)-cellulose (type III; Collaborative Research, Bedford, MA) and poly(A)$^+$ RNA is isolated as per the instructions of the manufacturer. The poly(A)$^+$ RNA is once again passed over a fresh column of oligo(dT)-cellulose to decrease the ribosomal fraction of the eluted poly(A)$^+$ RNA. We have routinely obtained 2–5 μg of poly(A)$^+$ RNA per gram tissue after two rounds of selection over the oligo(dT)-cellulose affinity column.

Construction of Short-Fragment cDNA Library

Poly(A)$^+$ RNA (5 μg) is primed with commercially available (Pharmacia, Piscataway, NJ) random hexanucleotide mixture [$p(dN)_6$], and the first strand is synthesized by Moloney murine leukemia virus (MMLV) reverse transcriptase in 40 μl of the first-strand mixture [50 m*M* Tris-HCl (pH 8.3), 75 m*M* KCl, 3.0 m*M* $MgCl_2$, 10 m*M* dithiothreitol (DTT), 562 μM each of dNTPs, 6.75 μg of $p(dN)_6$, 50 units (U) of RNasin, 450 U of MMLV reverse transcriptase]. The second strand is synthesized according to the Gubler and Hoffman method without adding ligase (17). Forty microliters of the first-strand synthesis reaction mix is used to construct the second strand in a total of 350 μl of second-strand reaction mix [25 m*M* Tris-HCl (pH 8.3), 100 m*M* KCl, 5 m*M* $MgCl_2$, 300 μM each of dNTPs, 5 m*M* DTT, 62.5 U of DNA polymerase I (pol I) of *Escherichia coli*, 2 U of RNase H]. *Eco*RI linkers are ligated after methylating all the *Eco*RI sites of the cDNA by *Eco*RI methylase. After overnight ligation at 15°C the cDNA is digested with *Eco*RI (20 U/μg of DNA). cDNA is selected in the range of approximately 400–2000 bp over low melting point agarose by phenol extraction (18). The purified cDNA is ligated to calf intestine alkaline phosphatase-treated λgt10 arms according to the protocol of the vendor (Stratagene, La Jolla, CA). Usually 3 to 6 million plaque-forming units (pfu) is plated to amplify a single library. Complex tissues such as brain or total fetus would require at least 7 to 10 million basic plaque-forming units for amplification purposes to obtain a good representation of rare messages.

λgt10 is the vector of choice for a cDNA library because (1) it is a well-documented insertion vector system, (2) being a phage, it lends itself to the amplification of up to 10^6 to 10^7 pfu, and (3) it lacks any plasmid vector sequences such as *lacZ* or *ampR*. This is particularly useful because the probes used in the evaluation of the quality of the library may be contaminated with parental plasmid sequences during their preparation and would generate false positives from the library.

Amplification of cDNA Inserts by Polymerase Chain Reaction

Three different sets of primers complementary to the sequence around the *Eco*RI cloning site of λgt10 are synthesized. These are sequentially named primer sets A, B, and C, with set A being the closest to the *Eco*RI site (see Fig. 1).

The cDNA inserts are amplified from the library by hot-start PCR (19) as described below using set C primers (outermost primers) (Fig. 1). Set B primers (Fig. 1) generate somewhat shorter products when compared to sets A and C. Hence, use of set C primers for the initial amplifications and set A primers for nested PCR amplifications is recommended. Originally, we reported PCR amplification on addition of all the components to one tube (4). However, it is now clear that hot-start PCR amplification considerably reduces nonspecific amplifications (11, 19). The method is briefly described below.

Hot-Start Polymerase Chain Reaction Amplification of Short-Fragment cDNA Library

Component	Volume (μl)	Final concentration
cDNA library [5 μg phage DNA/μl in TE buffer)	2	10 ng
Primer 1 (30 μM stock)	2	0.6 μM
Primer 2 (30 μM stock)	2	0.6 μM
10 × buffer [500 mM KCl, 100 mM Tris-HCl pH 8.3 at room temperature, 20 mM $MgCL_2$]	10	1 ×
AmpliTaq polymerase (5 U/μ1)(Perkin-Elmer Cetus, Norwalk, CT)	0.5	2.5 U
Sterile water	73.5	
	90.0 μl	

The samples are covered with mineral oil and the tubes placed in the PCR machine. The samples are denatured at 94°C for 2 min and then incubated at 80°C. At this stage, the machine is stopped and, without removing the tubes, 10 μl of 2.0 mM dNTP mixture (each 2 mM) is added by dipping the Pipetman tip almost to the bottom of the tube and releasing the solution with proper care (final dNTP concentration in PCR is 0.2 mM). The PCR cycles are continued using the following cycle conditions: (1) for the PCR 9600 machine (Perkin-Elmer, Norwalk, CT), use 94°C for 45 sec/50°C for 45 sec/72°C for 2 min for 30 cycles; (2) for the PCR 100 machine (Perkin-Elmer), use 94°C for 70 sec/50°C for 70 sec/70°C for 2 min for 30 cycles, with a final extension of 5 min at 72°C in either machine. The use of plugged (aerosol-resistant) Pipetman tips is recommended while setting up PCR reactions. The amplified samples are frozen at −20°C if not used or processed the same day. The amplified product is pooled into a single tube and extracted with a chloroform and isoamyl alcohol mixture (24:1), v/v) once and precipitated with ethanol. The cDNA between 400 and 2000 bp is fractionated by electrophoresis on a low melting point agarose gel and isolated by phenol–chloroform extraction and ethanol precipitation (18). This DNA is ready for reassociation experiments.

Hydroxyapatite Chromatography

Hydroxyapatite (HAP) chromatography is a well-documented method (20) for isolating single-stranded (ss) and double-stranded (ds) DNA fractions. Various factors that affect the separation are briefly described below.

Columns

Water-jacketed glass columns with sintered glass (G4 grade) bottoms are best suited for the purpose. The jacket covering the column should be extended at least 1 cm below the sintered glass bottom to ensure that the HAP material loaded in the column can be uniformly heated during the operation of the column.

Gel

Hydroxyapatite is available through various commercial organizations. This material is soaked in 0.01 *M* sodium phosphate buffer (pH 7.2), 0.1% (w/v) sodium dodecyl sulfate (SDS). Ten milliliters of the buffer is used to suspend 1 g of the gel and the fine particles are removed by aspiration after allowing the gel to settle for about 2–3 min. This process is repeated at least three times to remove most of the fine particles. Finally, the gel is suspended in 10 ml of 0.01 *M* sodium phosphate buffer (pH 7.2) and stored at 4°C for not more than a week.

Buffers

Because single-stranded and double-stranded fractions of DNA bind to hydroxyapatite at low salt concentrations and are differentially eluted with increasing salt concentrations, a series of concentrations of sodium phosphate buffers is prepared. To calibrate the HAP column, sodium phosphate buffers (pH 7.2) of 0.01, 0.05, and 0.1 to 0.4 *M* with 0.05 *M* increments of concentration are prepared as described earlier (4). All sodium phosphate buffers contain 0.1% SDS unless noted otherwise.

Tracer DNA

*Hae*III-digested M13 ssDNA is kinased with [γ-^{32}P]ATP (New England Nuclear, Boston, MA) using polynucleotide kinase and the double-stranded 1-kb DNA ladder [molecular weight markers (Bethesda Research Laboratories, Gaithersburg, MD) is nick-translated (18) to incorporate [α-^{32}P]dCTP (New England Nuclear, Boston, MA) to generate tracer ss- and dsDNA fractions. In addition, radiolabeled cDNA is prepared by incorporating 100 μCi of [α-^{32}P]dCTP (New England Nuclear) to the amplification reaction mixture. Amplification is performed by hot-start PCR as described earlier.

Calibration of Hydroxyapatite Columns

The column is connected to a circulation water bath (Neslab-endocal, Newington, NH) and maintained at 60°C. The gel material, resuspended in 0.01 *M* sodium phosphate buffer (pH 7.2) and prewarmed to 60°C in a separate water bath, is loaded on the column until the packed volume reaches a premarked 1.0-ml level. It is critical to maintain a constant DNA:gel ratio in all the experiments. The gel is washed with at least 10 bed volumes of 0.01 *M* sodium phosphate buffer (pH 7.2), 0.1% SDS. Tracer DNA (single or double stranded) in the range 0.1–2.0 μg is loaded in 0.01 *M* sodium phosphate buffer (pH 7.2) at flow rates of 8–10 ml/hr. The column is washed with 10 ml each of 0.01, 0.05, and 0.1–0.4 *M* (with 0.05 *M* increments) sodium phosphate buffer (pH 7.2), 0.1% SDS. Fractions of 1.0 ml are collected and counted in a scintillation counter. The elution profiles are monitored for ss- and dsDNA fractions by plotting radioactivity recovered against fraction number and the phosphate buffer concentration at the peak elution of ss- and dsDNA fractions is noted. It is essential to calibrate the HAP column in this way, because the conditions recommended by the manufacturer about the elution of ss- and dsDNA fractions could be considerably different from laboratory conditions. In addition, the storage mode and age of the gel may alter its elution properties.

Concentration of DNA Fractions

The ss- and dsDNA fractions isolated from the column are concentrated from an approximately 10-ml volume to 100 μl. Conventional butanol extraction procedures (18) result in poor recovery of the samples (<20%). Concentration of the samples by vacuum drying and further recovery by Geneclean (Bio 101, San Diego, CA) improve the recovery to 40–50%. However, concentration of the samples by Centricon (Amicon, Danvers, MA) centrifugation results in >90% recovery of the samples as measured by radioactivity. In our experience, Centricon 30 and Centricon 100 work equally well. Samples are extracted once with phenol and once with chloroform–isoamyl alcohol mixture (24:1, v/v), to reduce the concentration of SDS in the eluates. The DNA samples can be concentrated by Centricon filtration without any further treatment. Concentration is performed according to the protocol of the manufacturer.

Normalization Procedure

Reassociation Reactions

Reassociation of the DNA is dependent on the concentration of the reassociating species, salt concentration, temperature, and time of reassociation. Keeping the salt concentration at 0.3 *M*, the concentration of the DNA is varied from 1 to 20 μg/ml; reassociation is studied at 65 and 70°C; and the length of reassociation is varied from 12 to 288 hr. The ideal solution conditions for reassociation are 0.3 *M* sodium phosphate buffer (pH 7.0), 0.4 m*M* EDTA, 0.04% SDS, and 20 μg/ml of amplified DNA. The reaction volume is maintained at 50 μl in all reassociation reactions. The reaction mixture is overlaid with mineral oil and the DNA denatured by inserting the tube in a boiling water bath for 5 min. The tube is immediately transferred to a 65°C water bath and allowed to reassociate for the desired length of time. In general, cDNA from complex sources such as tissues or organs is reassociated for 72 hr, whereas cDNA from cell lines is reassociated for 36 hr, under the conditions stated above. Ideally, normalization may be performed in four 24-hr intervals. However, it is mandatory to evaluate the extent of normalization by using different probes that occur at various abundances in the native cDNA library.

Isolation of Single-Stranded and Double-Stranded DNA Fractions of cDNA

Reassociating DNA samples are mixed with 1.0 ml of 0.01 *M* sodium phosphate buffer (pH 7.2) containing 0.1% SDS at the end of a specific reassociation time and chilled on ice for 10 min. This diluted sample is briefly heated to 65°C and applied to an HAP column maintained at 60°C. Elution of the ss- and dsDNA fractions and concentration of the samples are performed as described above.

Reamplification of Single-Stranded and Double-Stranded DNA

As described above, the cDNA inserts amplified with set C primers retain the set A template region in the amplified cDNA (Fig. 1). Ten microliters of the concentrated ss- and dsDNA fractions are amplified using set A primers by the hot-start protocol. Usually, amplification in three tubes gives enough DNA to evaluate the extent of normalization and to construct cDNA libraries.

Evaluation of Normalization

Preparation of DNA Blots

The amplified material is pooled, and purified as described above. Using nylon membranes (Hybond-N; Amersham, Arlington Heights, IL), Southern blots are prepared from the amplified cDNA material, from both ss- and dsDNA fractions obtained at different time intervals (21). These blots are prehybridized in 1% crystalline-grade bovine serum albumin, 1 m*M* EDTA, 7% SDS (w/v), 0.5 *M* sodium phosphate buffer (pH 7.2) in a 65°C water bath or commercial hybridization oven. DNA probes are labeled by the random priming method (22) and added on denaturation to the prehybridization solution. After 10–16 hr of hybridization the filters are washed sequentially with 2× SSC [1× SSC is 0.15 *M* NaCl, 0.015 *M* sodium citrate (pH 7.0)], once at room temperature for 10 min and once at 65°C for 10 min; then once with 0.2× SSC at 65°C for 10 min and once with 0.1× SSC at 65°C for 10 min.

cDNA library is approximately 1 in 10,000 (4). Based on the probability equation (18), $N = \ln(1 - P)/\ln[1 - (1/n)]$, where N is the number of clones required to be screened, P is the probability desired (0.99), and n is the fractional abundance of the cDNA species, it is estimated that even rare messages may be identified in a normalized cDNA library by screening 45,000 to 60,000 clones. Such enriched libraries can be effectively used in subtraction and differential hybridization. Efforts are underway to use such libraries to understand protein–protein interactions by the contingent replication assay (25) and to select YAC-specific cDNA sequences (14).

Selection

ration of Reagents

Preparation of cDNA Library

A random hexamer-primed short-fragment cDNA library is prepared and cloned into λgt10 as described under Normalization (above). The size range of the cDNA is 0.4–1.5 kb. Usually a library containing at least 3 million independent clones is prepared. The cDNA inserts in the size range 400–2000 bp are amplified from the library by hot-start PCR as described above, using set C primers (outermost primers) (Fig. 1).

Preparation of Yeast Artificial Chromosome DNA

Electrophoretic Separation of Yeast Artificial Chromosomes

Yeast chromosomes are separated in 1% (w/v) SeaKem GTG agarose (FMC, Rockland, ME) in half-strength Tris–borate buffer [0.5× TBE (18)] run on a Bio-Rad (Richmond, CA) CHEF electrophoresis apparatus. If a YAC comigrates with a native yeast chromosome and is indistinguishable on the gel, it is identified by hybridization of Southern blot (from a single lane of the gel) to pBR322 or a related plasmid probe. During blotting, the remainder of the gel is wrapped in Saran wrap and stored at 4°C. After identification of the YAC, the band of interest is cut out and the agarose piece containing the YAC is stored in a capped 15-ml Falcon tube (Becton Dickinson, Oxnard, CA) at 4°C until further use.

Digestion of Yeast Artificial Chromosome DNA in Agarose Blocks with *Hin*dIII

Agarose pieces containing YACs are soaked in TE for 1 hr and then equilibriated with the *Hin*dIII enzyme buffer for 2 hr at 4°C with a change of buffer. An 8-cm-long agarose slice containing the YAC is immersed in 3–4 ml of *Hin*dIII buffer with 0.25–0.5 U of enzyme per microliter and incubated at 37°C for at least 6–8 hr. EDTA, pH 8, is added to a final concentration of 20 m*M* to stop the reaction. After 30 min it is replaced with 0.5× TBE and incubated another 30 min. *Hin*dIII may not be an ideal enzyme if the yield of YAC DNA is too low after pulsed-field

Choice of Probes

Ideally at least 10–12 different probes should be used to evaluate the extent of normalization. A detailed description of such probes is given elsewhere (4). A fair evaluation of the extent of normalization can be gained by use of the following probes.

Repetitive DNA Probes

Cloned *Alu* and LINE element repeats serve as excellent markers to monitor the depletion of repetitive DNA.

Ribosomal DNA Probes

Although poly$(A)^+$ RNA is twice selected over the oligo(dT)-cellulose column, a significant number (5–10%) of parent cDNA clones represent ribosomal DNA clones. Monitoring their abundance indicates the fate of *high-abundance* messages during normalization.

Actin Gene Probes/HLA Genes

Actin gene probes/HLA genes represent medium-abundance genes in cDNA libraries (0.1–0.15%). Because these probes recognize several members of the gene family, normalization patterns represent the behavior of *medium-abundance* genes.

Rare Probes

Housekeeping genes such as hypoxanthine phosphoribosyltransferase (*HPRT*) (23), or tissue/cell line-specific genes such as T cell receptor, c-*myc*, or T cell surface molecule CD_8, are present at 0.005% or less and serve as good markers to observe the normalization of *very rare messages*.

Identification of Optimal Normalization Conditions

Autoradiograms from Southern blots analyzed with various probes are compared to identify the ssDNA fraction showing almost identical intensities with all the probes at a specific C_0t value. cDNA material from such an ssDNA fraction is used to construct normalized cDNA libraries.

Construction of Normalized cDNA Libraries

Amplified cDNA from the best normalized ssDNA fraction (see above) is digested with *Eco*RI using 40 U/μg of DNA, for 4 hr at 37°C. The 500- to 2000-bp fraction is isolated from the low melting gel as described earlier (4). 40 ng of this DNA is cloned into λgt10 (Stratagene) as per the protocol of the vendor. Approximately

10^6–10^7 clones are obtained per microgram of the vector DNA. Amplification of 300,000–400,000 pfu is adequate to construct the normalized library.

Evaluation of cDNA Libraries

Although the cDNA libraries are constructed from the normalized cDNA material, quantitation of the extent of normalization of the libraries is useful. Plaques (100,000) are plated at a density of 20,000/plate and filters are lifted from each of the plates as described (18). These filters are probed with the same set of probes used in the Southern blots. An equal density of phage from the corresponding unnormalized library is also plated and the abundance of the probes is estimated by hybridization. The relative abundance of specific inserts in the libraries is compared to evaluate the extent of normalization.

Results

Hydroxyapatite Chromatography

Typically the ssDNA is eluted from the HAP column in the sodium phosphate buffer range 0.1–0.15 *M*. The dsDNA fraction is eluted at 0.3–0.4 *M* buffer concentration. However, dsDNA smaller than 250 bp coelutes with the ssDNA fractions. In order not to contaminate ssDNA fractions with low molecular weight dsDNA, cDNAs of a minimum of 400 bp length are used in reassociation experiments.

Reassociation Kinetics

High-abundance sequences, such as ribosomal DNA, are removed from the ssDNA fraction after a reassociation time as low as C_0t = 19 mol·sec/liter. On the other hand, very rare messages such as c-*myc* are partitioned into single-stranded as well as dsDNA fractions at the same C_0t value. Ideally very low-abundance messages should remain single stranded at short reassociation times, whereas high-abundance messages should have reassociated extensively. Empirically, however, we have observed that up to 30% of the scarce messages have apparently reassociated even after the short C_0t values mentioned above.

Evaluation of Normalization

In our previous studies (4), we observed a dramatic reduction (2500-fold) in the abundance of ribosomal DNA clones on normalization of a human adult thymus library. However, *Alu* repetitive DNA assessed by Blur 8 probe (4) decreased only by about fourfold. Similar results were also reported by Ko (5) in equalized libraries. The short length of the repeat sequences and the considerable divergence between the copies of the repeat (1) may be responsible for such empirical observations. The difference between the high-abundance (ribosomal DNA, 30%) and very low-

abundance (α-fodrin, 0.003%) cDNA messa
(4) is about 10,000-fold. The difference in t
normalization is only 2-fold, resulting in an
between the most abundant and the least abu
More than 500 random clones were sequen
Carrano, Lawrence Livermore National Labo
library. About 14% of the clones showed at
LINE, or other repetitive elements. About 7% o
homology to DNA or protein sequences in the
show no significant homology to the database se

cDNA

Prep

Discussion

Generation of uniform-abundance cDNA message
ization and cDNA selection methodologies. We h
in combination with PCR to facilitate the recov
stranded cDNA fractions. Short-fragment cDNA lib
the generation of normalized libraries because (1)
of the mRNA message, and (2) the library consis
400–2000 bp, which does not generate a significa
small fragments. However, the random oligonucleoti
tion/FPLC (fast protein liquid chromatography) puri
(Pharmacia) may not contain uniform representation o
a random hexanucleotide. A lack of all combinations
mRNA may result in the loss of information about som
Possibly, synthetic mixes or random hexamers would b
Although PCR is useful in amplifying small amounts
the amplification process itself may cause point mutatio
and generate a bias against the GC-rich fragments. Use
significantly reducing the amplifications of such hybrids.
Prolonged normalization may deplete high- and mediu
mixture of normalized DNA with different C_0t values of
generate a combined normalized library, it would adequ
rapid removal of medium-abundance messages. Setting u
reactions and analyzing the ss- and dsDNA fractions usi
considerable help in generating an ideal normalized library.
Prolonged incubation of cDNA at 65°C results in therma
detergents were shown to enhance the reassociation rates of
cies by about 10^4-fold (24). The use of such detergents may r
time considerably, thereby avoiding thermal degradation of t
loss in the efficiency of normalization.
The average frequency at which different cDNAs occur in

electrophoresis, *Hin*dIII digestion, and electroelution. It may be a good practice, in general, to make two preparations of YAC DNAs, each digested with a different enzyme, and then to pool them after electroelution and phenol extraction for cDNA selection.

Electroelution

About 10-cm-long (1.6-cm-diameter) dialysis tubing is prepared as described (18) and filled with 7–8 ml of 0.5× TBE and DNA containing agarose gel pieces. Electroelution is carried out at least overnight in 0.5× TBE. The polarity is reversed for 1 min before removal of samples.

Concentration

The electroeluted sample is concentrated in a Centricon (Amicon) to ~250 μl. The cDNA is ethanol precipitated after phenol–chloroform extraction in the presence of 0.3 *M* sodium acetate and 2 μg glycogen as carrier. The YAC DNA is dissolved in 10 μl TE and the concentration determined by ethidium bromide spot detection (18). The YAC DNA is adjusted to a concentration of 40–60 ng/μl.

Immobilization of Yeast Artificial Chromosome DNA onto Nylon Disks

Hybond-N nylon (Amersham) is cut into 2.5 × 2.5-mm squares. It is important that the batch of Hybond-N be tested and that it not give high background. The disks are marked gently with a pencil by putting small dots at the corners on the side that is not used for spotting DNA. In an Eppendorf tube, 2.5 μl *Hin*dIII-digested YAC DNA (40–60 ng/μl) is mixed with 2.5 μl *Hae*III-digested bacteriophage φX174 DNA (200 ng/μl in water). The DNA is denatured at 95–98°C for 3 min, chilled on ice immediately, and then briefly spun. An equal volume (5 μl) of 20× SSC, pH 7 (18), is added and mixed. The DNA is spotted in 0.5-μl aliquots onto a nylon disk on the side not marked with pencil marks and dried under a lamp; then additional 0.5-μl aliquots are loaded to immobilize 10–15 ng/disk. These conditions are for YACs of the size ~300–400 kb. When dealing with larger YACs, for example YACs of the size 500–1000 kb, it may be better to immobilize more DNA, for example 20 to 30 ng, on nylon disks of the same size as described here. The disks are placed on a small sheet of Whatman (Clifton, NJ) paper and transferred onto several sheets of Whatman 3MM soaked with 0.5 *M* NaOH, 1.5 *M* NaCl. The sheet of paper is transferred after 5 min to a dry paper to remove excess alkali and then placed on sheets of Whatman paper soaked with 0.5 *M* Tris (pH 7.5), 1.5 *M* NaCl (neutralizing solution) for 1 min; then it is transferred onto fresh Whatman paper soaked with fresh neutralizing solution for another 2–3 min, and finally to 10× SSC for 1 min. The disks are transferred onto a dry Whatman filter and ultraviolet (UV) cross-linked in a Stratalinker (Stratagene) at the autocross-link mode for 1 min while they are damp. The disks are then transferred into 0.5-ml Eppendorf tubes and the tubes are baked at 80°C with lids open in a vacuum oven for 0.5 hr.

The disks are stored dry in Eppendorf tubes with lids closed at 4°C until use. Baking alone is also successful if done for 2 hr at 80°C.

Quenching Materials

P15 Library

The P15 library is a human chromosome 15 genomic library [57740; American Type Culture Collection (ATCC), Rockville, MD] subcloned at the *Eco*RI site of the pTZ18 vector (Pharmacia). The size range is 0.8–6 kb (11). This may not be ideal for YACs from chromosome 15. In this case another chromosome library may be used.

XRLI

The XRLI library is a genomic repetitive sequence library prepared from the human X chromosome library (LAOXNLO1, ATCC 57750) by pooling 500 plaques that were selected because they hybridized to a probe prepared from total genomic DNA (11). This may be replaced by C_0t_1 DNA (Bethesda Research Laboratories), which is supplied as sonicated DNA.

pR7.3 and pR5.8

The pR7.3 and pR5.8 clones consist of human ribosomal RNA 45S precursor region *Eco*RI fragments (7.3 and 5.8 kb) cloned in pBR322 (11).

pK3535

The pK3535 clone is an ~7-kb *E. coli* ribosomal RNA operon *Bam*HI fragment in pBR322 (26). This is used if cDNA inserts for hybridization were generated by PCR from a crude phage DNA preparation rather than from a purified phage preparation.

Poly[(dI)·(dC)]

A Pharmacia product (Cat. No. 27-7875), poly[(dI)·(dC)] is dissolved in water and used as such without any further treatment.

Yeast DNA

Yeast DNA is prepared from the YAC host yeast strain AB 1380 by standard methods (18). Any other related yeast DNA should serve the same function, that is, to quench any coselection of cDNAs by contaminating yeast DNA in a YAC preparation. Addition of yeast ribosomal RNA may be necessary for large YACS (1–2 Mb) comigrating with large yeast chromosomes, known to harbor ~140 copies of ribosomal RNA.

Preparation of Quenching Materials

Plasmid DNA samples are prepared by the alkali–SDS–lithium chloride method as described (18). The plasmid DNA preparation is treated with RNase A and RNase T_1 and precipitated with ethanol and ammonium acetate. All the plasmid quenching agents and yeast DNA samples are digested with *Eco*RI and then sonicated to generate DNA of ~0.2–0.8 kb in size. Instead of sonication, one can use restriction enzymes with 4-base recognition sites such as *Sau*3A, *Rsa*I, and *Hin*fI. For this, a particular DNA sample can be digested in individual aliquots with different enzymes (single digests) and then pooled after the addition of EDTA for phenol extraction and ethanol precipitation. All the DNA samples are treated with mung bean nuclease (1 unit/μg DNA) for 30 min at room temperature and then extracted with phenol, phenol–chloroform, and chloroform, and ethanol precipitated. The DNAs are dissolved in a small quantity of TE for storage. After estimation of concentration, 20× stocks of each are made in water and stored at −20°C. The 20× stock concentration of each of the quenching reagents is as follows: p15, 1 μg/μl; XRLI or C_0t_1 DNA, 0.5 μg/μl; pR7.3 and pR5.8 (1:0.8 ratio) mixture, 0.8 μg/μl; pK3535, 0.5 μg/μl; yeast DNA, 0.5 μg/μl; poly[(dI)·(dC)], 0.4 μg/μl.

First Cycle of Selection

Prehybridization/Quenching

Stock	Volume/tube	Final concentration
Quenching reagents (20×)	2.5 μl of each (2.5 × 6 = 15 μl)	1×
Water	15 μl	—
The tube is placed in boiling water for 5 min and then chilled in ice. The following components are then added sequentially:		
20× SSPE (18)	12.5 μl	5× SSPE
50× Denhardt's (18)	5 μl	5× Denhardt's
The tube is placed in a 65°C water bath briefly just to warm the solution and then SDS is added:		
10% SDS	2.5 μl	0.5% SDS
	50 μl	

The above volumes are given for one tube, and the prehybridization mix is prepared in bulk depending on the number of tubes. Then 50 μl is quickly aliquoted into the tubes containing nylon disks with immobilized target DNA. The tubes are covered with mineral oil and incubated at 65°C for 24 hr with gentle rocking. The oil phase is removed and the disks are washed twice briefly with 5× SSPE, 0.1% SDS at room temperature, and then kept in this solution at room temperature until ready for hybridization.

Hybridization

Stock	Volume/tube	Final concentration
Quenching reagents (20×) except poly [(dI) · (dC)]	2 μl each (total, 10 μl)	1×
Amplified cDNA (with primer set C)	x μl	10μg/ml
Sterile water	y μl	
$x + y$	14 μl	
The tubes are put in boiling water for 5 min, chilled quickly, and briefly spun; then to each tube individually the following components are added:		
20× SSPE	10 μl	5× SSPE
50× Denhardt's	4 μl	5× Denhardt's
The tubes are warmed briefly at 65°C and then SDS is added:		
10% SDS	2 μl	0.5% SDS
	40 μl	

Quickly, 40 μl is aliquoted into fresh 0.5-ml tubes and the nylon disks are transferred individually from 5× SSPE solution with the help of Pipetman tips. Care is taken not to drag the DNA-immobilized side of the disk along the walls of the tube so that oil does not come in contact with the DNA. Enough hybridization mix is prepared for four tubes at a time. The tubes are covered with mineral oil and incubated at 65°C for 36–40 hr with gentle rocking.

Washing Disks

After removing the oil the disks are washed with 600 μl of wash solution, each time in 1.5-ml Eppendorf tubes. Using vacuum suction, most of the wash solution is removed, leaving behind enough to cover the disks. The last traces are removed with a Pipetman tip and care is taken not to allow the disks to dry at any stage. The wash solution is kept at 65°C between the washes. To maintain 65°C inside the Eppendorf tubes, the shaker water bath temperature is kept at 66–67°C. The wash schedule is as follows:

1. 2× SSC/0.1% SDS	Quick wash at room temperature	Three times
2. 2× SSC/0.1% SDS	At 65°C for 20 min each	Three times

The disks are transferred to new tubes and the wash continued:

3. 1× SSC/0.1% SDS	At 65°C for 20 min	Once
4. 0.2× SSC/0.1% SDS	At 65°C for 15–20 min	Once

The disks are transferred to new tubes and the wash continued:

5. 0.1× SSC/0.1% SDS	At 65°C for 20 min	Twice
6. 0.1× SSC alone	At room temperature	Twice

The tubes are transferred to new 0.5-ml PCR tubes containing 40 μl water while removing as much SSC as possible with a Pipetman tip. The tubes can be stored at 4°C overnight. The tubes with the disks are put in the PCR machine at 95°C for 5 min. Then they are chilled and, after a quick spin, 20 μl is taken for PCR (see the next section). The other half can be stored at −20°C as a backup sample.

First Round of Polymerase Chain Reaction

Hot-start PCR is carried out with 20 μl of eluate from above (see the previous section) in a 50-μl volume as described earlier using primer set C. The PCR is carried out for 30 cycles with the same cycling conditions as described in the earlier section. The concentration of primers is 0.3 μM in the final reaction mixture.

Second Round of Polymerase Chain Reaction

From the first round of PCR, 3 μl is taken out for hot-start PCR in a 100-μl volume with a 0.6 μM final concentration of primer set C as described earlier. An aliquot of 5 μl of the final product is analyzed on a 1% agarose gel. One should see a smear of an average size of ~0.3 kb. The remainder of the PCR product is extracted with a phenol–chloroform mixture and ethanol precipitated after the addition of EDTA to a final concentration of 5 mM, $\frac{1}{10}$ vol of 3 M sodium acetate, and 2 μg glycogen.

Size Selection of cDNA

Size fractionation of once-selected cDNA (the ethanol-precipitated material from the 100-μl PCR reaction described in the previous section) is carried out in 1% low melting point agarose. The agarose piece containing DNA greater than 0.35 kb is cut out and processed for DNA extraction and ethanol precipitation as before. If the volume of the solution before ethanol precipitation is ~600–700 μl then one can concentrate it to ~200–300 μl by Speed-Vac (Savant, Hicksville, NY); in the case of larger volumes, one can use a Centricon 30 or Centricon 100 (Amicon) before ethanol precipitation. After ethanol precipitation the DNA is dissolved in 50 μl TE and 20 μl is taken for spectrophotometric estimation, as above.

Second Cycle of Selection

Prehybridization/Quenching

The same steps are carried out as before using YAC DNA immobilized on a new nylon disk, and the same quenching materials and conditions are used as described in the first cycle of selection.

Hybridization

The same steps are carried out as before except that a lower concentration of cDNA is used than in the first cycle of the selection. The concentration of once-selected cDNA (see Size Selection of cDNA, above) in the final 40 μl of hybridization mixture is 0.25 μg/ml if only a single cDNA library was used in the first cycle of selection. This concentration can be increased to 0.5 μg/ml if a mixture of cDNA libraries was used in the first cycle of selection.

Washing Disks

Disks are washed in the same way as those in the first cycle.

First Round of Polymerase Chain Reaction Amplification

This round of PCR amplification is identical to that described in the first cycle except that primer set A is used (0.3 μ*M* in a 50-μl final volume).

Second Round of Polymerase Chain Reaction Amplification

Same as the second round of PCR of the first cycle of selection except that primer set A is used (0.6 μ*M* in a 100-μl volume). An aliquot of 5 μl is analyzed on an analytical agarose gel. To the rest of the PCR material, EDTA is added to a final concentration of 5 m*M*. The cDNA is extracted with phenol, phenol–chloroform, and chloroform alone, and then ethanol precipitated after the addition of ½ vol of 7.5 *M* ammonium acetate and 2 μg glycogen. The DNA is dissolved in 10 μl of TE.

Cloning and Analysis of Selected Library

Eco*RI Digestion and Size Selection of Selected cDNA*

The entire 10 μl of cDNA after the second round of amplification above is digested with 30 units of *Eco*RI in a 50-μl volume. The cDNA is extracted with phenol and chloroform and concentrated by ethanol precipitation after the addition of EDTA to 10 m*M* and 2 μg glycogen. The cDNA is size fractionated on 1% agarose and the agarose with DNA >0.3 kb is cut out. The DNA is extracted and ethanol precipitated in the presence of ammonium acetate and glycogen as described earlier. After dissolving DNA in 10 μl of 10 m*M* Tris-HCl (pH 7.5), 0.1 m*M* EDTA (TLE), a concentration of 2–4 μl is estimated in 0.5 ml TE at 260 nm. If the absorbance is too low, then the cDNA concentration can be determined by ethidium bromide spot detection under UV as earlier.

*Ligation of λgt10 Arms (*Eco*RI Cut and Dephosphorylated)*

*Eco*RI-cut and twice-selected cDNA is diluted in TLE to a concentration of 2.5 ng/μl. A ligation reaction with 200 units of T4 DNA ligase (New England BioLabs) is carried out in 5 μl with 0.5 μg of *Eco*RI-cut and dephosphorylated λgt10 arms (Promega or Stratagene) and 5 ng of twice-selected cDNA (see the previous section) in the presence of 50 m*M* Tris-HCl (pH 7.8), 10 m*M* $MgCl_2$, 1 m*M* ATP, and 5 m*M* DTT.

Packaging

The Gigapack Plus packaging kit (Stratagene) is used as per the instructions of the manufacturer. Out of the ligation mixes (see the previous section), 4 μl is used for *in vitro* packaging. The packaged DNA should be used within 1 week for amplifying the library (see below), as it is not very stable. The amplified library is stable at 4°C.

Infection into c600 hfl Cells and Titration

Escherichia coli c600 hfl cells are grown in NZYM medium (18) plus 0.2% maltose at 37°C until the OD_{600} is ~0.9–1.0. The cells are centrifuged and resuspended in a half volume of 10 m*M* magnesium sulfate. An aliquot of 5 μl (out of 500 μl of the packaging mix) is diluated to 50 μl with SM; two samples of 2 and 20 μl are titrated from this diluted sample for the plaque count. The total number of plaques in the original 500-μl packaging mix should be in the range of 100,000–3 million. If the number of plaques is too small, the cDNA should be purified by reprecipitation with ethanol and ammonium acetate as earlier.

Amplification of Selected Library

One confluent plate of plaques is made and phage lysate is harvested with 15 ml SM (18). This represents a high-titer selected library (titer, $\sim 10^{10}$).

Analysis of Selected cDNA Library

The selected and amplified library (above) is plated at low density to pick up random individual plaques. About 10–20 plaques are randomly picked and resuspended in 100 μl of SM. A drop of chloroform is added and the phage suspension is mixed, briefly spun (10 sec), and left at 4°C for at least 6–8 hr. The percentage of recombinants and the size distribution of their inserts are analyzed after PCR amplification of individual plaques. For this purpose it is not necessary to perform hot-start PCR; all the components of PCR, including dNTPs and enzyme, can be mixed and the tubes should be put into the machine when the temperature is at least 70°C. One can use either primer set A or C and 1 μl out of 100 μl phage suspension for a 50-μl PCR reaction. About 5 μl of the PCR product can be run on an agarose gel with a 1-kb size marker (Bethesda Research Laboratories). One should see distinct single bands ranging from 300 to 600 bp or more in size.

Screening Selected cDNA Library for New (Unknown) Clones

About 50 plaques for every 100 kb of YAC are picked at random, resuspended in 100 μl of SM, and amplified by PCR as described earlier. For example, one picks 150 plaques for a 300-kb YAC and amplifies all of them by PCR, using 1 μl in a 50-μl reaction and primer set C as described above. About 1 μl of the PCR product is dot-blotted from each sample in an array of 96 dots and multiple blots are made depending on the number of the probes to be used. One can, in general, probe the blots with ribosomal RNA probes, total genomic DNA for repeats, and any other probes known from that YAC. At this stage one can make pools of the PCR products from rows and columns of the 96-well grid plate and use them as probes to identify the overlapping clones. Both PCR amplification and dot blotting at this stage can be automated using a Biomek (Beckman, Fullerton, CA). After eliminating undesirable and/or overlapping clones, one can now sequence at least 150–250 bp of the PCR products rapidly by cycling PCR sequence (Bethesda Research Laboratories PCR-based cycling sequence kit). For sequencing individual PCR products, 2–3 μl is taken out of 50 μl of the PCR-amplified product (amplified with primer set C) and subjected to PCR-based cycling sequencing, without any intermediate purification, using one of the inner primers of set A as ^{32}P end-labeled sequencing oligonucleotide. These sequences can be analyzed for any sequence matches in the GENEMBL databank. Alternatively, the blotting is omitted and one can sequence directly and identify overlapping clones and some nonspecific ones such as ribosomal or repeat sequences; but this involves more sequencing than when blotting is performed. Whichever method one elects, it should be possible to identify unique sequences that are encoded in a YAC. Any unique clones from a selected library can be confirmed by Southern blots using total genomic DNA, YAC DNA, and DNA from a somatic hybrid cell line that is known to harbor a chromosome region corresponding to the YAC of interest. A full-length cDNA library can then be screened using clones derived from the selected library as probes.

Choice of Probes

Ideally at least 10–12 different probes should be used to evaluate the extent of normalization. A detailed description of such probes is given elsewhere (4). A fair evaluation of the extent of normalization can be gained by use of the following probes.

Repetitive DNA Probes

Cloned *Alu* and LINE element repeats serve as excellent markers to monitor the depletion of repetitive DNA.

Ribosomal DNA Probes

Although poly$(A)^+$ RNA is twice selected over the oligo(dT)-cellulose column, a significant number (5–10%) of parent cDNA clones represent ribosomal DNA clones. Monitoring their abundance indicates the fate of *high-abundance* messages during normalization.

Actin Gene Probes/HLA Genes

Actin gene probes/HLA genes represent medium-abundance genes in cDNA libraries (0.1–0.15%). Because these probes recognize several members of the gene family, normalization patterns represent the behavior of *medium-abundance* genes.

Rare Probes

Housekeeping genes such as hypoxanthine phosphoribosyltransferase (*HPRT*) (23), or tissue/cell line-specific genes such as T cell receptor, c-*myc*, or T cell surface molecule CD_8, are present at 0.005% or less and serve as good markers to observe the normalization of *very rare messages*.

Identification of Optimal Normalization Conditions

Autoradiograms from Southern blots analyzed with various probes are compared to identify the ssDNA fraction showing almost identical intensities with all the probes at a specific C_0t value. cDNA material from such an ssDNA fraction is used to construct normalized cDNA libraries.

Construction of Normalized cDNA Libraries

Amplified cDNA from the best normalized ssDNA fraction (see above) is digested with *Eco*RI using 40 U/μg of DNA, for 4 hr at 37°C. The 500- to 2000-bp fraction is isolated from the low melting gel as described earlier (4). 40 ng of this DNA is cloned into λgt10 (Stratagene) as per the protocol of the vendor. Approximately

10^6–10^7 clones are obtained per microgram of the vector DNA. Amplification of 300,000–400,000 pfu is adequate to construct the normalized library.

Evaluation of cDNA Libraries

Although the cDNA libraries are constructed from the normalized cDNA material, quantitation of the extent of normalization of the libraries is useful. Plaques (100,000) are plated at a density of 20,000/plate and filters are lifted from each of the plates as described (18). These filters are probed with the same set of probes used in the Southern blots. An equal density of phage from the corresponding un-normalized library is also plated and the abundance of the probes is estimated by hybridization. The relative abundance of specific inserts in the libraries is compared to evaluate the extent of normalization.

Results

Hydroxyapatite Chromatography

Typically the ssDNA is eluted from the HAP column in the sodium phosphate buffer range 0.1–0.15 *M*. The dsDNA fraction is eluted at 0.3–0.4 *M* buffer concentration. However, dsDNA smaller than 250 bp coelutes with the ssDNA fractions. In order not to contaminate ssDNA fractions with low molecular weight dsDNA, cDNAs of a minimum of 400 bp length are used in reassociation experiments.

Reassociation Kinetics

High-abundance sequences, such as ribosomal DNA, are removed from the ssDNA fraction after a reassociation time as low as C_0t = 19 mol·sec/liter. On the other hand, very rare messages such as c-*myc* are partitioned into single-stranded as well as dsDNA fractions at the same C_0t value. Ideally very low-abundance messages should remain single stranded at short reassociation times, whereas high-abundance messages should have reassociated extensively. Empirically, however, we have observed that up to 30% of the scarce messages have apparently reassociated even after the short C_0t values mentioned above.

Evaluation of Normalization

In our previous studies (4), we observed a dramatic reduction (2500-fold) in the abundance of ribosomal DNA clones on normalization of a human adult thymus library. However, *Alu* repetitive DNA assessed by Blur 8 probe (4) decreased only by about fourfold. Similar results were also reported by Ko (5) in equalized libraries. The short length of the repeat sequences and the considerable divergence between the copies of the repeat (1) may be responsible for such empirical observations. The difference between the high-abundance (ribosomal DNA, 30%) and very low-abundance (α-fodrin, 0.003%) cDNA messages in an unnormalized thymus library (4) is about 10,000-fold. The difference in the levels of various messages tested on normalization is only 2-fold, resulting in an ~5000-fold reduction in the variation between the most abundant and the least abundant cDNAs tested by us.

More than 500 random clones were sequenced (personal communication, Dr. A. Carrano, Lawrence Livermore National Laboratory) from the normalized thymus library. About 14% of the clones showed at least 60% homology to known *Alu*, LINE, or other repetitive elements. About 7% of the sequences showed at least 60% homology to DNA or protein sequences in the databases. The remaining sequences show no significant homology to the database sequences.

Discussion

Generation of uniform-abundance cDNA messages is useful in subtractive hybridization and cDNA selection methodologies. We have employed a kinetic approach in combination with PCR to facilitate the recovery of the single- and double-stranded cDNA fractions. Short-fragment cDNA libraries are particularly helpful in the generation of normalized libraries because (1) they represent the entire length of the mRNA message, and (2) the library consists of cDNAs in the size range 400–2000 bp, which does not generate a significant bias in the amplification of small fragments. However, the random oligonucleotide primer obtained by sonication/FPLC (fast protein liquid chromatography) purification of calf thymus DNA (Pharmacia) may not contain uniform representation of all possible combinations of a random hexanucleotide. A lack of all combinations of oligonucleotides to prime mRNA may result in the loss of information about some parts of mRNA sequences. Possibly, synthetic mixes or random hexamers would be preferable.

Although PCR is useful in amplifying small amounts of ss- and dsDNA fractions, the amplification process itself may cause point mutations in the amplified material and generate a bias against the GC-rich fragments. Use of hot-start PCR helps in significantly reducing the amplifications of such hybrids.

Prolonged normalization may deplete high- and medium-abundance cDNAs. If a mixture of normalized DNA with different C_0t values of reassociation is pooled to generate a combined normalized library, it would adequately compensate for the rapid removal of medium-abundance messages. Setting up multiple reassociation reactions and analyzing the ss- and dsDNA fractions using various probes is of considerable help in generating an ideal normalized library.

Prolonged incubation of cDNA at 65°C results in thermal degradation. Cationic detergents were shown to enhance the reassociation rates of homologous DNA species by about 10^4-fold (24). The use of such detergents may reduce the reassociation time considerably, thereby avoiding thermal degradation of the cDNA, without any loss in the efficiency of normalization.

The average frequency at which different cDNAs occur in a normalized thymus cDNA library is approximately 1 in 10,000 (4). Based on the probability equation (18), $N = \ln(1 - P)/\ln[1 - (1/n)]$, where N is the number of clones required to be screened, P is the probability desired (0.99), and n is the fractional abundance of the cDNA species, it is estimated that even rare messages may be identified in a normalized cDNA library by screening 45,000 to 60,000 clones. Such enriched libraries can be effectively used in subtraction and differential hybridization. Efforts are underway to use such libraries to understand protein–protein interactions by the contingent replication assay (25) and to select YAC-specific cDNA sequences (14).

cDNA Selection

Preparation of Reagents

Preparation of cDNA Library

A random hexamer-primed short-fragment cDNA library is prepared and cloned into λgt10 as described under Normalization (above). The size range of the cDNA is 0.4–1.5 kb. Usually a library containing at least 3 million independent clones is prepared. The cDNA inserts in the size range 400–2000 bp are amplified from the library by hot-start PCR as described above, using set C primers (outermost primers) (Fig. 1).

Preparation of Yeast Artificial Chromosome DNA

Electrophoretic Separation of Yeast Artificial Chromosomes

Yeast chromosomes are separated in 1% (w/v) SeaKem GTG agarose (FMC, Rockland, ME) in half-strength Tris–borate buffer [0.5× TBE (18)] run on a Bio-Rad (Richmond, CA) CHEF electrophoresis apparatus. If a YAC comigrates with a native yeast chromosome and is indistinguishable on the gel, it is identified by hybridization of Southern blot (from a single lane of the gel) to pBR322 or a related plasmid probe. During blotting, the remainder of the gel is wrapped in Saran wrap and stored at 4°C. After identification of the YAC, the band of interest is cut out and the agarose piece containing the YAC is stored in a capped 15-ml Falcon tube (Becton Dickinson, Oxnard, CA) at 4°C until further use.

Digestion of Yeast Artificial Chromosome DNA in Agarose Blocks with *Hin*dIII

Agarose pieces containing YACs are soaked in TE for 1 hr and then equilibriated with the *Hin*dIII enzyme buffer for 2 hr at 4°C with a change of buffer. An 8-cm-long agarose slice containing the YAC is immersed in 3–4 ml of *Hin*dIII buffer with 0.25–0.5 U of enzyme per microliter and incubated at 37°C for at least 6–8 hr. EDTA, pH 8, is added to a final concentration of 20 m*M* to stop the reaction. After 30 min it is replaced with 0.5× TBE and incubated another 30 min. *Hin*dIII may not be an ideal enzyme if the yield of YAC DNA is too low after pulsed-field

electrophoresis, *Hin*dIII digestion, and electroelution. It may be a good practice, in general, to make two preparations of YAC DNAs, each digested with a different enzyme, and then to pool them after electroelution and phenol extraction for cDNA selection.

Electroelution

About 10-cm-long (1.6-cm-diameter) dialysis tubing is prepared as described (18) and filled with 7–8 ml of 0.5× TBE and DNA containing agarose gel pieces. Electroelution is carried out at least overnight in 0.5× TBE. The polarity is reversed for 1 min before removal of samples.

Concentration

The electroeluted sample is concentrated in a Centricon (Amicon) to ~250 μl. The cDNA is ethanol precipitated after phenol–chloroform extraction in the presence of 0.3 *M* sodium acetate and 2 μg glycogen as carrier. The YAC DNA is dissolved in 10 μl TE and the concentration determined by ethidium bromide spot detection (18). The YAC DNA is adjusted to a concentration of 40–60 ng/μl.

Immobilization of Yeast Artificial Chromosome DNA onto Nylon Disks

Hybond-N nylon (Amersham) is cut into 2.5 × 2.5-mm squares. It is important that the batch of Hybond-N be tested and that it not give high background. The disks are marked gently with a pencil by putting small dots at the corners on the side that is not used for spotting DNA. In an Eppendorf tube, 2.5 μl *Hin*dIII-digested YAC DNA (40–60 ng/μl) is mixed with 2.5 μl *Hae*III-digested bacteriophage φX174 DNA (200 ng/μl in water). The DNA is denatured at 95–98°C for 3 min, chilled on ice immediately, and then briefly spun. An equal volume (5 μl) of 20× SSC, pH 7 (18), is added and mixed. The DNA is spotted in 0.5-μl aliquots onto a nylon disk on the side not marked with pencil marks and dried under a lamp; then additional 0.5-μl aliquots are loaded to immobilize 10–15 ng/disk. These conditions are for YACs of the size ~300–400 kb. When dealing with larger YACs, for example YACs of the size 500–1000 kb, it may be better to immobilize more DNA, for example 20 to 30 ng, on nylon disks of the same size as described here. The disks are placed on a small sheet of Whatman (Clifton, NJ) paper and transferred onto several sheets of Whatman 3MM soaked with 0.5 *M* NaOH, 1.5 *M* NaCl. The sheet of paper is transferred after 5 min to a dry paper to remove excess alkali and then placed on sheets of Whatman paper soaked with 0.5 *M* Tris (pH 7.5), 1.5 *M* NaCl (neutralizing solution) for 1 min; then it is transferred onto fresh Whatman paper soaked with fresh neutralizing solution for another 2–3 min, and finally to 10× SSC for 1 min. The disks are transferred onto a dry Whatman filter and ultraviolet (UV) cross-linked in a Stratalinker (Stratagene) at the autocross-link mode for 1 min while they are damp. The disks are then transferred into 0.5-ml Eppendorf tubes and the tubes are baked at 80°C with lids open in a vacuum oven for 0.5 hr.

The disks are stored dry in Eppendorf tubes with lids closed at 4°C until use. Baking alone is also successful if done for 2 hr at 80°C.

Quenching Materials

P15 Library

The P15 library is a human chromosome 15 genomic library [57740; American Type Culture Collection (ATCC), Rockville, MD] subcloned at the *Eco*RI site of the pTZ18 vector (Pharmacia). The size range is 0.8–6 kb (11). This may not be ideal for YACs from chromosome 15. In this case another chromosome library may be used.

XRLI

The XRLI library is a genomic repetitive sequence library prepared from the human X chromosome library (LAOXNLO1, ATCC 57750) by pooling 500 plaques that were selected because they hybridized to a probe prepared from total genomic DNA (11). This may be replaced by C_0t_1 DNA (Bethesda Research Laboratories), which is supplied as sonicated DNA.

pR7.3 and pR5.8

The pR7.3 and pR5.8 clones consist of human ribosomal RNA 45S precursor region *Eco*RI fragments (7.3 and 5.8 kb) cloned in pBR322 (11).

pK3535

The pK3535 clone is an ~7-kb *E. coli* ribosomal RNA operon *Bam*HI fragment in pBR322 (26). This is used if cDNA inserts for hybridization were generated by PCR from a crude phage DNA preparation rather than from a purified phage preparation.

Poly[(dI)·(dC)]

A Pharmacia product (Cat. No. 27-7875), poly[(dI)·(dC)] is dissolved in water and used as such without any further treatment.

Yeast DNA

Yeast DNA is prepared from the YAC host yeast strain AB 1380 by standard methods (18). Any other related yeast DNA should serve the same function, that is, to quench any coselection of cDNAs by contaminating yeast DNA in a YAC preparation. Addition of yeast ribosomal RNA may be necessary for large YACS (1–2 Mb) comigrating with large yeast chromosomes, known to harbor ~140 copies of ribosomal RNA.

Preparation of Quenching Materials

Plasmid DNA samples are prepared by the alkali–SDS–lithium chloride method as described (18). The plasmid DNA preparation is treated with RNase A and RNase T_1 and precipitated with ethanol and ammonium acetate. All the plasmid quenching agents and yeast DNA samples are digested with *Eco*RI and then sonicated to generate DNA of ~0.2–0.8 kb in size. Instead of sonication, one can use restriction enzymes with 4-base recognition sites such as *Sau*3A, *Rsa*I, and *Hin*fI. For this, a particular DNA sample can be digested in individual aliquots with different enzymes (single digests) and then pooled after the addition of EDTA for phenol extraction and ethanol precipitation. All the DNA samples are treated with mung bean nuclease (1 unit/μg DNA) for 30 min at room temperature and then extracted with phenol, phenol–chloroform, and chloroform, and ethanol precipitated. The DNAs are dissolved in a small quantity of TE for storage. After estimation of concentration, 20× stocks of each are made in water and stored at −20°C. The 20× stock concentration of each of the quenching reagents is as follows: p15, 1 μg/μl; XRLI or C_0t_1 DNA, 0.5 μg/μl; pR7.3 and pR5.8 (1:0.8 ratio) mixture, 0.8 μg/μl; pK3535, 0.5 μg/μl; yeast DNA, 0.5 μg/μl; poly[(dI)·(dC)], 0.4 μg/μl.

First Cycle of Selection

Prehybridization/Quenching

Stock	Volume/tube	Final concentration
Quenching reagents (20×)	2.5 μl of each (2.5 × 6 = 15 μl)	1×
Water	15 μl	—
The tube is placed in boiling water for 5 min and then chilled in ice. The following components are then added sequentially:		
20× SSPE (18)	12.5 μl	5× SSPE
50× Denhardt's (18)	5 μl	5× Denhardt's
The tube is placed in a 65°C water bath briefly just to warm the solution and then SDS is added:		
10% SDS	2.5 μl	0.5% SDS
	50 μl	

The above volumes are given for one tube, and the prehybridization mix is prepared in bulk depending on the number of tubes. Then 50 μl is quickly aliquoted into the tubes containing nylon disks with immobilized target DNA. The tubes are covered with mineral oil and incubated at 65°C for 24 hr with gentle rocking. The oil phase is removed and the disks are washed twice briefly with 5× SSPE, 0.1% SDS at room temperature, and then kept in this solution at room temperature until ready for hybridization.

Hybridization

Stock	Volume/tube	Final concentration
Quenching reagents (20×) except poly [(dI) · (dC)]	2 μl each (total, 10 μl)	1×
Amplified cDNA (with primer set C)	x μl	10μg/ml
Sterile water	y μl	
$x + y$	14 μl	
The tubes are put in boiling water for 5 min, chilled quickly, and briefly spun; then to each tube individually the following components are added:		
20× SSPE	10 μl	5× SSPE
50× Denhardt's	4 μl	5× Denhardt's
The tubes are warmed briefly at 65°C and then SDS is added:		
10% SDS	2 μl	0.5% SDS
	40 μl	

Quickly, 40 μl is aliquoted into fresh 0.5-ml tubes and the nylon disks are transferred individually from 5× SSPE solution with the help of Pipetman tips. Care is taken not to drag the DNA-immobilized side of the disk along the walls of the tube so that oil does not come in contact with the DNA. Enough hybridization mix is prepared for four tubes at a time. The tubes are covered with mineral oil and incubated at 65°C for 36–40 hr with gentle rocking.

Washing Disks

After removing the oil the disks are washed with 600 μl of wash solution, each time in 1.5-ml Eppendorf tubes. Using vacuum suction, most of the wash solution is removed, leaving behind enough to cover the disks. The last traces are removed with a Pipetman tip and care is taken not to allow the disks to dry at any stage. The

wash solution is kept at 65°C between the washes. To maintain 65°C inside the Eppendorf tubes, the shaker water bath temperature is kept at 66–67°C. The wash schedule is as follows:

1. 2× SSC/0.1% SDS	Quick wash at room temperature	Three times
2. 2× SSC/0.1% SDS	At 65°C for 20 min each	Three times

The disks are transferred to new tubes and the wash continued:

3. 1× SSC/0.1% SDS	At 65°C for 20 min	Once
4. 0.2× SSC/0.1% SDS	At 65°C for 15–20 min	Once

The disks are transferred to new tubes and the wash continued:

5. 0.1× SSC/0.1% SDS	At 65°C for 20 min	Twice
6. 0.1× SSC alone	At room temperature	Twice

The tubes are transferred to new 0.5-ml PCR tubes containing 40 μl water while removing as much SSC as possible with a Pipetman tip. The tubes can be stored at 4°C overnight. The tubes with the disks are put in the PCR machine at 95°C for 5 min. Then they are chilled and, after a quick spin, 20 μl is taken for PCR (see the next section). The other half can be stored at −20°C as a backup sample.

First Round of Polymerase Chain Reaction

Hot-start PCR is carried out with 20 μl of eluate from above (see the previous section) in a 50-μl volume as described earlier using primer set C. The PCR is carried out for 30 cycles with the same cycling conditions as described in the earlier section. The concentration of primers is 0.3 μM in the final reaction mixture.

Second Round of Polymerase Chain Reaction

From the first round of PCR, 3 μl is taken out for hot-start PCR in a 100-μl volume with a 0.6 μM final concentration of primer set C as described earlier. An aliquot of 5 μl of the final product is analyzed on a 1% agarose gel. One should see a smear of an average size of ~0.3 kb. The remainder of the PCR product is extracted with a phenol–chloroform mixture and ethanol precipitated after the addition of EDTA to a final concentration of 5 mM, $\frac{1}{10}$ vol of 3 M sodium acetate, and 2 μg glycogen.

Size Selection of cDNA

Size fractionation of once-selected cDNA (the ethanol-precipitated material from the 100-μl PCR reaction described in the previous section) is carried out in 1% low

melting point agarose. The agarose piece containing DNA greater than 0.35 kb is cut out and processed for DNA extraction and ethanol precipitation as before. If the volume of the solution before ethanol precipitation is ~600–700 μl then one can concentrate it to ~200–300 μl by Speed-Vac (Savant, Hicksville, NY); in the case of larger volumes, one can use a Centricon 30 or Centricon 100 (Amicon) before ethanol precipitation. After ethanol precipitation the DNA is dissolved in 50 μl TE and 20 μl is taken for spectrophotometric estimation, as above.

Second Cycle of Selection

Prehybridization/Quenching

The same steps are carried out as before using YAC DNA immobilized on a new nylon disk, and the same quenching materials and conditions are used as described in the first cycle of selection.

Hybridization

The same steps are carried out as before except that a lower concentration of cDNA is used than in the first cycle of the selection. The concentration of once-selected cDNA (see Size Selection of cDNA, above) in the final 40 μl of hybridization mixture is 0.25 μg/ml if only a single cDNA library was used in the first cycle of selection. This concentration can be increased to 0.5 μg/ml if a mixture of cDNA libraries was used in the first cycle of selection.

Washing Disks

Disks are washed in the same way as those in the first cycle.

First Round of Polymerase Chain Reaction Amplification

This round of PCR amplification is identical to that described in the first cycle except that primer set A is used (0.3 μ*M* in a 50-μl final volume).

Second Round of Polymerase Chain Reaction Amplification

Same as the second round of PCR of the first cycle of selection except that primer set A is used (0.6 μ*M* in a 100-μl volume). An aliquot of 5 μl is analyzed on an analytical agarose gel. To the rest of the PCR material, EDTA is added to a final concentration of 5 m*M*. The cDNA is extracted with phenol, phenol–chloroform, and chloroform alone, and then ethanol precipitated after the addition of ½ vol of 7.5 *M* ammonium acetate and 2 μg glycogen. The DNA is dissolved in 10 μl of TE.

Cloning and Analysis of Selected Library

Eco*RI Digestion and Size Selection of Selected cDNA*

The entire 10 μl of cDNA after the second round of amplification above is digested with 30 units of *Eco*RI in a 50-μl volume. The cDNA is extracted with phenol and chloroform and concentrated by ethanol precipitation after the addition of EDTA to 10 m*M* and 2 μg glycogen. The cDNA is size fractionated on 1% agarose and the agarose with DNA >0.3 kb is cut out. The DNA is extracted and ethanol precipitated in the presence of ammonium acetate and glycogen as described earlier. After dissolving DNA in 10 μl of 10 m*M* Tris-HCl (pH 7.5), 0.1 m*M* EDTA (TLE), a concentration of 2–4 μl is estimated in 0.5 ml TE at 260 nm. If the absorbance is too low, then the cDNA concentration can be determined by ethidium bromide spot detection under UV as earlier.

*Ligation of λgt10 Arms (*Eco*RI Cut and Dephosphorylated)*

*Eco*RI-cut and twice-selected cDNA is diluted in TLE to a concentration of 2.5 ng/μl. A ligation reaction with 200 units of T4 DNA ligase (New England BioLabs) is carried out in 5 μl with 0.5 μg of *Eco*RI-cut and dephosphorylated λgt10 arms (Promega or Stratagene) and 5 ng of twice-selected cDNA (see the previous section) in the presence of 50 m*M* Tris-HCl (pH 7.8), 10 m*M* $MgCl_2$, 1 m*M* ATP, and 5 m*M* DTT.

Packaging

The Gigapack Plus packaging kit (Stratagene) is used as per the instructions of the manufacturer. Out of the ligation mixes (see the previous section), 4 μl is used for *in vitro* packaging. The packaged DNA should be used within 1 week for amplifying the library (see below), as it is not very stable. The amplified library is stable at 4°C.

Infection into c600 hfl Cells and Titration

Escherichia coli c600 hfl cells are grown in NZYM medium (18) plus 0.2% maltose at 37°C until the OD_{600} is ~0.9–1.0. The cells are centrifuged and resuspended in a half volume of 10 m*M* magnesium sulfate. An aliquot of 5 μl (out of 500 μl of the packaging mix) is diluted to 50 μl with SM; two samples of 2 and 20 μl are titrated from this diluted sample for the plaque count. The total number of plaques in the original 500-μl packaging mix should be in the range of 100,000–3 million. If the number of plaques is too small, the cDNA should be purified by reprecipitation with ethanol and ammonium acetate as earlier.

Amplification of Selected Library

One confluent plate of plaques is made and phage lysate is harvested with 15 ml SM (18). This represents a high-titer selected library (titer, ~10^{10}).

Analysis of Selected cDNA Library

The selected and amplified library (above) is plated at low density to pick up random individual plaques. About 10–20 plaques are randomly picked and resuspended in 100 μl of SM. A drop of chloroform is added and the phage suspension is mixed, briefly spun (10 sec), and left at 4°C for at least 6–8 hr. The percentage of recombinants and the size distribution of their inserts are analyzed after PCR amplification of individual plaques. For this purpose it is not necessary to perform hot-start PCR; all the components of PCR, including dNTPs and enzyme, can be mixed and the tubes should be put into the machine when the temperature is at least 70°C. One can use either primer set A or C and 1 μl out of 100 μl phage suspension for a 50-μl PCR reaction. About 5 μl of the PCR product can be run on an agarose gel with a 1-kb size marker (Bethesda Research Laboratories). One should see distinct single bands ranging from 300 to 600 bp or more in size.

Screening Selected cDNA Library for New (Unknown) Clones

About 50 plaques for every 100 kb of YAC are picked at random, resuspended in 100 μl of SM, and amplified by PCR as described earlier. For example, one picks 150 plaques for a 300-kb YAC and amplifies all of them by PCR, using 1 μl in a 50-μl reaction and primer set C as described above. About 1 μl of the PCR product is dot-blotted from each sample in an array of 96 dots and multiple blots are made depending on the number of the probes to be used. One can, in general, probe the blots with ribosomal RNA probes, total genomic DNA for repeats, and any other probes known from that YAC. At this stage one can make pools of the PCR products from rows and columns of the 96-well grid plate and use them as probes to identify the overlapping clones. Both PCR amplification and dot blotting at this stage can be automated using a Biomek (Beckman, Fullerton, CA). After eliminating undesirable and/or overlapping clones, one can now sequence at least 150–250 bp of the PCR products rapidly by cycling PCR sequence (Bethesda Research Laboratories PCR-based cycling sequence kit). For sequencing individual PCR products, 2–3 μl is taken out of 50 μl of the PCR-amplified product (amplified with primer set C) and subjected to PCR-based cycling sequencing, without any intermediate purification, using one of the inner primers of set A as ^{32}P end-labeled sequencing oligonucleotide. These sequences can be analyzed for any sequence matches in the GENEMBL databank. Alternatively, the blotting is omitted and one can sequence directly and identify overlapping clones and some nonspecific ones such as ribosomal or repeat sequences; but this involves more sequencing than when blotting is performed. Whichever method one elects, it should be possible to identify unique sequences that are encoded in a YAC. Any unique clones from a selected library can be confirmed by Southern blots using total genomic DNA, YAC DNA, and DNA from a somatic hybrid cell line that is known to harbor a chromosome region corresponding to the YAC of interest. A full-length cDNA library can then be screened using clones derived from the selected library as probes.

Results

Sensitivity and Selectivity

Our previous data (11) indicated that as little as 1 pg of cosmid DNA was able to select is homologous cDNAs. By extrapolation, 10–20 pg of YAC DNA should be sufficient to select cDNAs encoded by a particular YAC. This sensitivity becomes particularly important if one wants to adapt the method to a chromosome-specific library or to a hybrid cell line. We have routinely used 10–15 ng of 300- to 400-kb YAC DNA as the target for generating cDNA libraries. At this target DNA concentration, material generated from a single tube after two rounds of PCR amplification is sufficient for cloning purposes. Our previous data with cosmids (11) indicated that increasing target concentration had no or very little effect on coselection of nonspecific cDNAs, whereas increasing cDNA concentration did cause coselection of a small proportion of contaminating cDNA after one round of selection. If one chooses to use a higher concentration of cDNA, it is probably possible to get rid of these undesirable sequences in the second round of selection, in which a much lower cDNA concentration is used. One can also increase the number of stringent washes or stringency of the washes if the need arises. In circumstances under which GC-rich regions pose a problem, one can use a 2.4 *M* tetraethylammonium chloride (Et_4NCl) (27) wash in the final stages of washing the disks. However, we have not so far encountered any problems with the several YACs tested and have not used this wash as a matter of routine. Detailed analysis of several randomly picked clones from the YAC-selected library showed that more than 85% of the clones do indeed come from the YAC after two rounds of selection. The number of clones with ribosomal RNA sequences or repetitive sequences (abundant type) was less than 7%. It has been possible to select a cDNA whose abundance in the original library is one in several hundred thousand.

Target DNA and Its Implications

The method described here has worked for cosmid DNA and YAC DNA. It worked much better with lysed bacterial colonies harboring cosmids than with lysed yeast colonies harboring YACs after one round of selection (data not shown). However, the results after two rounds of selection with bacterial or yeast colonies are unknown at this stage. It is conceivable that total yeast DNA rather than yeast colonies may be a better choice for the selection of cDNAs. Based on the experiments of minimum target concentration of cosmids, the amount of DNA needed for selection with genomic chromosomal specific libraries or hybrid cell lines is within manageable limits, but the precise conditions have yet to be worked out.

As a model experiment and approximation to a hybrid cell line in our previous experiments, we mixed various amounts of cosmid DNA (10–200 pg) with either 1 or 5 μg of hamster DNA and carried out cDNA selection (11). The data indicated that the yield of specific product was reduced when large amounts of hamster DNA were included with the cosmid DNA in comparison to cosmids alone. This reduction in yield could be due to competition between the large number of selected cDNAs during PCR amplification. A low level of ribosomal sequences in comparison to MHC class I was observed after one cycle of selection. However, this problem of coselection of ribosomal sequences and other conserved sequences can be circumvented by designing additional quenching reagents, and carrying out two cycles of selection.

Discussion

The approach described herein relies on the use of short-fragment random primed cDNAs that after hybrid selection are amplified by PCR. The advantage of using a short-fragment random primed library is that it is least affected by any bias introduced by PCR as a result of the size of cDNAs or GC-rich regions of a particular cDNA. Thus one can expect at least one, if not more, cDNA fragments to be recovered even in situations in which GC-rich or repetitive sequences occupy a part of a cDNA. The drawback of a short-fragment cDNA library, however, is that one must screen a full-length cDNA library with short-fragment probes to isolate full-length cDNAs. The development of *in vivo* recombination methods in *E. coli* (28) could facilitate the isolation of full-length cDNAs without the laborious method of screening libraries. Artifacts due to PCR such as point mutations are unavoidable but do not affect the feasibility of the method because it still identifies expressed sequence tags (ESTs); although their sequences cannot be used to deduce correct protein sequence, they still provide the primary function of generating probes for the isolation of cDNAs. The number of genes in a particular YAC may be overestimated if several short-fragment sequences originate from the same message without forming an overlap. This problem can be dealt with in several ways, such as by RNA blots or sequencing of full-length cDNAs. The generation of hybrid cDNA molecules as a result of coligations or PCR is again not a problem because they are either taken care of during the second round of selection or identified when sequencing full-length cDNAs. A limitation of the approach is that it requires a cDNA library representing all possible sequences. Although there are serious practical obstacles to obtaining such libraries for humans, the pooling of several libraries from multiple tissues and/or the use of pooled and normalized libraries makes it possible to approach such a complete library.

Some YACs may contain region-specific low-level repeats that are not competed out by the quenching cocktail. In such cases, a significant number of cDNAs with homologies to such repeats may be selected. One way of overcoming this problem is to hybridize the selected library plaques to such repeats and to remove them physically. Alternatively, one can make an additional quenching reagent from several plaques that hybridize to such repeats. The DNA from this quenching reagent can be readily prepared by PCR amplification of the plaques, followed by *Eco*RI digestion and mung bean nuclease treatment.

There are several advantages to using the present approach over other approaches for the identification of coding sequences. The selection procedure is insensitive to number and size of introns and to cryptic splice sites. The method detects 5′ and 3′ exons and genes that lack CpG-rich islands, as well as those genes that are not transcribed in hybrid cell lines or have highly diverged between species. The approach is much simpler than screening cDNA libraries with a series of unique probes derived from a YAC. Selection is particularly convenient for YACs because multiple YACs can be handled at the same time and several cDNA libraries can be screened for the identification of coding sequences in a relatively short period of time. The method has the potential for use with very complex DNA, and possibly for flow-sorted chromosomes or chromosome-specific libraries. The selection procedure can potentially be used to isolate polymorphic clones (such as sequences containing CA repeats) directly, to select linking or jumping clones corresponding to a large fragment, or to use linking clones to select jumping clones and vice versa for long-range physical mapping; thus the method can expedite the identification of disease genes in a defined region.

References

1. W. F. Bodmer, *Cold Spring Harbor Symp. Quant. Biol.* **51,** 1 (1986).
2. M. D. Adams, J. M. Kelley, J. D. Gocayne, M. Dubnick, M. H. Polymeropolous, H. Xiao, C. R. Meroid, A. Wu, B. Olde, R. F. Moreno, A. R. Koclarage, W. R. McCombie, and J. C. Venter, *Science* **252,** 1651 (1991).
3. J. O. Bishop, J. G. Morton, M. Rosebach, and M. Richardson, *Nature (London)* **250,** 199 (1974).
4. S. R. Patanjali, S. Parimoo, and S. M. Weissman, *Proc. Natl. Acad. Sci. U.S.A.* **88,** 1943 (1991).
5. M. S. H. Ko, *Nucleic Acids Res.* **18,** 5705 (1990).
6. J. M. Rommens, M. C. Ianuzzi, B.-S. Kerem, M. L. Drumm, G. Melmer, M. Dean, R. Rozmahel, J. L. Cole, D. Kennedy, N. Hidaka, M. Buchwald, J. R. Riordan, L.-C. Tsui, and F. S. Collins, *Science* **245,** 1059 (1989).
7. I. M. Hanson, A. Poutska, and J. Trowsdale, *Genomics* **10,** 417 (1991).
8. P. Elvin, G. Slynn, D. Black, A. Graham, R. Butler, J. Riley, R. Anand, and A. F. Markham, *Nucleic Acids Res.* **18,** 3913 (1990).
9. P. Liu, R. Legerski, and M. J. Siciliano, *Science* **246,** 813 (1989).
10. G. M. Dyuk, S. Kim, R. M. Myers, and D. R. Cox, *Proc. Natl. Acad. Sci. U.S.A.* **87,** 8995 (1990).
11. S. Parimoo, S. R. Patanjali, H. Shukla, D. Chaplin, and S. M. Weissman, *Proc. Natl. Acad. Sci. U.S.A.* **88,** 9623 (1991).

12. S. M. Weisman, *Mol. Biol. Med.* **4,** 133 (1987).
13. R. J. Britten, D. E. Graham, and B. R. Neufeld, *in* "Nucleic Acids and Protein Synthesis," Part E (L. Grossman and K. Moldave, eds.), Methods in Enzymology, Vol. 29, p. 363. Academic Press, New York, 1974.
14. R. K. Saiki, D. H. Gelfand, S. Stoffel, S. J. Scharf, R. Higuchi, G. T. Horn, K. B. Mullis, and H. A. Erlich, *Science* **239,** 487 (1988).
15. P. Chomczynski and N. Sacchi, *Anal. Biochem.* **162,** 156 (1987).
16. G. Cathala, J.-F. Sayouret, B. Mendez, B. L. West, M. Karin, J. A. Marctial, and J. D. Baxter, *DNA* **2,** 329 (1983).
17. U. Gubler and B. J. Hoffman, *Gene* **25,** 263 (1983).
18. J. Sambrook, E. F. Fritsch, and T. Maniatis, "Molecular Cloning: A Laboratory Manual," 2nd Ed. Cold Spring Harbor Lab. Press, Cold Spring Harbor, New York, 1989.
19. H. Erlich, D. Gelfand, and J. J. Sninsky, *Science* **252,** 1643 (1991).
20. G. A. Galau, W. H. Klein, R. J. Britten, and E. H. Davidson, *Arch. Biochem. Biophys.* **179,** 584 (1977).
21. E. Southern, *J. Mol. Biol.* **98,** 503 (1975).
22. A. P. Feinberg and B. Vogelstein, *Anal. Biochem.* **137,** 266 (1984).
23. D. Nelson, S. Ledbetter, L. Corbo, M. F. Victoria, R. R. Solis, T. D. Webster, D. H. Ledbetter, and T. C. Caskey, *Proc. Natl. Acad. Sci. U.S.A.* **86,** 6686 (1989).
24. B. W. Pontius and P. Berg, *Proc. Natl. Acad. Sci. U.S.A.* **88,** 8237 (1991).
25. H. Vasavada, S. Ganguly, S. Settleman, D. DiMaio, and S. M. Weissman, *Indian J. Biochem. Biophys.* **25,** 488 (1988).
26. J. Brosius, A. Ulrich, M. A. Raker, A. Gray, T. J. Dull, R. R. Gutell, and H. F. Noller, *Plasmid* **6,** 112 (1981).
27. W. I. Wood, J. Gitschier, L. A. Lasky, and R. M. Lawn, *Proc. Natl. Acad. Sci. U.S.A.* **82,** 1585 (1985).
28. D. M. Kurnit and B. Seed, *Proc. Natl. Acad. Sci. U.S.A.* **87,** 3166 (1990).

[3] Identification of Genes Differentially Expressed during Mammalian Embryogenesis: Subtraction Hybridization of Embryonic cDNA Libraries

Ann M. Lawler, Corey H. Mjaatvedt, and John D. Gearhart

Introduction

One of the challenges in identifying novel genes involved in mammalian embryogenesis is the small size of the embryos and the limited quantities of mRNA available for experiments. For example, in the study of gastrulation in the mouse, it is necessary to examine changes in gene expression between pregastrula, egg cylinder-stage embryos at 6.5 days postcoitum (dpc) and gastrula stage embryos at 7.5 dpc. However, 6.5-dpc embryos contain fewer than 2000 cells with an estimated 0.1 μg of total RNA and 1 ng of mRNA per embryo. Due to the extensive and rapid cell proliferation that occurs during gastrulation, a 7.5-dpc embryo consists of approximately 14,000 cells and contains roughly 0.6 μg of total RNA and 14 ng of mRNA. To prepare RNA for a single Northern blot, spanning these developmental stages, it would be necessary to extract RNA from 500 6.5-dpc embryos and 40 7.5-dpc embryos. Each day of gestation brings dramatic increases in the size of the embryo, and by 8.5 dpc the heart, neural tube, and somites have begun to form and the embryo has reached 100,000 cells. Once the body plan is established and the organs have begun to form, it is possible to focus on the development of a particular organ or tissue, requiring the careful dissection of numerous embryos to obtain sufficient RNA.

To establish a reference source for studies of gene expression during embryogenesis, we have built a number of λ phage libraries from embryos and tissues from various stages. The libraries have proved useful for isolating homologs to known genes, for roughly estimating mRNA abundance, and for ± differential screens to identify relatively high-abundance messages (greater than 0.1% of messages) with levels of expression that vary between stages (1). However, development is expected to be accompanied by altered levels of expression of numerous genes, many of moderate (0.01 to 0.1% of messages) to low abundance (less than 0.01% of messages). Therefore it is necessary to turn to subtraction hybridization methods, which are designed to enrich for lower abundance messages that are differentially expressed. For a review of the theoretical aspects of subtraction hybridization, refer to Sargent (2).

The goal of subtraction hybridization is to produce a library that is enriched in the messages expressed at a higher level in one developmental stage relative to

Methods in Molecular Genetics, Volume 1

TABLE I Strategies for Generating Subtracted Libraries

Target	Driver	Subtraction method	Vector	Library size	Percentage differential	Refs.
cDNA	poly(A)$^+$ RNA	Hydroxyapatite	pBR322	150,000	50	(3)
cDNA	poly(A)$^+$ RNA	Hydroxyapatite	Bluescribe	20,000	13	(6)
cDNA/cRNA[a]	poly(A)$^+$ RNA	Hydroxyapatite	λSWAJ	300,000	48	(9)
cDNA/cRNA[a]	cRNA	Biotin–streptavidin	λZapII	nr[b]	0.6	(12)
ssDNA	ssDNA	Biotin–streptavidin	λZapII	nr[b]	13	(7)

[a] cDNA reverse transcribed from cRNA from the library.
[b] nt, Not reported.

another. For example, subtraction hybridization could be used to identify genes whose expression increases during gastrulation in the mouse by removing the common messages expressed in 6.5-dpc embryos from those messages expressed in 7.5-dpc embryos. Genes that are expressed at higher levels in 7.5-dpc embryos will be enriched in the subtracted library. The method can also be used to study the development of an organ- or tissue-specific cell type. For example, genes differentially expressed during the normal development of the septa-forming cardiac mesenchyme can be identified by subtraction of mRNA from embryos that lack cardiac mesenchyme (8.5 dpc) and embryonic hearts in which relatively large numbers of the mesenchyme are present (10.5 dpc). To perform subtraction hybridization, a population of molecules containing the desired messages (target) is mixed with a 10- to 30-fold excess of molecules representing messages to be removed (driver) under conditions that favor hybridization of homologous sequences. Commonly expressed molecules hybridize to each other and are physically removed. The remaining molecules are isolated and cloned to form a subtracted library. To illustrate the diversity of subtraction hybridization strategies and techniques that have been successfully employed in the study of development, several experiments will be briefly summarized below (Table I).

Subtraction of cDNA with mRNA

Sargent and Dawid (3) first applied subtraction hybridization to the study of development in their work on gastrulation in *Xenopus laevis*. A subtracted library of 150,000 clones was generated by hybridizing 7.5 μg of cDNA derived from gastrula stage embryos to 270 μg of poly(A)$^+$ RNA from unfertilized eggs in the ovary. The cDNA–RNA hybrids were removed by hydroxyapatite chromatography and the remaining cDNA was cloned into pBR322. A differential screen was performed on 84 randomly selected clones by using cDNA probes from eggs, blastulas, gastrulas, and tadpoles. Nearly half of the clones represented low-abundance messages with no detectable hybridization. Most of the remaining clones represented genes that were expressed at higher levels in the gastrula than in unfertilized eggs, blastulas, or tadpoles, although some of these clones were redundant. One clone, later described as the *Xpo* gene, was found to be highly expressed in the posterior region of early tailbud embryos (4).

This subtracted library was also used by Rosa (5) to identify the *Mix.1* homeobox gene by a differential screen using cDNA probes isolated from uninduced animal caps and animal caps induced to form mesoderm by fibroblast growth factor or mesoderm-inducing factor. In this case, the probes were also enriched by hybridization to ovarian poly(A)$^+$ RNA and subtraction over hydroxyapatite prior to radioactive labeling. In addition to *Mix.1*, 21 other clones were differentially expressed in a screen of 1000 clones.

Another subtracted library was used by Sato and Sargent (6) to obtain molecular probes for dorsoanterior development in *X. laevis*. cDNA derived from stage 24–26 embryos was hybridized to 100 μg of poly(A)$^+$ RNA from completely ventralized embryos plus 20 μg of muscle poly(A)$^+$ RNA, and subtracted and cloned as described above. Fifty-three clones were randomly selected and the inserts were used to probe Northern blots from normal and ventralized embryos. Seven clones displayed a 10-fold increase in expression in normal embryos versus ventralized embryos, whereas over half of the clones represented low-abundance messages.

Subtraction with Material Generated from Libraries

Although subtraction hybridization using poly(A)$^+$ RNA directly is possible for experiments in which large amounts of tissue are available, it would be prohibitive for experiments on mammalian embryos. However, several methods have been devised to perform subtraction hybridization between libraries or reagents derived from libraries.

Schweinfest *et al.* (7) described a method using subtraction between λZapII libraries (Stratagene, La Jolla, CA) from colon carcinoma and normal mucosa. Using helper phage and the f1 origin of replication in the vector, single-stranded DNA (ssDNA) was produced from each library. One pool of ssDNA was biotinylated with a photoactivatable analog of biotin (8) and hybridized in 10-fold excess to the target ssDNA. Hybrids were removed with streptavidin followed by phenol–chloroform extraction rather than hydroxyapatite chromatography. The remaining ssDNA was converted to double-stranded DNA and transformed into bacteria. An initial screen of 14 randomly selected clones with 2 cDNA probes showed that 3 clones displayed differential expression.

An alternative cloning procedure was described by Palazzolo *et al.* (9) that allowed the production of RNA directly from the library, obviating the need to collect large amounts of material for the subtraction hybridization. The goal was to isolate cDNA clones specifically expressed in the *Drosophila* head but not in embryos. Initially, a λ cDNA library was built from adult head poly(A)$^+$ RNA using an oligo(dT) primer with an *Xho*I linker, permitting the cDNA molecules to be inserted into the vector in one orientation. The T7 RNA polymerase promoter within the vector was used to transcribe cRNA from the library and this was converted into cDNA with reverse transcriptase. The cDNA was subsequently hybridized to 162 μg of poly(A)$^+$ RNA from *Drosophila* embryos and the hybrids were removed by hydroxyapatite chromatography. The remaining cDNA was cloned directionally into the same vector to yield a library of 300,000 clones. An initial screen using embryo cDNA eliminated about 25% of the clones that were not removed by the subtraction. As a secondary screen, cRNA probes were transcribed from 741 individual clones and hybridized to Northern blots of head and embryo RNA. Fifty percent of the clones displayed differential expression between the two tissues. More extensive Northern blot analysis revealed 39 clones that were specific to the visual system, including a homolog of human arrestin (10) and a novel G_α molecule (11).

A further development in the use of directional cDNA libraries, described by Klar *et al.* (12), uses two separate libraries to generate the cDNA and RNA for hybridization. Directional cDNA libraries were constructed from floor plate mRNA and from dorsal spinal cord mRNA. Library DNA was used to transcribe cRNA copies of the inserts. The floor plate cRNA was reverse transcribed into cDNA and hybridized to dorsal cord cRNA that had been synthesized with UTP-biotin. The cDNA–RNA hybrids were removed with streptavidin and the remaining cDNA was amplified by the polymerase chain reaction (PCR) and cloned. A differential screen of 4000 plaques of the subtracted library with dorsal spinal cord and floor plate cDNA probes revealed 24 clones specifically expressed in the floor plate. One of the clones corresponded to F-spondin, a novel protein involved in cell adhesion and neurite extension.

These three techniques demonstrate the use of subtraction hybridization when sufficient material is available for the construction of libraries. However, application of PCR methods to subtraction hybridization may eventually permit comparison of gene expression patterns between individual cells. Papers describing techniques that use PCR to amplify cDNA molecules for generation of probes and construction of libraries from varying amounts of starting material include Belyavsky *et al.* (13), 10 cells; Hla and Maciag (14), 200 ng poly(A)$^+$ RNA; Brunet *et al.* (15), 1 μg total RNA; and Welsh *et al.* (16), 17.5 ng total RNA from 50 mouse oocytes. However, problems can arise during the amplification and cloning process, including preferential amplification of small inserts, cloning of primer artifacts, a high background frequency of clones with no inserts, and amplification and cloning of trace nucleic acids present in the cDNA synthesis reagents and in the carrier molecules used for precipitations.

Several papers have also described methods that combine PCR amplification with subtraction hybridization (17–19); however, relatively large amounts of RNA

[1.5–10 μg poly(A)$^+$ RNA] were used to generate the initial cDNA for amplification. One advantage of these techniques is the possibility of generating sufficient material to perform multiple rounds of hybridization and subtraction. Duguid and Dinauer (18) performed a total of six rounds of hybridization and subtraction with PCR amplification after every two rounds. Screening of the subtracted library of 20,000 clones with normal and subtracted probes revealed 7 differential recombinants, 2 of which represent low-abundance clones. To minimize biases due to differential amplification of small inserts, Wang and Brown (19) included a restriction enzyme digestion prior to ligation of linkers and amplification. The sizes of the amplified products ranged from 0.2 to 2 kilobase pairs (kbp). This method also used several rounds of hybridization and subtraction with 20-hr hybridization times to remove moderate- and low-abundance messages and short, 2-hr hybridizations to remove abundant messages. With this procedure, Wang and Brown (19) were able to isolate 16 of the estimated 30 genes judged to be differentially expressed between normal tadpole tails and tails induced with thyroid hormone. Besides allowing the isolation of differentially expressed genes, this method also provides a means for estimating the number of differential genes; however, validation of this process in different experimental systems remains to be demonstrated.

Considerations in Planning Subtraction Hybridization Experiments

A number of factors must be considered when planning a subtraction hybridization experiment. The initial investment of time, energy, and materials needed for the construction of libraries from a particular developmental stage or tissue is generally necessary to establish a foundation for future work. Polymerase chain reaction methods have been successful, as mentioned above; however, the resulting clones are usually small and isolation of a full-length clone would require a library. Although the efficiency of building unidirectional libraries is generally lower than that of random orientation libraries, the advantages of unidirectional libraries for generating cRNA for subtraction and for sense or antisense probes and expression studies outweigh the extra effort required for the initial construction. An estimation of the number of differences expected between the libraries and the relative abundance of the differential clones will indicate the type of library to be used and the strategy of the screen. If the differential genes are expected to represent a large fraction of the clones in the subtracted library, it is not necessary to screen large numbers of clones and it is possible to work with plasmid libraries. However, to find rare differential genes, it is necessary to screen large numbers of clones, which is simpler with λ libraries. But for the most part, the decision to use plasmid libraries or λ libraries is a matter of personal preference. Because some type of screen is generally required to identify differential clones in the subtracted library, the abundance of the desired genes should be considered. Moderate- to high-abundance messages can be identified with cDNA probes, but detection of low-abundance messages would require the use of subtracted probes or the use of probes made from individual clones to screen Northern blots or dot blots. Screens of a subtracted library to isolate homologs of known genes are also possible. A measure of the effectiveness of subtraction is helpful and is most easily accomplished by monitoring the clone frequency of a gene, such as actin or tubulin, that is expected to be removed during subtraction. A positive control is also helpful if a gene is known that is preferentially expressed in the target tissue. Its clone frequency should increase in the subtracted library relative to the original library. The most critical aspect of planning a subtraction hybridization experiment is to have established specific criteria and methodology to select differential clones for further study quickly and efficiently. These methods can range from *in situ* hybridization histochemistry to detect tissue-specific or stage-specific expression, partial sequencing to identify homologs of known genes, or functional assays, such as the use of antisense oligonucleotides, if an appropriate system is available.

Methods

The purpose of this article is to discuss techniques involved in performing subtraction between libraries and not to cover in detail the construction of cDNA libraries, because an extensive literature on library construction already exists. Also, our experience has been that the success of any of these techniques varies substantially between laboratories and between individuals within a laboratory. However, several general comments are worth noting. Standard methods for extraction of RNA and construction of the libraries can be used, but care should be taken to optimize each step in the process. It is necessary to establish a successful method of library construction before committing precious embryonic tissues to the endeavor. Embryonic material can be stored in guanidinium isothiocyanate lysis buffer (20) at $-80°C$ for months until sufficient material is accumulated to begin library construction. Isolation of total RNA by the acid–phenol method (20) is simple and effective and can be easily scaled up or down to accommodate varying amounts of tissue. Purification of poly(A)$^+$ RNA can be accomplished by oligo(dT)-cellulose chromatography. Several kits for isolation of poly(A)$^+$ RNA have also been used that employ oligo(dT)-cellulose and quick-spin columns or oligo(dT) attached to magnetic beads; however, it is preferable to store embryonic tissue in guanidinium isothiocyanate and perform one acid–phenol extraction before using the procedures in the kits. A Beckman (Fullerton, CA) TL100 ultracentrifuge facilitates precipitation of RNA and small amounts of cDNA by spinning at 100,000 rpm for 1 h at 4°C (1). Libraries can be built using Gubler–Hoffman methods (21) or for unidirectional libraries the Uni-ZapII cDNA synthesis kit (Stratagene) may be used. Several substitutions for reagents in the kit have improved results, including Superscript reverse

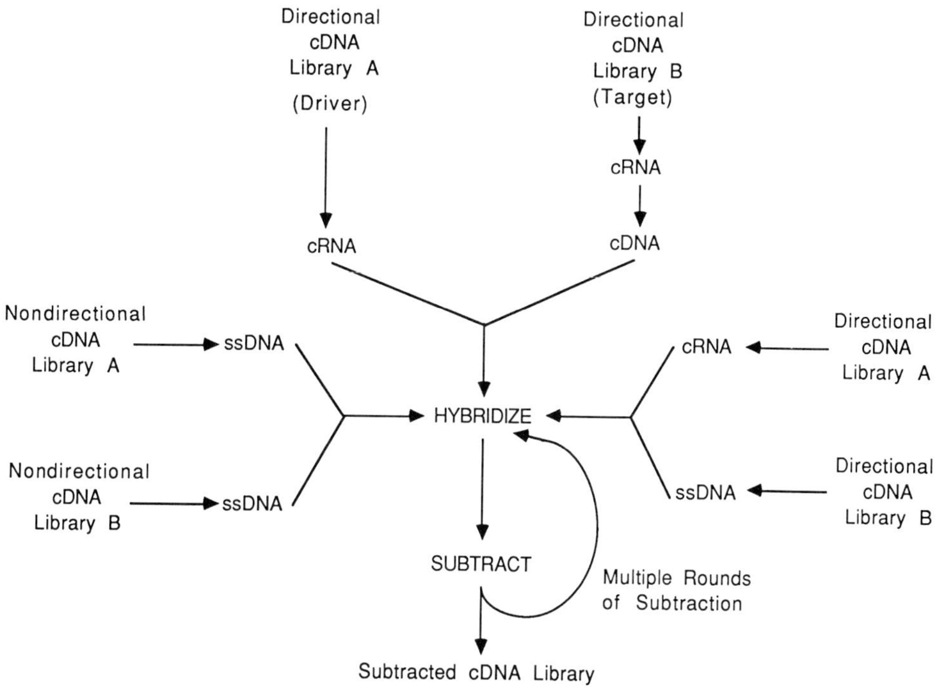

FIG. 1 Alternative strategies for construction of a subtracted cDNA library. See text for further details.

transcriptase (8053SA; GIBCO–Bethesda Research Laboratories, Gaithersburg, MD) and T4 DNA ligase (202S; New England BioLabs, Beverly, MA).

The following methods describe procedures to perform subtraction hybridization between random orientation or unidirectional λZapII libraries. As shown in Fig. 1, random orientation libraries have been subtracted using ssDNA, whereas it is possible to use either ssDNA or cRNA from unidirectional libraries. Although theoretically possible, the combination of ssDNA and cRNA in a subtraction hybridization has not been described. One advantage to using cRNA as a driver rather than ssDNA is that it can be completely removed by alkaline hydrolysis after the subtraction and there is no chance of cloning some of the driver material in the subtracted library. Our experience with subtraction of random orientation libraries using ssDNA has been directed toward identifying genes uniquely expressed during murine gastrulation, neurulation, and heart septation. Several potentially exciting genes that have been identified in these subtracted libraries appear to be unique and expressed at the appropriate stage of development. For example, of 31 clones randomly picked from 3 separate subtracted heart libraries, 16 appear to be unique and 9 have significant homologies with known genes, including a high-mobility group

DNA-binding protein, an inositol phosphate pathway enzyme, and a GTP-binding protein. The remaining clones were identified as ribosomal RNA (four clones), and two contained no inserts. Preliminary sequencing of six 8.5-dpc minus 7.5-dpc clones revealed genes homologous to two ribosomal proteins, mitochondrial ATP synthase proteolipid, lactate dehydrogenase, a hamster proline-rich protein, and a novel gene product, 7-12, that is expressed at low but increasing levels during gastrulation (0.002–0.004% of the messages in the cDNA libraries).

Generation of Single-Stranded DNA from λZapII Libraries

This procedure is modified from Schweinfest *et al.* (7).

1. Mix the following in a 50-ml conical tube: 250 μl XL1-Blue cells (Stratagene) at OD_{600} = 1.0 2 × 10^6 λZapII library, 1 μl R408 helper phage [10^8 plaque-forming units (pfu), Stratagene]. Incubate at 37°C for 15 min.
2. Add 5 ml 2× YT [16 g Bacto-tryptone (Difco, Detroit, MI), 10 g yeast extract, 10 g NaCl in 1 liter H_2O]. Shake at 37°C for 5 hr.
3. Heat tubes at 70°C for 20 min.
4. Centrifuge at 3000 rpm for 5 min to pellet cells, and recover the supernatant containing single-stranded phage and helper phage.
5. Mix the following: 1 ml XL1-Blue cells grown overnight in 2× YT with 50 μg/ml tetracycline and 200 μl phage supernatant from step 4. Incubate at 37°C for 15 min.
6. Add to 80 ml 2× YT with 50 μg/ml tetracycline and 50 μg/ml ampicillin, in a 250-ml flask. Shake at 250 rpm at 37°C for 5–6 hr.
7. Transfer cultures to two 50-ml conical tubes and centrifuge for 15 min at 3000 rpm to pellet bacteria. Save 200 μl of supernatant for determining the titer of single-stranded phage and helper phage.
8. Transfer supernatants to a clean 50-ml conical tube and heat to 70°C for 20 min.
9. Add 10 ml 20% (w/v) polyethylene glycol, 3.5 *M* ammonium acetate. Mix and incubate at room temperature for 30 min.
10. Spin at 3000 rpm for 45 min to pellet phage. Pour off the supernatant and drain the tube on a paper towel for 5 min.
11. Resuspend the phage pellet in 1 ml SM buffer [5.8 g NaCl, 2 g $MgSO_4 \cdot H_2O$, 50 mM Tris-HCl pH 7.5), 0.1% (w/v) gelatin] and transfer to a 1.5-ml microfuge tube. Spin at 14,000 rpm for 30 sec to remove cellular debris. Transfer the supernatant to a clean tube containing 250 μl 20% polyethylene glycol, 3.5 *M* ammonium acetate. Incubate at room temperature for 30 min.
12. Pellet phage by spinning at 14,000 rpm for 3 min. Remove the supernatant. Respin the tube for 5 sec and remove traces of supernatant with a capillary tube.
13. Resuspend the phage pellet in 300 μl TE [10 m*M* Tris-HCl, (pH 8.0), 1

mM ethylenediaminetetraacetic acid (EDTA), pH 8.0]. Extract with equal volumes of neutralized phenol, phenol–chloroform, and chloroform. Add 300 μl 7.5 M ammonium acetate. Mix. Add 600 μl 100% ethanol. Mix and incubate at -20°C for 1 hr.

14. Recover DNA by spinning at 14,000 rpm for 5 min. Remove the supernatant and wash the pellet with 70% ethanol. Dry briefly and resuspend the pellet in 100 μl TE. The yield is approximately 100 μg DNA containing equal amounts of recombinant phage and helper phage.

Generation of cRNA from λZapII Libraries

The first step in generating cRNA from λZapII libraries is to isolate DNA from plate lysates. Typically, 1×10^6 plaques is plated on 20 plates at a density of 50,000 plaques/150-mm plate using 0.7% (w/v) agarose in NZY (5 g NaCl, 2 g $MgSO_4 \cdot H_2O$, 5 g yeast extract, 10 g NZ-amine, casein hydrolysate per liter). Agarose in the plating layer is essential as agar contains contaminants that will inhibit digestion of λ DNA.

1. Flood each plate with 8 ml of SM. Gently swirl overnight at 4°C.
2. Recover and pool SM containing phage from each plate. Add 5 ml SM to each plate and continue to elute phage for an additional 2 hr at 4°C. Pool eluted phage in SM in 50-ml conical centrifuge tubes.
3. To each 50-ml conical tube of phage lysate add DNase I and RNase A to 1 μg/ml. Incubate at room temperature for 30 min.
4. Add 2.92 g NaCl and dissolve. Incubate on ice for 1 hr. Centrifuge at 3000 rpm for 10 min at 4°C to pellet cell debris. Transfer the supernatant to a clean tube.
5. Add 5 g polyethylene glycol 8000 and mix thoroughly until dissolved. Incubate on ice for 1 hr. Centrifuge for 10 min at 3000 rpm at 4°C to pellet phage. Drain off the supernatant.
6. Resuspend pellet in 400 μl SM and transfer to a 1.5-ml microfuge tube. Add 2 μl 500 mM EDTA (pH 8.0), 2 μl proteinase K (20 mg/ml), 25 μl 10% (w/v) sodium dodecyl sulfate (SDS). Incubate at 55°C for 1 hr.
7. Extract with neutralized phenol, phenol–chloroform, and chloroform. Add 45 μl 3 M sodium acetate (pH 7), 1 ml 100% ethanol. Mix to precipitate the DNA. Incubate on ice for 10 min. Centrifuge at 14,000 rpm for 5 min. Wash the pellet with 70% ethanol. Dry the pellet briefly and resuspend in 200 μl TE. The average yield is 500 μg DNA from 50 ml of phage lysate.

Prior to transcription, the DNA template must be digested to completion with an enzyme that cleaves on the 3′ side of the insert and must be treated to remove RNA and RNase.

1. Digest 50 μg λ DNA with 300 units of *Xho*I in a 500-μl reaction for 3 hr at 37°C.

2. Add 2 μl RNase A (20 mg/ml). Incubate at 37°C for 15 min.

3. Add 5 μl proteinase K (20 mg/ml), 25 μl 10% SDS. Incubate at 55°C for 30 min.

4. Using RNase-free reagents (22), extract with neutralized phenol, phenol–chloroform, and chloroform. Add 55 μl 3 *M* sodium acetate (pH 7), 1 ml 100% ethanol. Mix to precipitate the DNA. Incubate on ice for 10 min. Centrifuge at 14,000 rpm for 5 min. Wash the pellet with 70% ethanol. Dry the pellet briefly and resuspend in 40 μl TE.

5. T3 RNA polymerase transcription reaction: Mix the following:

Digested DNA template	40 μl
Transcription buffer (10×) [400 m*M* Tris-HCl (pH 8.0), 80 m*M* $MgCl_2$, 50 m*M* NaCl, 20 m*M* spermidine, 300 m*M* dithiothreitol (DTT)	10 μl
ATP (75 m*M*)	10 μl
UTP (75 m*M*)	10 μl
CTP (75 m*M*)	10 μl
GTP (75 m*M*)	10 μl
RNase inhibitor (Cat. No. 799017; Boehringer Mannheim, Indianapolis, IN)	5 μl
T3 RNA polymerase	500 units

Remove 5 μl to a separate tube containing 2 μl [α-^{32}P]UTP (3000 Ci/mmol) to measure the percentage incorporation. Incubate both tubes at 37°C for 2 hr.

6. Add 5 μl RNase-free DNase I. Incubate at 37°C for 15 min. Add 30 μl diethyl pyrocarbonate (DEPC)-treated H_2O, 15μl 5 *M* ammonium acetate. Mix thoroughly. Extract with neutralized phenol, phenol–chloroform, and chloroform. Add 450 μl 100% ethanol. Mix to precipitate the RNA. Incubate on ice for 20 min. Centrifuge at 14,000 rpm for 10 min at 4°C. Wash the pellet with 70% ethanol. Dry the pellet briefly and resuspend in 40 μl DEPC-treated H_2O.

Biotinylation of Single-Stranded DNA and cRNA for Subtraction

This method is described by Sive and St John (23) and can be used for any nucleic acid, although all reagents and plasticware must be maintained RNase free when working with RNA.

1. In a darkroom with subdued light, mix equal volumes of up to 50 μl of photobiotin (1 mg/ml in H_2O) (Cat. No. 0695P-1; Research Organics, Cleveland, OH)

and nucleic acid (1 mg/ml up to 45 μg) in a microfuge tube. Place on ice 10 cm below a sun lamp (275 W).

2. Irradiate with the cap off for 15 min.
3. Add 10 μl 1 *M* Tris-HCl (pH 9.0), and adjust the volume to 100 μl with H_2O.
4. Extract with water-saturated 2-butanol four or five times to remove unreacted photobiotin. Extract once with chloroform. Add 11 μl 3 *M* sodium acetate (pH 4.8), and 250 μl 100% ethanol. Precipitate nucleic acids on ice for 15 min and recover by spinning at 14,000 rpm for 10 min at 4°C. Rinse the pellet with 70% ethanol and dry briefly. Resuspend the pellet in the starting volume of H_2O.
5. Repeat biotinylation as above.

Hybridization and Subtraction of Biotin Complexes

1. Mix and precipitate target cDNA or ssDNA and a 10- to 30-fold excess of biotinylated driver RNA or ssDNA. Resuspend the pellet in 8 μl DEPC-treated H_2O. Add 2 μl 5× hybridization buffer [250 m*M* *N*-2-hydroxyethylpiperazine-*N*′-2-ethanesulfonic acid (HEPES) (pH 7.6), 1% SDS, 10 m*M* EDTA, 1.25 *M* NaCl]. Overlay with 20 μl paraffin oil. Heat at 95°C for 5 min. Cool on ice for 1 min. Hybridize at 65°C for the desired length of time.
2. Add 50 μl dilution buffer [50 m*M* HEPES (pH 7.6), 2 m*M* EDTA, 0.25 *M* NaCl]. Mix in 10 μl streptavidin (Cat. No. 5532LB; GIBCO-Bethesda Research Laboratories), 10 mg/ml in 50% glycerol. Incubate at room temperature for 5 min.
3. Extract with 65 μl phenol–chloroform, buffered with dilution buffer. Remove the aqueous phase to a clean tube. Back extract the organic phase with 50 μl dilution buffer and combine with the first aqueous phase.
4. Add 10 μl streptavidin and incubate at room temperature for 5 min.
5. Extract with 125 μl phenol–chloroform, buffered with dilution buffer. Remove the aqueous phase to a clean tube. Back extract the organic phase with 50 μl dilution buffer and combine with the aqueous phase.
6. Add 10 μl streptavidin and incubate at room temperature for 5 min.
7. Extract with 185 μl phenol–chloroform, buffered with dilution buffer. Remove the aqueous phase to a clean tube. Back extract the organic phase with 50 μl dilution buffer and combine with the aqueous phase. Extract with 200 μl chloroform. Add 1 ml 100% ethanol to precipitate remaining material and repeat hybridization or use to construct a subtracted library.

Construction of Subtracted Libraries

Single-stranded DNA remaining after subtraction can be cloned simply and efficiently by replicating with the Klenow fragment of DNA polymerase and directly transforming bacteria. The result is a plasmid library rather than a λ phage library and it must be plated, screened, and stored accordingly (22).

1. Mix 5 μl ssDNA in TE with 5 μl SK primer (2 ng/μl; Stratagene). Heat to 95°C for 5 min and cool slowly to room temperature.
2. Add the following:

Reaction buffer (10×) [500 m*M* Tris-HCl (pH 8.0), 100 m*M* $MgCl_2$]	4 μl
dNTP mix (10 m*M*) (10 m*M* dATP, 10 m*M* dCTP, 10 m*M* dGTP, 10 m*M* dTTP)	1 μl
Dithiothreitol (100 m*M*)	0.5 μl
H_2O	32 μl
Klenow fragment (6 U/μl)	2 μl

Incubate at 37°C for 1 hr. Precipitate with 80 μl 0.3 *M* sodium acetate, and 300 μl 100% ethanol.
3. Resuspend the DNA in 10 μl TE and transform 1 μl by electroporation (22) into XL1-Blue cells or any appropriate host bacteria.

Cloning the cDNA remaining after subtraction is more challenging because several additional steps are involved and care must be taken to recover the minute amounts of cDNA at each step. The process involves converting the single-stranded cDNA to double-stranded cDNA, using the SK primer and Sequenase DNA polymerase (United States Biochemicals, Cleveland, OH). The *Xho*I site must be regenerated to clone the molecule. The cDNA is digested with *Xho*I and *Eco*RI, purified from the primers and ends, and ligated into λ arms for packaging (22). Trace amounts of [^{32}P]dCTP in the polymerase reaction can aid in following the recovery of the cDNA in subsequent steps.

1. Mix 10 μl cDNA in TE with 5 μl SK primer (2 ng/μl; Stratagene). Heat to 95°C for 5 min and cool slowly to room temperature.
2. Add the following:

Reaction buffer (5×) [200 m*M* Tris-HCl (pH 7.5), 100 m*M* $MgCl_2$, 250 m*M* NaCl]	6 μl
dNTP mix (10 m*M*) (10 m*M* dATP, 10 m*M* dCTP, 10 m*M* dGTP, 10 m*M* dTTP)	1 μl
Dithiothreitol (100 m*M*)	3 μl
H_2O	5 μl
Sequenase (undiluted)	1 μl

Incubate at 37°C for 1 hr.

3. Extract with phenol–chloroform and chloroform and precipitate by adding 3.5 μl 3 *M* sodium acetate (pH 4.8), 100 μl 100% ethanol. Incubate overnight at -20°C. Recover the DNA by spinning at 100,000 rpm in a Beckman TL100 ultracentrifuge for 1 hr at 4°C. Wash the pellet with 70% ethanol and dry briefly. Resuspend in 40 μl TE.

4. Add the following:

*Xho*I buffer (10×) [500 m*M* Tris-HCl (pH 8.0), 100 m*M* $MgCl_2$, 500 m*M* NaCl]	5 μl
*Eco*RI (10 U/μl	2.5 μl
*Xho*I (10 U/μl	2.5 μl

Incubate at 37°C for 1 hr. Extract with phenol–chloroform, then chloroform.

5. Purify on a Sepharose CL-4B column (22). Coprecipitate with 1 μg λZapII arms digested with *Xho*I and *Eco*RI.

6. Resuspend the DNA pellet in 4 μl H_2O. Add 0.5 μl 10× ligase buffer [500 m*M* Tris-HCl (pH 7.5), 100 m*M* $MgCl_2$, 200 m*M* dithiothreitol, 10 m*M* ATP], 0.5 μl T4 DNA ligase (200 U/μl) (Cat. No. 202S; New England BioLabs). Incubate at 16°C for 12 hr. or more.

7. Package with high-efficiency packaging extracts, following the instructions of the manufacturer.

In cases in which the amount of cDNA is small, PCR can be used to amplify the material prior to digestion with *Eco*RI and *Xho*I. The SK primer and *Xho*I/oligo(dT) linker–primer (Stratagene) are used for the amplification in 20 cycles of 48°C annealing for 1 min, 72°C extension for 1 min, and 95°C denaturation (12).

Screening Subtracted Libraries

Since the process of subtraction hybridization is not quantitative, a screen of the subtracted library is generally necessary to identify clones of interest. A variety of strategies have been used for screens, depending on the object of the search, the number of clones in the library, the suspected frequency of desired clones in the library, and the availability of reagents for the screen. A typical method of screening involves probing replica filters of the library with two cDNA probes. The probes can either be synthesized directly from poly(A)$^+$ RNA or can be enriched by subtraction hybridization.

Synthesis of cDNA Probes

1. Mix 1 μg poly(A)$^+$ RNA with 2 μg oligo(dT) (Cat. No. 27-7858-01; Pharmacia, Piscataway, NJ) in a volume of 5 μl. Heat at 70°C for 5 min. Cool on ice.

2. Dry 250–500 μCi [α-^{32}P]dCTP (3000 Ci/mmol). Resuspend with:

DTT (0.1 *M*)	1 μl
Reverse transcriptase buffer (5×) [250 m*M* Tris-HCl (pH 8.3), 375 m*M* KCl, 50 m*M* DTT, 15 m*M* $MgCl_2$]	3 μl
dATP (10 m*M*)	1 μl
dGTP (10 m*M*)	1 μl
dTTP (10 m*M*)	1 μl

3. Add the following:

RNase inhibitor (Cat. No. 799017; Boehringer Mannheim)	1 μl
RNA/oligo(dT) mix	5 μl
Superscript reverse transcriptase (Cat. No. 8025SA; GIBCO-Bethesda Research Laboratories)	1 μl

Incubate at 37°C for 15 min. Add the following:

dCTP (10m*M*)	1 μl
Superscript reverse transcriptase	1 μl

Incubate at 37°C for 1 hr.

4. Add 2 μl 250 m*M* EDTA. Remove 2 μl to measure the percentage incorporation. Add 2 μl 1 *M* NaOH. Heat at 70°C for 20 min to hydrolyze the RNA. Neutralize with 2 μl 1 *M* Tris-HCl (pH 7.5), 2 μl 1 *M* HCl. Purify the cDNA probe from unincorporated nucleotides by a standard method (22).

Subtracted probes can be synthesized as described above and submitted to two rounds of hybridization and subtraction with biotinylated RNA. The remaining cDNA is sufficiently radioactive to be used directly on filters. Alternatively, low specific activity cDNA probes can be synthesized, then subtracted, and finally labeled to high specific activity with random oligonucleotides and the Klenow fragment of DNA polymerase (24) as described below.

1. Mix 1 μg poly$(A)^+$ RNA with 2 μg oligo(dT) in a volume of 5 μl. Heat at 70°C for 5 min. Cool on ice.
2. Add the following:

Reverse transcriptase buffer (5×) [250 m*M* Tris-HCl (pH 8.3), 375 m*M* KCl, 50 m*M* DTT, 15 m*M* $MgCl_2$]	3 μl
DTT (0.1 *M*)	1 μl
dATP (10 m*M*)	1 μl

dGTP (10 mM)	1 μl
dTTP (10 mM)	1 μl
dCTP (10 mM)	1 μl
RNase inhibitor	1 μl
[α-^{32}P]dCTP (3000 Ci/mmol)	1 μl
Superscript reverse transcriptase (Cat. No. 8025SA; GIBCO-Bethesda Research Laboratories)	1 μl

Incubate at 37°C for 1 hr.

3. Add 2 μl 250 mM EDTA. Remove 2 μl to measure the percentage incorporation. Add 2 μl 1 M NaOH. Heat at 70°C for 20 min to hydrolyze the RNA. Neutralize with 2 μl 1 M Tris-HCl (pH 7.5), 2 μl 1 M HCl. Purify the cDNA probe from unincorporated nucleotides by a standard method.

The labeled cDNA can be enriched by two rounds of subtraction hybridization with 1 μg biotinylated RNA and labeled to high specific activity as follows.

1. Resuspend the subtracted cDNA in 4 μl H_2O. Add the following:

[α-^{32}P]dCTP (3000 Ci/mmol)	100 μCi
Primer extension buffer (10×) [60 mM Tris-HCl (pH 7.5), 500 mM NaCl, 60 mM $MgCl_2$, 2 mM DTT]	2.5 μl
p(dN)$_6$, 90 OD U/ml (Cat. No. 27-2166-01; Pharmacia)	3 μl
dATP (10 mM)	1 μl
dTTP (10 mM)	1 μl
dGTP (10 mM)	1 μl
Klenow fragment DNA polymerase (6 U/μl) (Cat. No. 8012SA; GIBCO-Bethesda Research Laboratories)	1 μl

Incubate at room temperature for 2 hr. Measure the percentage incorporation and purify labeled cDNA from unincorporated nucleotides using standard methods (22).

Conclusions

The concept of subtraction hybridization as a method to identify genes whose expression changes during development and differentiation continues to attract the efforts of a number of workers. The wide variety of techniques employed suggests, at present, that the process has not matured to a point at which a single method can be successfully applied to any biological system. Nonetheless, a review of the literature reveals that most subtraction hybridization experiments identify several important genes displaying differential expression, but probably do not identify all of the differentially expressed genes. Further advances in the combination of PCR technology and subtraction hybridization may be able to build on the work of Wang and Brown (19) and others to improve the percentage of differentially expressed genes that can be identified. At that point, workers can more fully pursue the exciting task of determining the functional significance of each differentially expressed gene.

Acknowledgments

The authors gratefully acknowledge the advice and expertise of David Weng, Developmental Genetics Laboratory, and Anthony Lanahan and Kyu-Ho Lee, Department of Molecular Biology (Johns Hopkins University School of Medicine). This work was supported by Public Health Service Grant HD25863.

References

1. D. Weng, Ph.D. Thesis, Johns Hopkins Univ., Baltimore, 1992.
2. T. D. Sargent, *in* "Guide to Molecular Cloning Techniques" (S. Berger and A. Kimmel, eds.), Methods in Enzymology, Vol. 152, p. 423. Academic Press, Orlando, Florida, 1987.
3. T. D. Sargent and I. B. Dawid, *Science* **222,** 135 (1983).
4. S. M. Sato and T. D. Sargent, *Development* **112,** 747 (1991).
5. F. M. Rosa, *Cell* **57,** 965 (1989).
6. S. M. Sato and T. D. Sargent, *Dev. Biol.* **137,** 135 (1990).
7. C. W. Schweinfest, K. W. Henderson, J.-R. Gu, S. D. Kottaridis, S. Besbeas, E. Panotopoulou, and T. S. Papas, *Genet. Analysis Technol. Appl.* **7,** 64 (1990).
8. A. C. Forster, J. L. McInnes, D. C. Skingle, and R. H. Symons, *Nucleic Acids Res.* **13,** 745 (1985).
9. M. J. Palazzolo, D. R. Hyde, K. VijayRaghavan, K. Mecklenburg, S. Benzer, and E. Meyerowitz, *Neuron* **3,** 527 (1989).
10. D. R. Hyde, K. L. Mecklenburg, J. A. Pollock, T. S. Vihtelic, and S. Benzer, *Proc. Natl. Acad. Sci. U.S.A.* **87,** 1008 (1990).
11. Y.-J. Lee, M. B. Dobbs, M. L. Verardi, and D. R. Hyde, *Neuron* **5,** 889 (1990).
12. A. Klar, M. Baldassare, and T. M. Jessell, *Cell* **69,** 95 (1992).
13. A. Belyavsky, T. Vinogradova, and K. Rajewsky, *Nucl. Acids Res.* **17,** 2919 (1989).
14. T. Hla and T. Maciag, *Biochem. Biophys. Res. Commun.* **167,** 637 (1990).
15. J.-F. Brunet, E. Shapiro, S. A. Foster, E. R. Kandel, and Y. Iino, *Science* **252,** 856 (1991).
16. J. Welsh, J.-P. Liu, and A. Efstradiatis, *Genet. Analysis Technol. Appl.* **7,** 5 (1990).

17. C. Timblin, J. Battey, and W. M. Kuehl, *Nucleic Acids Res.* **18,** 1587 (1990).
18. J. R. Duguid and M. C. Dinauer, *Nucleic Acids Res.* **18,** 2789 (1990).
19. Z. Wang and D. D. Brown, *Proc. Natl. Acad. Sci. U.S.A.* **88,** 11505 (1991).
20. P. Chomczynski and N. Sacchi, *Anal. Biochem.* **162,** 156 (1987).
21. U. Gubler and B. J. Hoffman, *Gene* **25,** 263 (1983).
22. J. Sambrook, E. F. Fritsch, and T. Maniatis, "Molecular Cloning: A Laboratory Manual," 2nd Ed. Cold Spring Harbor Lab. Press, Cold Spring Harbor, New York, 1989.
23. H. L. Sive and T. St John, *Nucleic Acids Res.* **16,** 10937 (1988).
24. A. P. Feinberg and B. Vogelstein, *Anal. Biochem.* **132,** 6 (1983).

[4] Identification and Cloning of Genes Involved in Progression of Transformed Phenotype

P. Gopal Reddy, Zao-Zhong Su, and Paul B. Fisher

Introduction

Based on both epidemiological evidence and direct experimental approaches, it is now well established that the carcinogenic process is multifactor with respect to its etiology and multistep in terms of its evolution (reviewed in Refs. 1–3). Although progress has been made in defining a number of the critical processes contributing to carcinogenesis, the specific biochemical and molecular mechanisms underlying many of these complex phenomena still remain to be elucidated (reviewed in Refs. 3–5). Operationally, the neoplastic process has been divided into three distinct, yet many times overlapping, phases termed *initiation, promotion,* and *progression* (1–5). Of the three phases of carcinogenesis, the least understood phenomenon, on both a biochemical and molecular level, is the mechanism by which transformed cells acquire a new set of transformation-specific phenotypic traits or a further elaboration of transformation-related properties, a process termed *progression* (1–3). Studies support the hypothesis that the development of cancer often involves the activation of protooncogenes (which function as oncogenes) and the inactivation of genes that function as inhibitors of carcinogenesis (tumor suppressor or antioncogenes) (1–5). It is not known whether progression involves a further change in the expression of the same oncogenes and/or tumor suppressor genes that are involved in the initiation and promotion phases of cancer, or if this process involves the modification of additional genetic elements (tumor progression-inducing and/or tumor progression-suppressing genes) in the evolving tumor genome. It is clear, however, that an understanding of the mechanism of action of oncogenes, tumor suppressor genes, and tumor progression-inducing and tumor progression-suppressing genes (once identified and characterized) will be mandatory to define the neoplastic process mechanistically. A more precise understanding of the biochemical and molecular events mediating neoplastic development should eventually result in the development of improved strategies for identifying specific stages in the carcinogenic process and for the design of improved modalities for cancer prevention and treatment (reviewed in Ref. 6).

As indicated above, the molecular basis of tumor cell progression remains to be defined. Possible factors that may mediate progression include (1) an inherent genetic instability of the tumor genome resulting in a constantly evolving heterogeneous tumor cell population displaying unique or enhanced oncogenic properties;

Methods in Molecular Genetics, Volume 1

(2) host-selective pressures exerted on the developing neoplasm resulting in the selection of tumor cells with a more aggressive phenotype (i.e., an enhanced growth rate, the production of proteins facilitating metastatic tumor spread, the ability to secrete proteins stimulating angiogenesis, etc.) and/or a reduced ability to be recognized and destroyed by the immune system; (3) mutational activation or changes in the expression of specific subsets of oncogenes and/or progression-inducing genes that then function as positive regulators of progression; (4) inactivation of specific tumor suppressor and/or tumor progression suppressor genes resulting in the absence of appropriate negative regulators that normally inhibit the progression phenotype; and/or (5) a combination of the above processes (see references cited in Ref. 7).

Advances in molecular biology have resulted in improved strategies for the identification and cloning of genetic elements displaying differential patterns of gene expression (8, 9). A potentially powerful approach for identifying genes that are differentially expressed between closely related cell populations employs a technique termed subtractive hybridization (10–12). The basic substractive hybridization procedure involves constructing cDNA libraries from mRNAs isolated from the cell type expressing the desired phenotype (or induced to produce the desired phenotype); hybridizing these first-strand cDNAs with an excess (of at least 10-fold) of mRNAs from a matched cell type not expressing the desired phenotype; isolating single-stranded unhybridized subtracted cDNAs from the RNA/cDNA hybrids; and screening cDNA libraries for the identification of differentially expressed or induced sequences (10, 13). This approach has been successfully employed to identify genes that are specifically expressed at growth arrest (14), induced by specific DNA damaging agents (15), expressed during specific stages of B cell development (16), expressed at specific stages of malignant progression (17), and associated with programmed cell death (18). By appropriate modifications, the basic subtractive hybridization technique has been improved for the identification of rare transcripts (16, 19–21), or transcripts exhibiting small differences in expression between two cell types (15, 19). A further refinement in subtractive hybridization methodologies, resulting in increased sensitivity for the detection of differentially expressed transcripts, involves the use of polymerase chain reaction (PCR) amplification (18–23). In addition, studies have described the construction of uniform-abundance (normalized) cDNA libraries capable of detecting and isolating cDNAs corresponding to species of mRNA present at a low level in only a small proportion of cells in a complex tissue (23). Although it is beyond the scope of this article to describe these refinements in detail, several other modifications have been reported that facilitate the separation of the hybridized from the subtracted cDNAs. These either employ an oligo(dT) primer, conjugated to a small latex bead, to initiate cDNA synthesis (24) or utilize photoactivatable biotin and a phenol extraction (25).

In the present article we describe a model cell culture system that is ideally suited for the identification of gene(s) that function as both positive regulators (progression inducing) and negative regulators (progression suppression) of the progression phenotype. Strategies for the construction of cDNA libraries and the application of subtractive hybridization procedures to identify and isolate appropriate progression-related genes are also described. The basic procedure described here utilizes commercially available cDNA synthesis and RNA transcription kits (Stratagene, La Jolla, CA) with modifications appropriate for subtractive hybridization applications.

Analysis of Progression of Transformed Phenotype Using Adenovirus Type 5-Transformed Rat Embryo Cells

Transformation of rat embryo (RE) cells by adenovirus type 5 (Ad5) is often a multistep process that is affected by diverse factors, including chemical carcinogens, hormones, and tumor promoters (26–29). In this model system, morphological transformation can precede anchorage-independent growth and acquisition of oncogenic potential (27, 28). By isolating cells following anchorage-independent growth in agar or after tumor induction in nude mice, a series of Ad5-transformed Sprague-Dawley RE clones has been obtained that stably displays an increased expression of the transformed state, that is, the *progression* phenotype (27, 28, 30, 31). The progression phenotype is associated with enhanced anchorage-independent growth and decreased latency periods for tumor induction in nude mice in comparison with their respective parental unprogressed Ad5-transformed RE clone (Table I) (27, 28, 30–32). The process of progression in Ad5-transformed cells was shown to be reversible by the single application of the demethylating agent 5-azacytidine (Aza) (31). In these studies, a high percentage of Aza-treated Ad5-transformed RE clones (≥ 95%) stably reacquired the unprogressed phenotype of the original parental Ad5-transformed RE clone following a single 24-hr exposure to Aza and then growth in the absence of Aza (31). In addition, by selecting Aza-treated unprogressed Ad5-transformed cells for colony formation in agar or after tumor induction in nude mice, the progression phenotype was stably reacquired by these Ad5-transformed cells (31) (Table I). These observations have led to the hypothesis that progression in this system is a reversible process that may be the consequence of inactivation of a progression suppressor gene(s) in Ad5-transformed RE cells. The identity of this progression suppressor gene(s) remains to be determined, but based on the reversibility of the progression phenotype by Aza, the transcriptional activity of the progression suppressor gene(s) may be correlated with the status of methylation of its promoter region. Alternatively, progression in Ad5-transformed RE cells could involve the induction or enhanced expression of a specific progression-inducing gene(s) whose effect on cellular phenotype is mediated in a coordi-

TABLE I Genetic Control of Progression Phenotype Investigated by Somatic Cell Hybridization

Cell Line[a]	Growth in agar (%)[b]	Tumorigenicity in nude mice[c]	Progression phenotype[d]
Rat embryo	<0.001	0/5	Neg
CREF	<0.001	0/14	Neg
CREF^{cap}	<0.001	0/6	Neg
E11	2.9 ± 0.3	8/8 (36)	Neg
E11-NMT	34.3 ± 4.1	12/12 (20)	**Pos**
E11-NMT Aza 10-11	2.2 ± 0.3	6/6 (48)	Neg
E11-NMT Aza 10-11 R5	33.2 ± 4.6	3/3 (17)	**Pos**
E11^{hyg}	3.9 ± 0.6	3/3 (45)	Neg
E11-NMT^{neo}	39.0 ± 3.7	3/3 (19)	**Pos**
Somatic cell hybrids			
CREF^{cap} X E11-NMT^{neo}			
R1	72.2 ± 9.4	3/3 (17)	**Pos***
R2	57.4 ± 6.9	3/3 (17)	**Pos***
F1	2.0 ± 0.3	0/3	Neg
F2	1.5 ± 0.1	0/3	Neg
E11^{hyg} X E11-NMT^{neo}			
IIId	5.6 ± 0.7	3/3 (56)	Neg
3b	1.5 ± 0.2	3/3 (31)	Neg
IIa	30.1 ± 3.5	3/3 (12)	**Pos**

[a] Cell lines: rat embryo, 2° Sprague-Dawley rat embryo cell culture (26); CREF, specific clone of Fischer rat embryo fibroblast cells (35, 36); CREF^{cap}, chloramphenicol-resistant CREF cell clone (32); E11, H5ts125-transformed 2° Sprague-Dawley rat embryo clone (26); E11-NMT, subclone of E11 cells isolated from a tumor induced in a nude mouse by E11 cells (31); E11-NMT Aza 10-11, subclone of E11-NMT cells isolated after a single treatment with $10 \mu M$ Aza (31); E11-NMT Aza 10-11 R5, subclone of E11-NMT Aza 10-11 cells isolated after growth in 0.4% agar (31); E11^{hyg}, hygromycin-resistant E11 subclone obtained by transfection with pRSV1.1 (which contains a cloned bacterial hygromycin gene) and selection in hygromycin (Z.-Z. Su and P. B. Fischer, unpublished observations, 1991); E11-NMT^{neo}, G418 resistant E11-NMT subclone obtained by transfection with pSV2*neo* (which contains a cloned bacterial neomycin gene) and selection in G418 (32); R1 and R2, hybrid clones with a round morphology formed by fusing CREF^{cap} and E11-NMT^{neo} cells and selecting cells in chloramphenicol and G418 (*, enhanced expression of the progression phenotype) (32); F1 and F2, hybrid clones with a flat morphology formed by fusing CREF^{cap} and E11-NMT^{neo} cells and selecting cells in chloramphenicol and G418; IIId, 3b, and IIa, hybrid clones formed by fusing E11^{hyg} and E11-NMT^{neo} cells and selecting cells in hygromycin and G418.

[b] Percentage cloning efficiency in agar was determined by seeding cells at 1×10^3, 5×10^3, 5×10^4, or 1×10^5 in medium (Dulbecco's modified Eagle's medium supplemented with 10% fetal bovine serum) containing 0.4% (w/v) Noble agar on a 0.8% (w/v) Noble agar/medium base layer. Cultures were fed every 4 days with medium containing 0.4% (w/v) Noble agar and colonies >0.1 mm were determined after 21 days of growth. Further details can be found in Refs. 27 and 28.

[c] Four-week-old BALB/c nude mice were injected with 1×10^6 cells. Results indicate number of animals with tumors per number of animals injected and numbers in parentheses indicate average latency time in days, that is, first appearance of a palpable tumor.

[d] Neg, Absence of the progression phenotype; Pos, expression of the progression phenotype.

nated manner by the inactivation of the progression suppressor gene(s). If either or both of the above genetic changes are involved in progression and if progression is regulated transcriptionally (as is suggested by the Aza studies), then by employing appropriate strategies it should be possible to identify and isolate both progression-suppressing and progression-inducing genes from stage-specific Ad5-transformed cells.

The identity of the genetic elements that control the progression phenotype in Ad5-transformed RE cells are not presently known. Detailed studies of this phenomenon have employed a series of Ad5-transformed Sprague-Dawley RE cells displaying stable differences in their expression of the progression phenotype (1, 7, 26–31) (Table I), including E11 (*unprogressed* parental Ad5-transformed RE clone), E11-NMT (a subclone of E11 cells derived from a nude mouse tumor that displays a *progressed* phenotype), Aza 10-11 (a single-cell clone of E11-NMT cells isolated after a single 24-hr exposure to 10 μM Aza that displays an *unprogressed* phenotype), and Aza 10-11 R5 (a subclone of Aza 10-11 cells reisolated following growth in 0.4% (w/v) agar medium and displaying a *progressed* phenotype). In prior studies, progression of the transformed phenotype in Ad5-transformed RE cells was shown not to involve an alteration in the pattern of integration of the Ad5 transforming gene(s) (1, 30, 31). Progression was also shown not to be associated with alterations in the steady-state levels of mRNA produced by the viral transforming genes, E1A and E1B, or the cellular gene c-*myc* (7) (Fig. 1).

The tumor-promoting agent, 12-*O*-tetradecanoylphorbol 13-acetate (TPA), which has been shown to induce a progressed phenotype in Ad5-transformed RE cells (1, 27, 28, 30), did not alter the transcription rates of the Ad5 *E1A* or *E1B* genes, the TPA-inducible gene *TPA-S1*, or the TPA-responsive genes *Pro1* or protein kinase C (*PKC*) (Fig. 2) (7). The agent TPA did, however, increase within 1 hr the steady-state level of c-*fos* mRNA to an equivalent extent in both progressed E11-NMT and unprogressed E11 cells (7). Progression also did not involve a change in the RNA transcription rate of a number of cellular and viral oncogenes, including actin, c-Ha-*ras*, c-*myc*, v-*fos*, *erbB*, TGF-α (transforming growth factor α), TGF-β, Pro2, transin, TPA-R1, v-*myb*, and c-*mos*, or other adenovirus genes in addition to E1A and E1B, including E2A and E4 (Fig. 2) (7). Employing E1B polyclonal antibody, which recognizes an M_r 21,000 polypeptide encoded by E1B, and immunoblotting analysis, it was further demonstrated that progression was not associated with a change in the levels of this Ad5-encoded transforming protein (7). Similarly, immunoprecipitation analysis with an Ad2 E1A monoclonal antibody indicated similar levels of the M_r 55,000 and 48,000 E1A polypeptides, as well as coprecipitated proteins of M_r 300,000, 107,000, and 105,000 [which is the retinoblastoma (Rb) protein] in progressed E11-NMT and unprogressed E11 cells (Fig. 3).

Immunoprecipitation of cell lysates with a monoclonal antibody specific for the

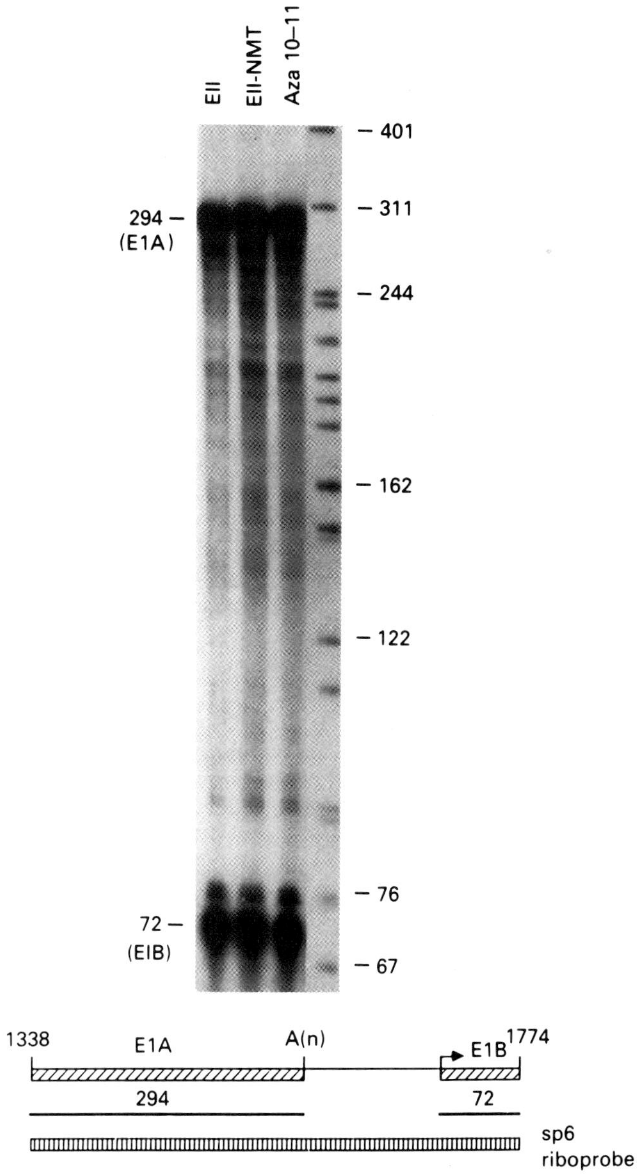

FIG. 1 Steady state mRNA levels of the viral E1A and E1B mRNAs in cells displaying different stages in the transformation progression phenotype. *Top:* Cytoplasmic RNA was isolated from each of the cell lines shown. Twenty micrograms of RNA was hybridized to a uniformly ^{32}P-labeled sp6-generated antisense RNA probe [a total of 7.5×10^5 counts per minute (cpm)] representing the E1A second exon and the E1B gene cap site (see diagram at the bottom). Following RNase T2 digestion and gel electrophoresis, protected RNAs of 294 (E1A) and 72 (E1B) nucleotides are observed. *Bottom:* Exonic sequences are depicted by hatched boxes. [Reprinted with permission from Duigou *et al.* (7).]

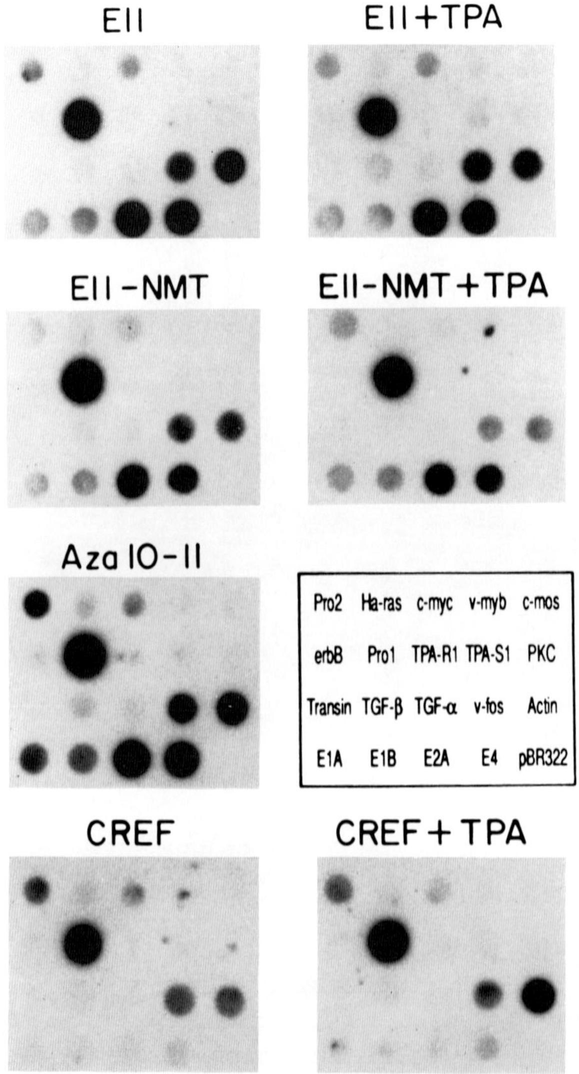

FIG. 2 Analysis of viral and cellular gene transcription rates for cell lines stably displaying different stages in the transformation progression lineage in the presence and absence of 12-*O*-tetradecanoylphorbol 13-acetate (TPA). Nuclei from 1×10^8 cells were isolated from each of the cell lines shown, and nuclear RNA was labeled *in vitro* and subsequently hybridized to denatured DNA probes on nitrocellulose filters. For each hybridization reaction, an equal number of total counts representing a similar number of cell nuclei was used, so that comparative rates of transcription could be obtained. Cells treated with TPA (100 ng/ml) were serum starved for 48 hr prior to treatment. Further details can be found in Ref. 7. [Reprinted with permission from Duigou *et al.* (7).]

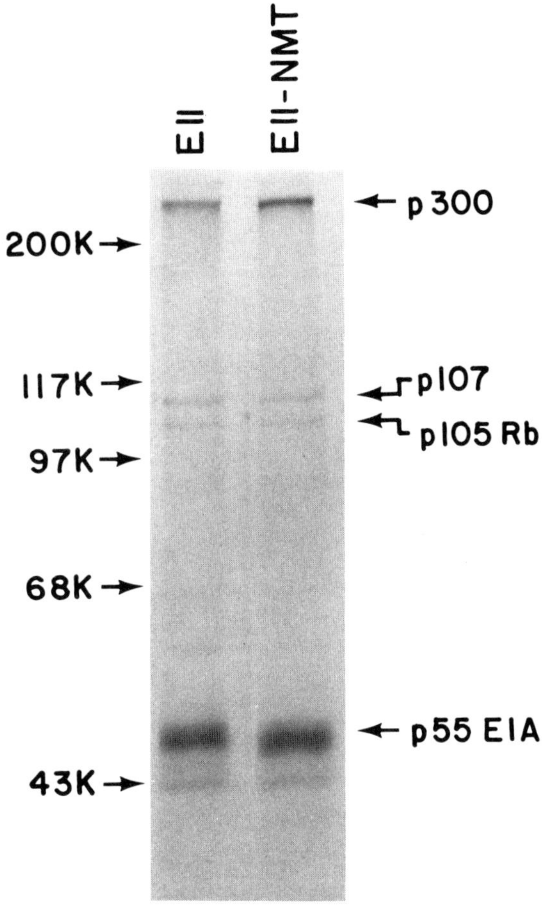

FIG. 3 Expression of E1A-encoded polypeptides in cells displaying different stages in the progression phenotype. E11 and E11-NMT cell lysates were immunoprecipitated with an Ad2 E1A monoclonal antibody (Ad2 E1A Ab-1) as described in Ref. 7. Molecular weight markers are indicated on the left-hand side. Coprecipitated p300, p107, and hypophosphorylated p105 Rb proteins and the p55 E1A proteins are indicated on the right-hand side. [Reprinted with permission from Duigou *et al.* (7).]

M_r 105,000 Rb protein also demonstrated that progression was not associated with a change in the level or state of phosphorylation of the Rb protein (7). However, transfection of a human *Rb* gene (also containing a neomycin resistance gene) into Ad5-transformed RE cells was more inhibitory, with respect to formation of G418-resistant colonies, in unprogressed than in progressed Ad5-transformed RE cells. These findings suggest that progression of the transformed phenotype in

Ad5-transformed RE cells does not involve the altered expression of virus-encoded transforming genes. Moreover, progression of these cells is not associated with an alteration of genes previously shown to be associated with enhanced expression of the transformed cell phenotype in other progression model cell culture systems. These results indicate that progression of the transformed phenotype in Ad5-transformed RE cells is controlled by a cellular-encoded gene(s) that modulates stages of progression without directly altering expression of viral genes or the *Rb* tumor suppressor gene, which is associated with expression of the transformed phenotype.

In addition to occurring spontaneously following selection for tumor formation in nude mice (*spontaneous progression*), the progression phenotype can also be induced in E11 cells by transfection with and expression of the Ha-*ras* oncogene (E11-*ras*) (referred to as *oncogene-induced progression*) or the β_1 *PKC* gene (E11-PKC) (referred to as *growth factor-related, gene-induced progression*) (33, 34). As was observed for spontaneous progression, both oncogene-induced and growth factor-related, gene-induced progression also appear to be regulated by the expression of a progression suppressor gene that is responsive to Aza demethylation. A single exposure of E11-*ras* or E11-*PKC* cells (plated at clonal densities) to Aza for 24 hr, followed by growth in the absence of Aza, results in clones that stably display a reversion in the progression phenotype. These findings support the hypothesis that the putative progression suppressor gene(s) when expressed in E11 cells can directly modulate the progression phenotype independent of the mode of induction of progression, that is, spontaneously, as a result of Ha-*ras* expression or as a consequence of overexpression of β_1 *PKC*.

The ability to analyze the molecular basis of adenovirus transformation critically has been increased by the isolation of a specific cloned rat embryo fibroblast cell line, CREF (Fig. 4) (35, 36). CREF cells were initially identified on the basis of their increased susceptibility to transformation by adenovirus (35, 36). CREF cells are transformed >100- to 200-fold more efficiently than early-passage rat embryo cells or other established rat embryo cell lines by Ad5 (36). In addition, CREF cells are contact inhibited, they are anchorage dependent, they do not undergo spontaneous transformation, they are resistant to transformation induced by both chemical and physical carcinogens, and they are nontumorigenic in both nude mice and syngenic Fischer rats (37, 38). With the exception of being immortal, CREF cells display all of the properties normally associated with normal cells. As will be discussed below, CREF cells also appear to express a dominant-acting suppressor gene that can suppress the progression phenotype of E11-NMT cells without altering Ad5 gene expression (32).

To analyze the genetic regulation of progression we have also employed somatic cell hybridization techniques (32). Hybrids were constructed between neomycin-resistant E11-NMT^{neo} cells and an untransformed chloramphenicol-resistant rat embryo fibroblast cell line, CREF^{cap}. As observed with neomycin-sensitive E11-NMT cells, E11-NMT^{neo} cells grew efficiently in agar and displayed a reduced latency

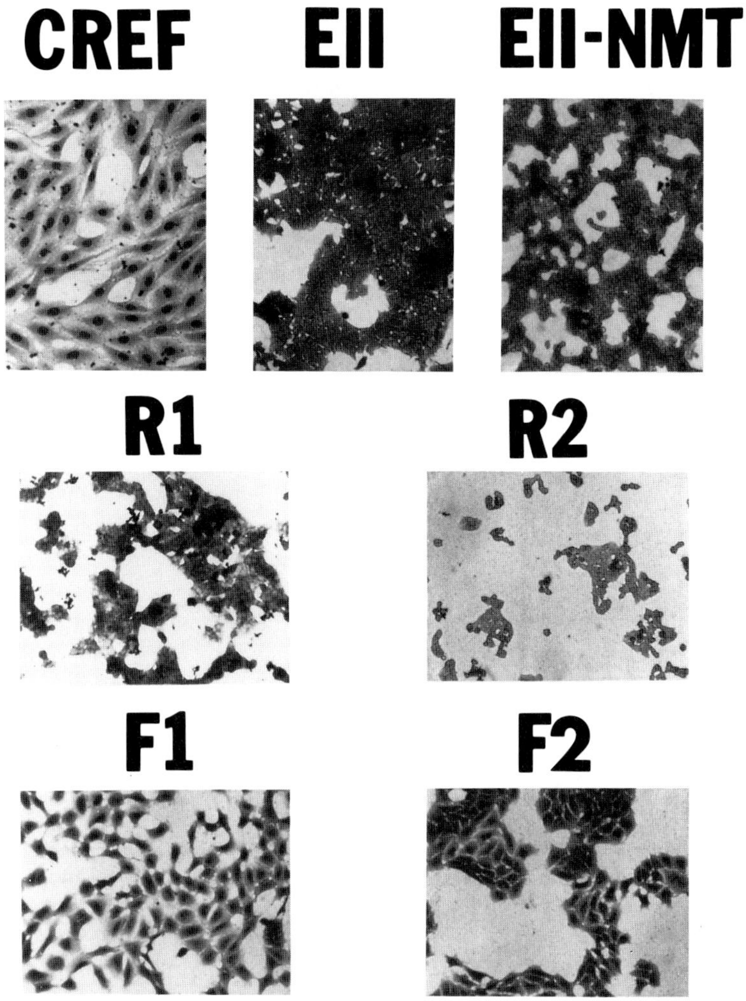

FIG. 4 Morphology of $CREF^{cap}$, E11, E11-NMT^{neo}, and $CREF^{cap}$/E11-NMT^{neo} hybrid R1, R2, F1, and F2 cells. Description of cells can be found in the legend to Table I and Ref. 32. Cells were grown in 35-mm dishes, fixed with 4% formaldehyde, and stained with Giemsa stain. Magnification: ca. ×200. [Reprinted with permission from Duigou *et al.* (32).]

time versus the time required by E11 cells for tumor formation in nude mice, that is, the progressed phenotype. A high proportion (84%) of the presumptive hybrid cell types obtained after fusion and genetic selection (G418 and chloramphenicol) displayed a flat morphological phenotype intermediate between $CREF^{cap}$ and E11-NMT^{neo} cells, suggesting a trans-dominant extinction phenomenon had occurred

(Fig. 4). Two hybrids with a round morphology (R), which still exhibited the progressed phenotype, and two hybrids with a flat morphology (F), which had lost the progressed phenotype, were employed for detailed analysis (32). The F hybrids displayed a reduced growth in agar (similar to unprogressed E11 cells), but they were completely devoid of tumorigenic potential in nude mice (Table I). This observation suggests that growth in agar and tumorigenic capacity were dissociatable phenomena in the F hybrids. An analysis of the viral DNA integration patterns and the rates of transcription, steady-state mRNA accumulation, and relative levels of Ad5 *E1A* and *E1B* gene products revealed no differences among the parental and hybrid cells (Figs. 5 and 6). These studies indicate that normal CREF cells may contain a suppressor gene(s) that can inhibit the expression of specific traits of the progression phenotype in Ad5-transformed cells and that this suppression is not associated with changes in the expression of Ad5 transforming genes.

To determine if the progressed phenotype of E11-NMT was dominant or recessive to the unprogressed phenotype of E11 cells, an additional set of somatic cell hybrids has been constructed between hygromycin-resistant E11 cells, $E11^{hyg}$ and $E11\text{-}NMT^{neo}$ cells (Table I). Hybrid clones have been isolated after growth in medium containing hygromycin and G418, and they have been confirmed as hybrids by karyotype analysis. Two types of hybrids have been identified, some that display the unprogressed E11 phenotype (such as IIId and 3b) and others that display the progressed E11-NMT phenotype (such as IIa). These studies indicate that the progression suppressor gene function of E11 cells can inhibit the progression phenotype in E11-NMT cells. The mechanism by which this trans-dominant suppressor function of E11 cells exerts its effect on E11-NMT cells remains to be determined.

The various Ad5-transformed RE clones and somatic cell hybrids described above and in Table I represent valuable experimental tools for analyzing the genetic basis of the progression phenotype, and they represent ideal systems for the identification and cloning of genes involved in progression. In addition, the various cell strains will also prove useful in defining the relevance of specific gene expression changes in mediating the progression phenotype. As one approach to assigning functional roles to progression-suppressing and progression-inducing genes (once identified and cloned), one could employ sense (5′ to 3′) and antisense (3′ to 5′) expression vector constructs containing appropriate progression cDNAs. By using inducible promoters, such as the Moloney murine tumor virus (MMTV) promoter region, which is responsive to dexamethasone, to control expression of the sense and antisense genes, a direct relationship between expression of a specific construct and the progression phenotype could be obtained (39). A direct effect of a specific negative-acting progression-suppressing (suppressor) cDNA in modulating progression would be indicated by the ability of the sense construct to induce the unprogressed phenotype in E11-NMT cells and the antisense construct to induce the progression phenotype in E11 cells. A direct effect of a specific positive-acting progression-inducing cDNA in modulating progression would be indicated by the

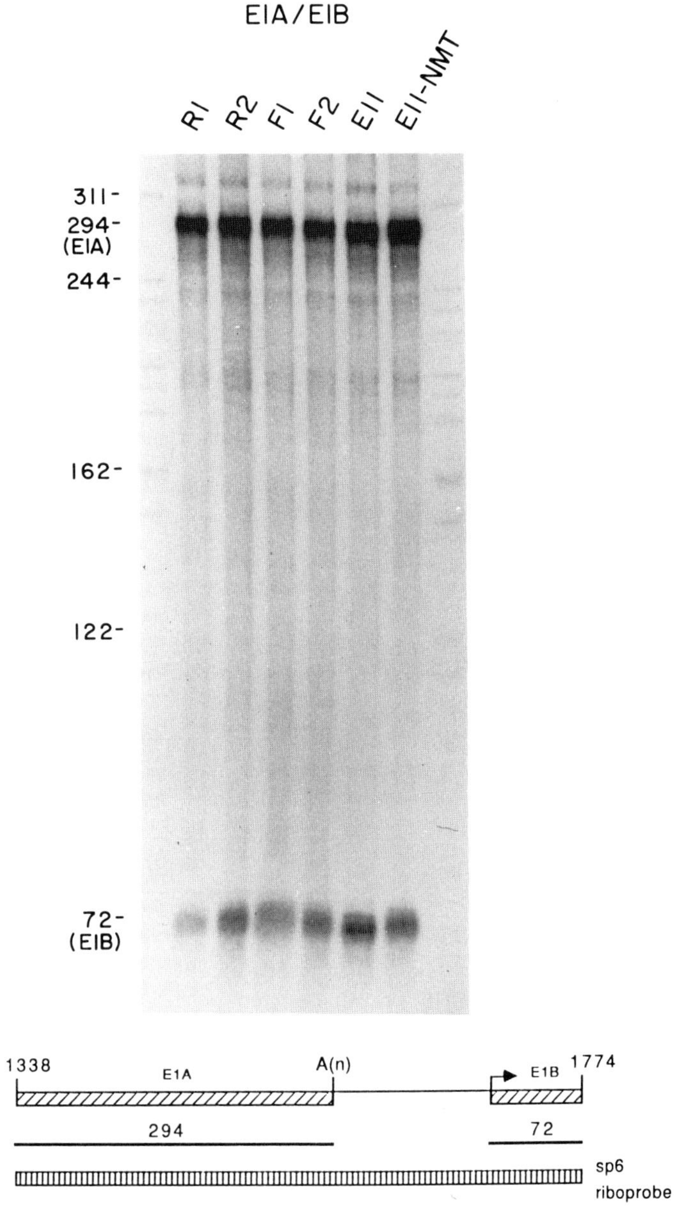

FIG. 5 Analysis of steady-state levels of Ad5 *E1A* and *E1B* RNA. Total cytoplasmic RNA (12.5 μg) extracted from each cell line was hybridized for 24 hr to antisense ^{32}P-labeled RNA probes containing sequences from the *E1A* and *E1B* transcription units of Ad5. RNase T_2-resistant hybrids were fractionated through denaturing 5% polyacrylamide gels, and the gels were autoradiographed. RNase-resistant fragments of predicted sizes of 294 nt (E1A specific) and 72 nt (E1B specific) were localized by using a labeled size standard run in lanes flanking analyzed RNAs. [Reprinted with permission from Duigou *et al.* (32).]

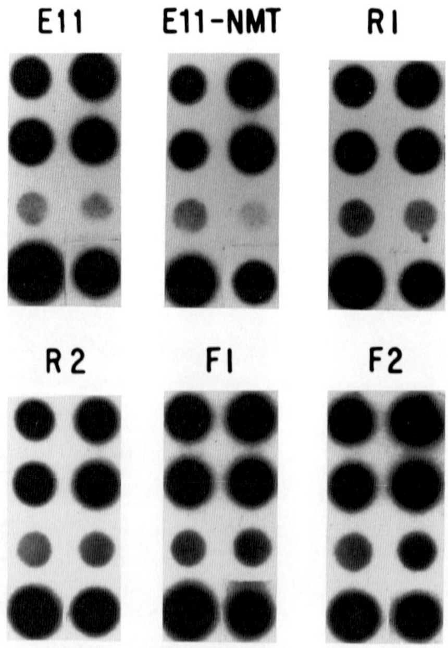

E1A	tubulin
E1B	actin
pBR322	ligandin
E4	E2

FIG. 6 Analysis of transcription in Ad5-transformed E11, E11-NMT^{neo}, and hybrid cells. [^{32}P]UTP-Labeled RNA was extracted from isolated nuclei after chain elongation of nascent transcripts and used as a hybridization probe of various cloned DNAs immobilized as dots on nitrocellulose filters. Each dot contained 7 μg of DNA that was hybridized to 10^7 cpm of labeled RNA from each cell line, hybrids were digested with RNase, and filters were subjected to autoradiography. The types and locations of cloned DNAs immobilized in each dot are shown in the box under the autoradiograph. pBR322 DNA was included as a control to determine the amount of nonspecific hybridization in this experiment. Further details can be found in Ref. 30. [Reprinted with permission from Duigou *et al.* (32).]

ability of the sense construct to induce the progression phenotype in E11 cells and the antisense construct to induce the unprogressed phenotype in E11-NMT cells. Strategies for the identification of appropriate progression suppressor and progression inducer cDNAs from E11 and E11-NMT cells, respectively, will be described in detail below.

Construction of cDNA Libraries

Although originally difficult, the ability to synthesize large cDNA libraries with a high proportion of full-length inserts is now readily achievable using commercially available reagents (reviewed in Refs. 8 and 9). A complete cDNA library should contain at least one cDNA corresponding to each mRNA in a cell. It is estimated that a given mammalian cell will contain approximately 34,000 different types of mRNA and approximately 500,000 mRNA molecules will be present per mammalian cell. A complete library containing at least 1 copy of every mRNA molecule should, therefore, consist of approximately 500,000 to 1 million independent cDNA clones. For our initial cDNA library constructions we have employed the ZAP-cDNA synthesis kit available from Stratagene, which utilizes many of the common cDNA synthesis strategies (9, 40). The basic scheme for the construction of E11 and E11-NMT cDNA libraries in the λZAP phage vector is shown in Fig. 7. The reasons we choose the λZAP vector and the Stratagene-cDNA synthesis kit include the following: (1) the use of methyl-dCTP in the initial step of the first-strand cDNA synthesis eliminates the need for *Eco*RI methylase for the internal methylation of the *Eco*RI sites; (2) the λZAP system permits the construction of unidirectional cDNA libraries by employing an *Xho*I primer/linker; (3) the cDNA insert can easily be converted into sense or antisense RNA by T3 or T7 RNA polymerase, which can then be used as a probe or in the construction of subtracted cDNA libraries or for *in situ* hybridization; (4) the cDNA insert in the λZAP vector can be excised directly into a pBluescript phagemid (Stratagene, La Jolla, CA) without having to reclone the insert into another plasmid vector; and (5) the sense RNA copy of the cDNA can be used for *in vitro* translation to identify its encoded protein.

A key component in the development of cDNA libraries is the isolation of undegraded RNA of high quality. Suitable total cellular RNA preparations for cDNA library construction can be obtained using the guanidinium thiocyanate technique (41), the Fast Track lysis procedure (Invitrogen, San Diego, CA), or the Stratagene RNA isolation kit. Because RNA is easily degraded by RNase A and T1 (which are resistant to inactivation), it is imperative that all equipment, reagents, and manipulations (because oils present on skin contain RNase) be handled appropriately. Thus, distilled water (dH_2O) and all reagents that may come in contact with RNA should be treated with 0.1% (v/v) of the irreversible enzyme inhibitor diethyl pyrocarbonate (DEPC), followed by autoclaving to destroy the inhibitor (DEPC treatment). Some reagents (such as nucleotide triphosphates) that would be destroyed by autoclaving should be dissolved in DEPC-treated dH_2O and then passed through a 0.22-μm sterile filter to remove any possible bacterial contamination. Additionally, reusable glassware should have their final rinse with DEPC-containing dH_2O and subsequently should be autoclaved or extensively baked. Disposable materials (such as microfuge tubes and pipette tips) should always be handled with gloves and, while they could be DEPC treated, usually an autoclaving is sufficient.

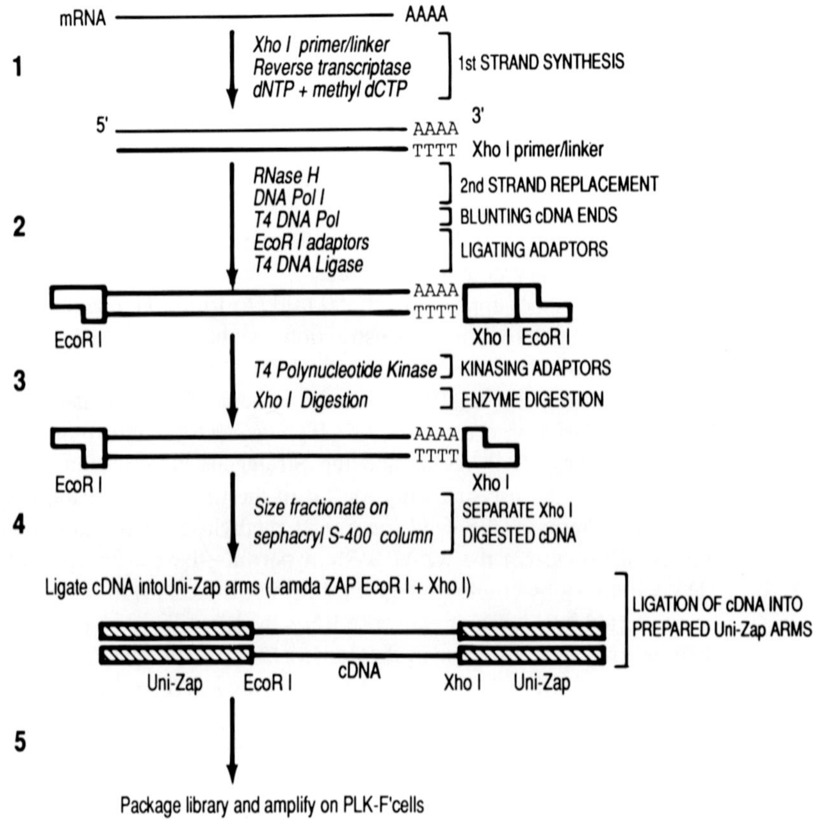

FIG. 7 Schematic diagram for the construction of E11 and E11-NMT cDNA libraries. The basic sequence for cDNA library construction is shown in this diagram, including: (1) First-strand synthesis; (2) second-strand replacement, blunt ending cDNA, and ligating cDNA to adaptors; (3) kinasing adaptors and *Xho*I enzyme digestion; (4) separation of *Xho*I-digested cDNA; and (5) ligation of cDNA into prepared Uni-Zap arms of the λZAP vector.

If total RNA is to be obtained from paired samples (for the construction of libraries to be used in differential and subtractive screening procedures), it is important that the cells be grown under similar conditions. This will reduce and hopefully avoid gene expression differences based on growth or treatment conditions and will result in representative RNAs reflecting inherent constitutive changes in the two cell populations. In the case of E11 and E11-NMT cells, cultures are grown to 75% confluence in 100-cm tissue culture plates at 37°C in a humidified 5% CO_2/95% air incubator in Dulbecco's modified Eagle's medium supplemented with 5% fetal bovine serum (DMEM-5), and 24 hr prior to RNA isolation the medium is changed.

The isolation of total RNA after shorter periods following the medium change increases the risk of isolating differentially expressed genes responding to growth factors present in the medium.

From the total RNA population, which contains rRNA and tRNA in addition to the mRNA transcripts, the poly(A)$^+$ RNA fraction is selected by passage through an oligo (dT)-cellulose column (9). For the preparation of cDNA libraries, a total of 5 μg of poly(A)$^+$ RNA from E11 or E11-NMT cells is used. As indicated above, the procedures we have employed for constructing cDNA libraries and cloning cDNAs into the Uni-Zap arms of the λZAP vector were as described in the ZAP-cDNA synthesis kit protocol outlined by Stratagene. The first step in cDNA synthesis involves first-strand synthesis (Fig. 7). In this procedure mRNA (the template) is converted into a double-stranded RNA/DNA hybrid molecule by catalysis with reverse transcriptase in the presence of nucleotides (dNTPs: dATP, dGTP, dTTP, and 5-methyl dCTP), an appropriate buffer, and a primer (*Xho*I primer/linker). The *Xho*I primer/linker incorporated in the ZAP-cDNA synthesis kit is a 50-base oligomer (5′ GAGAGAGAGAGAGAGAGAGAACTAGTCTCGAG TTTTTTTTTTTTTTTTTT 3′). The underlined sequence encodes the *Xho*I restriction enzyme recognition site used in the cloning procedure, and the presence of the "GAGA" sequence at the 5′ end ensures survival of the full cleavage site throughout the procedures employed. The 18-base poly(dT) sequence at the 3′ end of the oligomer permits hybridization with the poly(A) tail on the mRNA molecule. This primer/linker allows the completed cDNA to be inserted into the λZAP vector (42) in a sense orientation (*Eco*RI–*Xho*I) with respect to the *LacZ* promoter. By employing high concentrations of Moloney murine leukemia virus reverse transcriptase (M-MuLVRT) and dNTPs, synthesis of full-length cDNAs is facilitated. In addition, by employing the dNTP analog 5-methyl-dCTP in the first-strand synthesis, the methyl group on each cytosine will protect the cDNA from cleavage by the restriction enzymes used in later cloning steps.

Step 1: First-Strand Synthesis

In a 1.5-ml sterile and RNase-free microcentrifuge tube, the following materials (supplied in the ZAP-cDNA synthesis kit; Stratagene) are added in the indicated sequence: 5μl 10× first-strand buffer, 2.5 μl first-strand methyl nucleotide mixture (10 m*M* dATP, dGTP, and dTTP and 5 mM 5-methyl-dCTP), 2 μl linker–primer (1.4 μg/μl), 32 μl DEPC-treated H_2O, and 5 μl RNase Block II (Stratagene, La Jolla) (1 U/μl). The tube is then vortexed and 5 μl (containing 5 μg) of E11 or E11-NMT poly(A)$^+$ mRNA is added and the tube is again gently vortexed. The tube is then left at room temperature for 10 min to permit the primer to anneal to the template mRNA. To this mixture 2.5 μl of M-MuLVRT (18 U/μl) is added (total volume, 50 μl). The tube is then vortexed briefly, the contents in the microcentrifuge tube are spun down, and 5 μl is aliquoted to a separate microcentrifuge tube containing 0.5 μl of [α-^{32}P]dATP (specific activity, 800 Ci/mmol; Amersham, Arlington Heights, IL). This control sample is utilized to determine the efficiency and quality (by running a gel) of the first-strand synthesis. Both the control (5.5 μl) and experimental (45 μl) samples are then incubated at 37°C for 1 hr. After this point the control labeled sample is placed at −20°C (for subsequent analysis) and the experimental sample is placed at 4°C.

Subsequent steps in cDNA library construction include second-strand cDNA synthesis, blunting cDNA ends, and ligating *Eco*RI adaptors (Fig. 7). A major factor in the success of the second-strand replacement is the quality of the first-strand cDNA (total yield, length, and purity) (43). It is also important to follow directly, using labeled dNTP, the synthesis of the second strand (44). For second-strand synthesis, the RNA bound to the first-strand cDNA is nicked with RNase H, resulting in the production of a large number of fragments that can be employed as primers for *Escherichia coli* DNA polymerase I (DNA Pol I). The RNA fragments are synthesized into second-strand cDNAs by a nick translation-type reaction catalyzed by DNA Pol I. To reduce the possibility of 5-methyl-dCTP incorporation into the second strand, the second-strand nucleotide mixture contains an increased concentration of dCTP. This modification is important to increase the probability that the appropriate restriction sites in the linker/primer are responsive to cleavage with the suitable restriction endonucleases. Because the double-stranded cDNAs usually can contain uneven terminal regions, the termini are enzymatically removed or new nucleotides are incorporated by using T4 DNA polymerase (T4 DNA Pol). Following this procedure, *Eco*RI adaptors [consisting of a 9-base oligomer (3′ GCCGTGCTC 5′) and a 13-base oligomer (5′ AATTCGGCACGAG 3′)] are then able to be ligated to the blunt ends of the cDNAs.

Step 2: Second-Strand Replacement

For second-strand synthesis it is important that all reagents be kept at temperatures less than 16°C when DNA Pol I is added. This is critical because temperatures above 16°C will result in the development of hairpin secondary structures that will interfere with successful incorporation of the cDNAs into the vector. To the 45 μl of nonradioactive first-strand cDNA mixture the following materials (as indicated in the Stratagene ZAP-cDNA synthesis kit protocol) are added in the following sequence: 40 μl of 10× second-strand buffers, 6 μl of second-strand nucleotide mixture (10 m*M* dATP, dGTP, and dTTP and 36 m*M* dCTP), 297 μl of sterile dH_2O (it is not necessary at this stage to use DEPC-treated dH_2O), and 2 μl of

[α-^{32}P]dATP (specific activity, 800 Ci/mmol; Amersham). To the sides of the microcentrifuge tube add 0.8 μl of RNase H (4 U/μl) and 9 μl of DNA Pol I (11 U/μl). Quickly vortex the mixture, briefly spin the tube at maximum speed for 1 to 2 min at room temperature in a microcentrifuge to pellet the reaction mix, and incubate for 2.5 hr at 16°C. *Caution:* As indicated above, make sure the incubation temperature does not exceed 16°C. After second-strand synthesis add 400 μl of phenol : chloroform (1 : 1, v/v). Microfuge the reaction mixture for 2 to 3 min at maximum speed at room temperature and then remove the upper aqueous layer (containing the cDNA) into a new microcentrifuge tube. To the aqueous layer add an equal volume of chloroform, vortex, microfuge as above, and collect the upper aqueous layer into a new microcentrifuge tube. The reaction mixture is then precipitated overnight by the addition of 33.3 μl of 3 *M* sodium acetate (final concentration, 0.1 *M*) and 867 μl of 100% ethanol (2.2 vol). To orient the direction in which the precipitate will form it is advisable to mark the microcentrifuge tube. The cDNAs are pelleted by microcentrifuging the mixture at 4°C or room temperature for 20 to 30 min at maximum speed. Remove the radioactive supernatant (which contains unincorporated counts) and discard. Gently wash the pellet by adding 1 ml of 70% ethanol (away from the precipitate) and microcentrifuge the sample for 2 min at room temperature at maximum speed. Aspirate the ethanol and dry the pellet in a Speed-Vac (Savant, Hicksville, NY) (or other suitable apparatus) until completely dry. Resuspend the pellet in 43.5 μl of sterile dH_2O and remove 4.5 μl and freeze at −20°C (for analysis of the quality of the second-strand synthesis).

Step 3: Blunting cDNA Termini

The cDNA product from above is converted to blunt ends by the addition of T4 DNA Pol. To the 39-μl cDNA sample from above the following reagents are added from the Stratagene ZAP-cDNA synthesis kit in the following order: 5 μl of 10× T4 DNA Pol buffer; 2.5 μl of 2.5 m*M* dNTP mix; and 3.5 μl of T4 DNA Pol (2.9 U/μl). The mixture is then incubated at 37°C for 30 min. This step is followed by the addition of 50 μl of dH_2O (to bring the final volume to 100 μl), extraction with 100 μl of phenol–chloroform, and then a second extraction with an equal volume of chloroform. The supernatant is then transferred to a new microcentrifuge tube and precipitated (for a minimum of 30 min at −70°C) by the addition of 7 μl of 3 *M* sodium acetate and 226 μl of 100% ethanol. Using the appropriate orientation mark the reaction mixture is microcentrifuged at maximum speed for 20 to 30 min at either room temperature or 4°C. If no pellet forms, then it is advisable to precipitate the supernatant overnight at −20°C. Wash the pellet with 150 μl of 70% ethanol, remove the wash solution, and dry the pellet as before until completely dry.

Step 4: Ligating Eco*RI Adaptors*

The pellet from above is resuspended by vortexing in 14 μl of *Eco*RI adaptors (0.4 μg/μl) and 2 μl of 10× ligase buffer (contained in the Stratagene ZAP-cDNA synthesis kit). A Geiger counter is utilized to be certain that the cDNA is resuspended. The mixture is incubated for 10 min at 37°C and 2 μl of 10 m*M* ATP and 2 μl of T4 DNA ligase is then added. The reaction mixture is first vortexed, then briefly spun in a microcentrifuge and incubated overnight at 4°C. (*Note:* Adaptor ligation efficiency is increased by longer ligation times.) The following day T4 DNA ligase is inactivated by heating at 70°C for 30 min.

The *Eco*RI adaptor oligomers are partially complementary, thus one end of the adapter is blunt and is supplied with a phosphate on the 5′ end of the 9-mer oligomer (3′ GCCGTGCTC 5′). This permits the 9-mer to ligate to additional blunt-end termini available from the cDNA and other adaptors. In contrast, the 13-mer oligomer (5′ AATTCGGCACGAG 3′), being longer, supplies an *Eco*RI cohesive overhang with its four 5′ bases. This oligomer is supplied in a dephosphorylated form, preventing it from ligating to additional cohesive ends. Thus, extensive ligation creating multiple adaptor forms are prevented. Following adaptor/cDNA ligation, the ligase is heat inactivated and the 13-mer is kinased, permitting its ligation into the dephosphorylated Uni-Zap vector arms.

Step 5: Kinasing Adaptors and Enzyme Digestion

After the ligase is heat inactivated, the reaction mixture is microcentrifuged briefly (10 sec) at maximum speed. The reaction is then allowed to cool at room temperature for 5 to 10 min. The adaptor ends are then kinased by adding in the order indicated the following reagents (supplied by Stratagene in the ZAP-cDNA synthesis kit): 2 μl of 10× ligation buffer, 4 μl of 10 m*M* ATP, 12 μl of sterile dH_2O, and 2 μl of T4 polynucleotide kinase (10 U/μl). The reaction mixture is then incubated for 30 min at 37°C. The kinase can then be heat inactivated by placing the microcentrifuge tube at 70°C for 30 min. The reaction mixture is briefly spun at maximum speed for 10 sec and left at room temperature for 5 to 10 min (procedure 1). Alternatively (procedure 2), the kinase enzyme is removed after heat inactivation by phenol–chloroform and then chloroform extraction, as described above. The cDNA is precipitated with ethanol, spun, and dried, again as described previously.

For enzyme digestion following procedure 1, 28 μl of 10× *Xho*I buffer supplement and 2 μl of *Xho*I (45 U/μl) (supplied by Stratagene) are combined with the reaction mixture and incubated for at least 2 hr at 37°C and then allowed to cool at room temperature. To this reaction mixture 5 μl of 10× STE [1× STE is 10 m*M*

Tris-HCl, 10 m*M* NaCl, 1 m*M* ethylenediaminetetraacetic acid (EDTA)] is added. For enzyme digestion following procedure 2, the dried sample is resuspended in 43 μl of 1× TE [10 m*M* Tris (pH 8.0), and 1 m*M* EDTA], 5 μl of 10× *Xho*I restriction enzyme buffer and 2 μl of *Xho*I (45 U/μl). The reaction mixture is incubated for at least 2 hr at 37°C and then allowed to cool at room temperature.

Following *Xho*I digestion, the cDNA is separated from *Eco*RI adaptor fragments and residual linker/primer by employing a Sephacryl S-400 column. The appropriately sized cDNA can then be precipitated and finally ligated to the Uni-Zap vector arms derived from the Uni-ZAP XR vector.

Step 6: Separating XhoI*-Digested cDNA and Ligation of cDNA into Prepared Uni-Zap Arms*

The *Xho*I-digested cDNA is separated from small fragments by passage through a Sephacryl spin column (as described by Stratagene). The spin column is prepared from a 1-ml plastic syringe, the plunger is discarded, and a small cotton plug (equivalent to approximately one-half the cotton from a 1-ml pipette) is inserted into the syringe. Employing a sterile Pasteur pipette the cotton plug is pushed to the bottom of the syringe. The syringe is completely filled with the supplied Sephacryl S-400 slurry, placed in a 15-ml conical centrifuge tube, and spun for 2 to 4 min at 600 to 800 *g*. This procedure is repeated an additional time and the column is subsequently washed twice with 300 μl of 1× STE. The spin column is now ready for collecting cDNA fractions by placing it in a 15-ml conical centrifuge tube with the addition of a 1.5-ml microcentrifuge tube (without its lid) under the plastic syringe (spin column). The cDNA is carefully loaded onto the prepared spin column by Gilson Pipetman (Rainin Instrument Co., Woburn, MA) and the centrifuge is spun at 600 *g* for 2 min at room temperature. The effluent is removed and placed in a new microcentrifuge tube (fraction 1). The procedure is repeated a second time using 50 μl of 1× STE and fraction 2 is collected into a new microcentrifuge tube. Further fractions can also be collected, but the size range of the cDNA that will be present in these fractions will decrease. By using a sizing gel, the fractions can be evaluated and if they contain cDNA of appropriate size they can be pooled.

The pooled cDNA fractions are extracted with an equal volume of phenol–chloroform (1:1, v/v). After vortexing, the sample is microcentrifuged for 4 min at room temperature and the upper aqueous layer is transferred into a new microcentrifuge tube. To this sample an equal volume of chloroform is added and the extraction procedure is repeated. The sample in the aqueous layer is then precipitated overnight at −20°C and the cDNA library is concentrated in the pellet by microcentrifugation for 60 min. To the extracted cDNA library 2.2 times the volume of 100% ethanol is added, and the cDNA library is precipitated either on ice for 60 min or overnight at -20°C. Microcentrifuge the sample at maximum speed for 60 min at 4°C, remove the supernatant, and verify that the majority of counts are present in the pellet. If this has occurred, wash the pellet with an equivalent volume of 70% ethanol, remove the ethanol, and vacuum evaporate the sample until completely dry (usually 5 to 10 min). The cDNA can then be resuspended in 10 μl of sterile H_2O.

To ligate the cDNA into the λ Uni-Zap arms (supplied by Stratagene as pre-*Eco*RI and *Xho*I cut, as well as dephosphorylated) the following reagents are added in the order indicated (described in the Stratagene ZAP-cDNA synthesis kit): 5 μl resuspended cDNA, 1 μl of 10× ligation buffer, 1 μl of 10 m*M* ATP, 2 μl of Uni-ZAP XR vector (1 μg/μl), and 1 μl of T4 DNA ligase (4 Weiss U/μl). The mixture is then incubated for 24 hr at 12°C (or alternatively for up to 2 days at 4°C) and for 2 hr at room temperature. Under these conditions the ligated cDNA/vector will tend to form large concatamers (which can be destroyed if handled roughly or with freezing, which will result in decreased efficiency).

Step 7: Packaging the cDNA Library and Amplifying on PLK-F′ Cells

After ligation, 2 μl of the ligated cDNA/vector is packaged into phage by using the Gigapack II Gold packaging extract (as described by Stratagene). Packaging involves combining 2 μl of DNA to the freeze/thaw extract (on ice), adding 15 μl of sonic extract, and mixing (avoid air bubbles). To be certain that the reaction mixture is on the bottom of the tube, the mixture can be spun for 3 to 5 sec at maximum speed. The reaction mixture is then incubated 22°C for 2 hr. Five hundred microliters of phage dilution buffer is added, followed by 20 μl of chloroform, and the reactants are gently mixed. The culture is then spun briefly to sediment debris and the supernatant is ready to be titered (store at 4°C) (9). It is recommended that a positive control for the Gigapack packaging extracts, such as wild-type *c*I857 *S*am7 plated on the host strain VCS257, be used to determine efficiency of packaging. Both a diluted library (1:10) and an undiluted library should be tested for the efficiency of the packaged ligation products by plating on PLK-F′ cells. The use of the PLK-F′ bacterial cell is recommended by Stratagene because PLK-F′ cells are *recA*-, *mcrA*-, and *mcrB*- and they will not destroy DNA that is methylated, which will occur when using wild-type bacterial strains. In more recent protocols from Stratagene, the SURE bacterial cell is used in place of the PLK-F′ bacterial cell. Because primary libraries can be unstable, it is also recommended that amplification of Uni-ZAP XR libraries be performed relatively soon after library construction. This will result in a higher titer and a more stable stock of the library. The protocol for amplification is as described by Stratagene (ZAP-cDNA synthesis kit) and in Ref. 9.

A primary reason the Uni-ZAP XR vector system was chosen for the construction of cDNA libraries was the ability to excise and recircularize *in vivo* the cloned cDNA insert (located in the λ vector), resulting in the formation of phagemid containing the cloned insert. This procedure is possible because the λ phage genome has been constructed to contain DNA sequences that respond to specific proteins (including proteins derived from the f1 bacteriophage) permitting *in vivo* excision. The Uni-ZAP XR vector contains two positive-strand origins (one site for initiation and one site for termination of DNA synthesis) that have been subcloned at different locations. By simultaneously infecting the appropriate strain of *Escherichia coli* with both the λ vector and the f1 bacteriophage, the λ phage is made receptive to the f1-derived proteins. This results in nicking in one of the two DNA strands followed by DNA synthesis in the λ vector downstream of the nicked DNA. Once a termination sequence is reached 3′ of the initiation sequence, DNA synthesis is stopped. The presence of a gene II product from the f1 bacteriophage results in the circularization of the single-stranded DNA molecule that contains all of the sequences cloned between the initiator and terminator sequence of the Uni-ZAP XR vector [i.e., the pBluescript SK(−) phagemid plus any cloned insert]. Present in the terminator DNA sequence (the f1 terminator origin) are the required signals necessary for the packaging of the phagemid that is then secreted from *E. coli.* To generate double-stranded DNA, the packaged pBluescript DNA is combined with a new culture of *E. coli* and layered onto LB/ampicillin (GIBCO, Grand Island, NY) plates to generate colonies. The colonies can then be used to produce DNA minipreparations for the analysis of the insert DNA by sequencing, subcloning, mapping, and expression. In addition, the minipreparation DNA can be utilized to produce single-stranded DNA that can be used for dideoxy sequencing and site-specific mutagenesis. The phagemid rescue protocol, and procedures for increasing the efficiency of this process, are performed as described in the Stratagene ZAP-cDNA synthesis kit protocol (*in vivo* excision protocol).

It is important to evaluate cDNA libraries to ensure that a negative screening result is not a consequence of technical difficulties encountered during library construction. If a high cDNA cloning efficiency is obtained, this usually indicates that the procedure has been performed appropriately. However, if on screening the cDNA library one obtains a negative result, this might indicate problems with the library construction, the probe, the screening method, and/or the source of the RNA used for the probing. To ensure that library construction is not the problem, several approaches can be utilized to evaluate the cDNA library, including (1) screening libraries for expression of an abundant mRNA sequence, such as actin or GAPDH, which is known to be present in the library, (2) screening libraries with a combination of total mixed cDNA probes, or (3) picking 15 or more randomly chosen cDNA clones and analyzing their inserts. The library can be appraised by screening duplicate filters from a single plating of 10,000 to 60,000 clones (from a plasmid or phage library) with a probe recognizing an abundant mRNA (such as actin). Each positive signal on both filters is considered positive, and in the case of actin can range from 0.02 to 0.06% (liver) to 1 to 4% (fibroblast cell line) of the clones screened, depending on the source of mRNA used to construct the cDNA library. An alternate approach is to screen the library with the same source of mRNA used to construct the library. Employing this approach, approximately 50 to 90% of the plated clones should generate positive signals. A final approach involves the random isolation of 15 single recombinant clones and the production of phage DNA minipreparations. These preparations are then digested with the appropriate restriction endonuclease(s) and electrophorsed through a 1.5% (w/v) agarose gel in TBE [89 m*M* Tris, 89 m*M* boric acid, and 2 m*M* EDTA] buffer. A minimum of 12 of the 15 recombinant clones should contain inserts of average sizes of ~1 kb or more, which can be visualized by their fluorescence after ethidium bromide staining.

Subtractive Hybridization Procedures for Identification and Cloning of Progression-Suppressing and Progression-Inducing Genes

A powerful technique for identifying and cloning differentially expressed genes from closely related cell types is subtractive hybridization (10–24). Unlike a complete cDNA library, subtractive libraries contain cDNA clones representing mRNAs present in the positive cDNA library (target) that are not expressed or that are expressed at greatly reduced levels in the negative subtractor cell mRNA (driver). One approach for subtraction hybridization involves the construction of two separate cDNA libraries prepared in either phagemid vectors (45, 46) or vectors containing a specific bacteriophage promoter (such as SP6, T3, or T7) that will allow the *in vitro* generation of single-stranded RNA molecules. This subtractive hybridization approach can be applied with a related pair of cells, such as E11 and E11-NMT, for the identification of differentially expressed genes that may function as suppressors of the progression phenotype or inducers of the progression phenotype. To obtain cDNA clones representing mRNAs that may encode proteins that are important in mediating suppression of the progression phenotype, the E11 parental cell cDNA library is used as the target and mRNA from the E11-NMT cell line is used as driver (Fig. 8). In contrast, to obtain cDNA clones representing mRNAs that may encode proteins that contribute to expression of the progression phenotype, the E11-NMT cell cDNA library is used as the target and mRNA from the E11 parental cell line is used as driver (Fig. 9).

The protocol used to obtain subtracted E11 cDNA libraries is outlined in Fig. 8. A similar approach is used to generate subtracted E11-NMT cDNA libraries as outlined in Fig. 9. Approximately 10 μg of E11 (for constructing E11 subtracted libraries) or E11-NMT (for constructing E11-NMT subtracted libraries) cDNA library in the λ Uni-Zap vector (Fig. 7) is digested with *Xho*I and transcribed to an E11 sense cRNA or an E11-NMT sense cRNA, respectively, as described below.

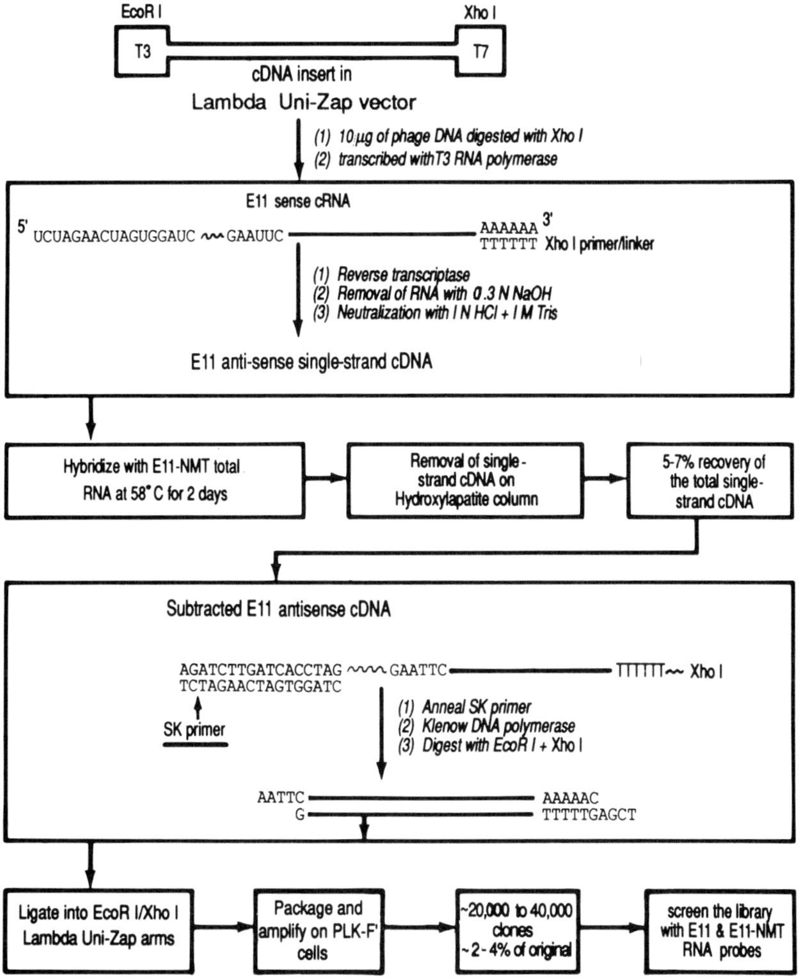

FIG. 8 Schematic diagram for the construction of subtracted E11 cDNA libraries. The essential steps in this process are indicated, including conversion of E11 cDNA into an E11 sense cRNA; conversion of E11 sense cRNA into an E11 antisense single-strand cDNA; hybridizing of E11 antisense single-strand cDNA with an excess of E11-NMT total RNA; removal of single-strand cDNA on a hydroxylapatite column; recovery of total single-strand-subtracted E11 antisense cDNA; conversion of this single-strand-subtracted E11 antisense cDNA into a double-stranded E11 cDNA; ligating the subtracted E11 cDNA into *Eco*RI/*Xho*I-prepared λ Uni-Zap arms; packaging and amplifying on PLK-F′ cells; and isolating phage and screening library with E11 and E11-NMT RNA probes.

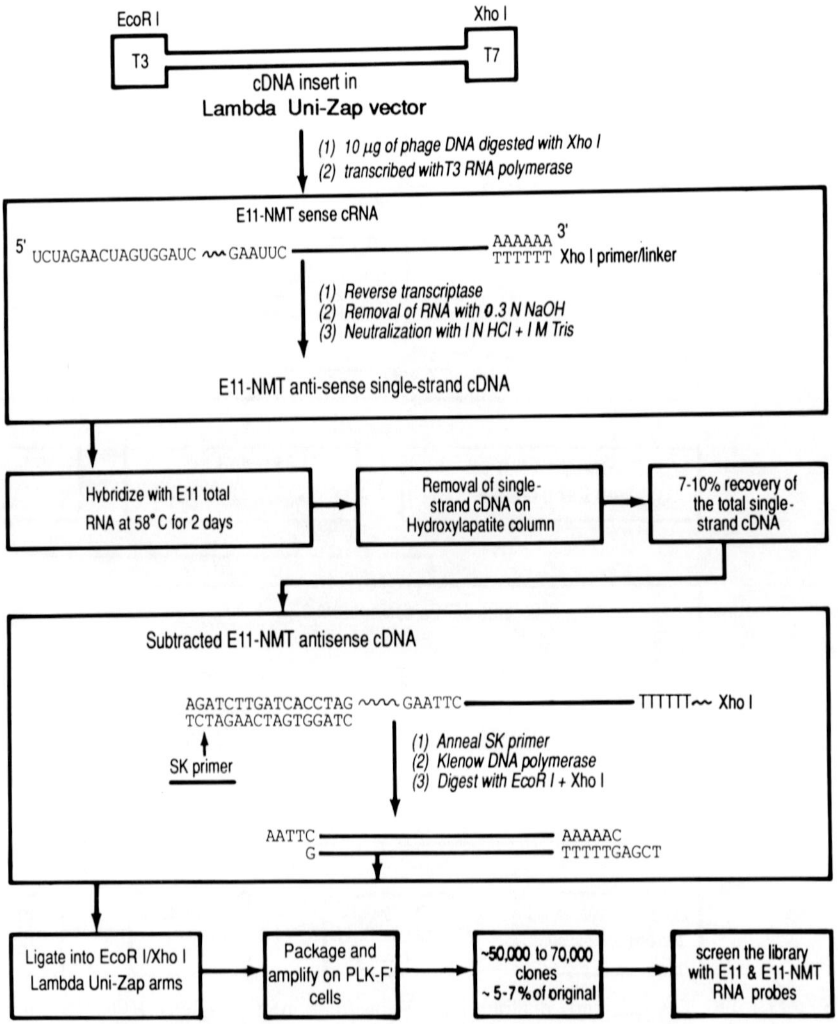

FIG. 9 Schematic diagram for the construction of subtracted E11-NMT cDNA libraries. The essential steps in this process are indicated, including conversion of E11-NMT cDNA into an E11-NMT sense cRNA; conversion of E11-NMT sense cRNA into an E11-NMT antisense single-strand cDNA; hybridizing of E11-NMT antisense single-strand cDNA with an excess of E11 total RNA; removal of single-strand cDNA on a hydroxylapatite column; recovery of total single-strand-subtracted E11-NMT antisense cDNA; conversion of this single-strand-subtracted E11-NMT antisense cDNA into a double-stranded E11-NMT cDNA; ligating the subtracted E11-NMT cDNA into *Eco*RI/*Xho*I-prepared λ Uni-Zap arms; packaging and amplifying on PLK-F′ cells; and isolating phage and screening library with E11 and E11-NMT RNA probes.

Step 8: Production of Sense cRNA

For a total volume of 50 μl, combine the following in the indicated sequence: 10 μl of phage DNA (1 μg/μl), 5 μl of 10× *Xho*I restriction enzyme buffer, 32 μl sterile H_2O, and 2 μl of *Xho*I (10 U/μl). Incubate the mixture for 2 hr at 37°C and extract with equal volumes of phenol–chloroform and then chloroform. After extraction add 250 μl of cold ethanol and place the reaction mixture at −70°C for 30 min. Precipitate and dry the DNA and resuspend it in 10 μl of 1× TE. The phage DNA is then transcribed into sense cRNA by adding the following materials in the sequence described (as outlined by Stratagene in their transcription kit): 10 μl of 5× transcription buffer, 10 μl *Xho*I-digested phage DNA, 2 μl 10 m*M* ATP, 2 μl 10 m*M* CTP, 2 μl 10 m*M* GTP, 2 μl 10 m*M* UTP, 1 μl RNase Block II (1 U/μl); 17 μl of DEPC-treated sterile H_2O, and 2 μl of T3 RNA polymerase (50 U/μl). Incubate the mixture for 30 min at 37°C. To remove DNA add 2 μl of RNase-free DNase (10 U/μl) and incubate for 15 min at 37°C. The cRNA (aqueous phase) is extracted with phenol–chloroform and then chloroform and the cRNA is precipitated, dried, and resuspended in 10 μl of 1× TE.

Step 9: Production of Antisense Single-Strand cDNA

First-strand synthesis is performed as described previously for cDNA library construction, using reagents provided by Stratagene with the exception of the random primer, which was obtained from Bethesda Research Laboratories, Gaithersburg, MD. The following materials are added, in the order indicated, for reverse transcription of the sense cRNA into an antisense single-strand cDNA: 5 μl of 10× first-strand buffer, 2 μl of *Xho*I primer/linker (1.4 μg/μl), 1 μl of random primer (3 μg/μl), 2.5 μl of 10 m*M* first-strand methyl-NTP mix, 1 μl of RNase block II (1 U/μl), 5 μl of [α-^{32}P]dATP (specific activity, 800 Ci/mmol; Amersham), 10 μl of cRNA, and 21 μl of DEPC-treated dH_2O. The mixture is briefly vortexed and primers are annealed by incubation at room temperature for 10 min. To this mixture 2.5 μl of M-MuLVRT (18 U/μl) is added and the microfuge tube is incubated for 1 hr at 37°C. The mixture is then extracted first with 50 μl phenol–chloroform followed by the same amount of chloroform, and the cRNA/cDNA hybrid is precipitated with 250 μl of cold ethanol at −70°C for 30 min. The contents of the tube are spun down, the ethanol is aspirated off, and the pellet is dried and then resuspended in 20 μl of 1× TE.

Removal of the RNA from the duplex is achieved by alkaline hydrolysis. To the 20 μl of solution in 1× TE buffer containing the sense cRNA: antisense cDNA, 140 μl of 0.3 *N* NaOH in 2 m*M* EDTA is added and the solution is boiled for 1.5 min. The solution is then neutralized by the addition of 20 μl of 1 *M* Tris (pH 7.5) and 20 μl of 1 *N* HCl. The mixture is again extracted with an equal volume of phenol–

chloroform and chloroform, and antisense single-strand cDNA is precipitated by the addition of 800 μl of cold ethanol at −70°C for 30 min. The contents of the tube are spun down, the ethanol is aspirated off, and the pellet is dried and then resuspended in 20 μl of 1× TE.

Step 10: Hybridization and Separation of Cell-Specific Messages

To produce a subtracted E11 antisense cDNA, the E11 antisense cDNA is hybridized with an excess of E11-NMT total RNA [with or without the addition of E11-NMT poly(A)$^+$ RNA] and single-strand unhybridized antisense cDNA is removed by fractionation through a hydroxylapatite column, based on the different affinities of single- and double-stranded DNA for the hydroxylapatite resin. In a similar manner, to produce a subtracted E11-NMT antisense cDNA, the antisense E11-NMT cDNA is hybridized with an excess of E11 total RNA [with or without the addition of E11 poly(A)$^+$ RNA] and single-strand unhybridized antisense cDNA is removed by fractionation through a hydroxylapatite column. [*Note:* A significant improvement in subtraction efficiency can be obtained by using a 5- to 10-fold excess of subtractor poly(A)$^+$ RNA, as opposed to using total subtractor RNA.]

To remove cDNAs that are expressed by both cell types, 20 μl of antisense cDNA is mixed with 200 μl of total subtractor RNA (300 μg in 200 μl of 0.3 *M* phosphate buffer, pH 7.0). To increase the subtraction process further an additional 3 to 5 μg of subtractor poly(A)$^+$ RNA can be added. Hybridization is then achieved by incubating the reactants for 2 days at 58°C. To prevent excessive fluid evaporation, a 150 to 200 μl overlay of mineral oil (Sigma, St. Louis, MO) is applied.

To separate single-stranded cDNA from double-stranded cDNA/RNA hybrid complexes, the mixture is fractionated by passage through a hydroxylapatite column that is maintained at 60°C. The hydroxylapatite column is prepared by first hydrating the hydroxylapatite powder (HTP DNA grade; Bio-Rad, Richmond, CA) with 0.12 *M* sodium phosphate at pH 6.8 (at about 2 to 3 ml/g). One milliliter of the slurry is added to a siliconized Pasteur pipette plugged with glass wool and maintained at 60°C by placing it in a 50-ml centrifuge tube that is immersed in a 60°C water bath. The column is then equilibrated with 10 ml of the 0.12 *M* phosphate buffer, prewarmed to 60°C. The hybridization mixture, containing the single- and double-stranded nucleic acids, is adjusted to about 0.12 *M* phosphate by the addition of 330 μl of DEPC-treated dH_2O. The mixture is layered onto the hydroxylapatite column and, because the unhybridized antisense cDNA should not bind to the column resin under these conditions, 500-μl fractions are immediately collected. The column is then washed six times with 500 μl of 0.12 *M* phosphate buffer also prewarmed to 60°C. Fractions of approximately 500 μl are continually collected and monitored by Geiger counter for radioactivity. The majority of antisense cDNA elutes in fractions 2 to 6, which are pooled and dialyzed extensively against 1× TE at 4°C. The volume of the cDNA fractions can be reduced by extraction with *sec*-(2)-butanol until it is about 200 μl (the aqueous phase is now the lower phase in these extractions). Residual *sec*-butanol is removed by chloroform extraction (the aqueous phase is back on top) and the single-stranded cDNA can be precipitated with 500 μl of ethanol at −70°C for 30 min. The cDNA is pelleted, washed with 70% ethanol, repelleted, and dissolved in 10 μl of 1× TE.

To the subtracted antisense cDNA an SK primer is annealed, a second-strand sense cDNA is synthesized, and the double-stranded complex is digested with *Eco*RI and *Xho*I and ligated onto Uni-Zap arms. Using a primer to direct second-strand synthesis requires modifying the procedures described in step 2 above. The exonuclease (proofreading) activity of DNA Pol I would interfere with primer-directed synthesis, thus a modified (genetically engineered) DNA polymerase is used. In this description, Sequenase (a modified T7 DNA polymerase; United States Biochemical Corporation, Cleveland, OH) is used; however, the Klenow fragment of DNA Pol I is equally efficient in this application.

Step 11: Second-Strand Synthesis and Ligation of the Subtracted cDNA onto the Uni-ZAP Arms

The SK primer (from Stratagene) complements a region of the pBluescript plasmid directly upstream of the inserted cDNA and is used to initiate synthesis of the second or sense cDNA strand. The primer is annealed to the subtracted antisense cDNA by adding the following reagents in the order indicated in a total volume of 37 μl: 10 μl of subtracted antisense cDNA, 2.5 μl SK primer (1 μg/ml from Stratagene), 10 μl of 5× Sequenase buffer (from the Sequenase kit obtained from United States Biochemical Corporation), and 14.5 μl of sterile dH_2O. Annealing is performed by incubating the mixture at 60°C for 45 min. After annealing, the following reagents are added in the order given: 0.5 μl of 1 m*M* dATP, 1.5 μl of 10 m*M* dCTP, 1.5 μl of 10 m*M* dGTP, 1.5 μl of 10 m*M* dTTP, 2 μl of [α-^{32}P]dATP (specific activity, 800 Ci/mmol; Amersham, Arlington Heights, IL), 5 μl of 0.1 *M* DDT, and 1.5 μl of Sequenase enzyme (from the Sequenase kit obtained from United States Biochemical Corporation). The reaction mixture is incubated at 37°C for 30 min, extracted with 50 μl each of phenol–chloroform and chloroform, and precipitated by adding 250 μl of cold ethanol at −70°C for 30 min. The cDNA is pelleted, washed with 70% ethanol, dried, and dissolved in 50 μl of 1× TE. Because the cDNA already has *Eco*RI and *Xho*I restriction enzyme sites, it is not necessary to add *Eco*RI linkers (however, this procedure is optional). The subtracted antisense cDNA is then digested with the restriction enzymes *Xho*I and *Eco*RI by adding the following materials in the order indicated: 50 μl of antisense cDNA, 6 μl of 10× *Xho*I enzyme buffer, 2 μl of sterile dH_2O, and 2 μl of *Xho*I enzyme (10 U/μl). The mixture is then incubated for 2 hr at 37°C followed by the addition of 3μl of 10× *Eco*RI enzyme buffer and 2 μl of *Eco*RI enzyme (10 U/μl). The mixture is extracted with an equal volume of phenol-chloroform and then chloroform and cDNA is ethanol precipitated at −70°C for 30 min. The cDNA is collected, washed, dried, and then resuspended in 10 μl of 1× TE.

The subtracted cDNA is then ligated into the λ Uni-Zap arms prepared by treating the Uni-ZAP vector with *Eco*RI and *Xho*I. To ligate the cDNA into the λ Uni-Zap arms the following reagents are added in the order indicated (basically as described in the Stratagene ZAP-cDNA synthesis kit): 5 μl subtracted cDNA, 1 μl of 10× ligation buffer, 1 μl of 10 m*M* ATP, 2 μl of Uni-ZAP XR vector (1 μg/μl), and 1 μl of T4 DNA ligase (4 Weiss U/μl). The mixture is then incubated for 24 hr at 12°C (or alternatively for up to 2 days at 4°C) and for 2 hr at room temperature. The ligated product is packaged with Gigapack II Gold packaging extract (Stratagene) and amplified on a PLK-F′ bacterial strain for demethylation of the cDNA as described in Step 7 of Construction of cDNA Libraries. The subtracted library is then screened using the XL-1 blue strain as described by Stratagene. Subtracted E11 libraries consisting of 20,000 to 40,000 independent clones (2 to 4% of the original cDNA clones) and subtracted E11-NMT libraries consisting of 50,000 to 70,000 independent clones (5 to 7% of the original cDNA clones) have been obtained using the procedures described in this article.

The subtracted libraries are screened with E11 and E11-NMT RNA probes to identify differentially expressed cDNAs. XL-1 blue bacteria are infected with approximately 5000 phage and plated on 150-mm plates. Duplicate nitrocellulose filters (Nitrocellulose HATF, 137 mm; obtained from Millipore, Inc., Bedford, MA) are applied to plates and the filters are processed by washing once with sterile 1× SSC, denaturing for 30 sec in 1.5 *M* NaCl plus 0.5 *M* NaOH, neutralizing by incubation for 8 min in 1 *M* Tris (pH 8.0) plus 1.5 *M* NaCl, washing with 2× SSC, and baking (under vacuum) for 2 hr at 80°C. Filters are then prehybridized and hybridized in 5× SSC, 1× Denhardt's solution, 47% (v/v) formamide (deionized), 0.1% (w/v) sodium dodecyl sulfate (SDS), and 100 μg/ml of denatured salmon sperm DNA, at 42°C, with agitation. RNA probes are prepared from E11 and E11-NMT cDNA and used to screen the library as described in Ref. 9. We have employed T7 RNA polymerase for RNA synthesis because the T7 RNA probes are flanked by only an additional 26 bp of DNA preceding the cDNA. In contrast, if T3 RNA polymerase is employed it would result in an additional 77 bp of DNA prior to the cDNA. This extra 51 bp of DNA could complicate the screening procedure. RNA probes are prepared by adding the following reagents in the order indicated: 10 μg of each phage library DNA [digested with 10 μl (1 μg/μl) of the restriction enzyme *Eco*RI], 10 μl of 5× reaction buffer (Stratagene), 2 μl of 10 m*M* ATP, 2 μl of 10 m*M* CTP, 2 μl of 10 m*M* GTP, 2 μl of 10 m*M* UTP, 5 μl of

[α-^{32}P]UTP (Amersham), 2 μl of 0.75 *M* DTT, 13 μl of DEPC-treated dH_2O, and 2 μl T7 RNA polymerase enzyme (50 U/μl). The mixture is incubated for 30 min at 37°C and processed as described above. For hybridization, each filter is incubated for 4 hr in 15 ml of hybridization solution at 45°C (prehybridization); 3×10^6 cpm of probe is added and incubation at 45°C is continued for an additional 16 hr; the filters are washed with 2× SSC in 0.2% (w/v) SDS for 30 min, and dried and exposed to X-ray film for autoradiography. Positive phages are isolated and excised *in vitro* to pBluescript with the addition of the helper phage as described by Stratagene. The same hybridization conditions are employed to screen Northern blots, except the probe is labeled with a multiprime labeling kit (α-^{32}P random primer labeled) (Amersham) using [α-^{32}P]dCTP.

Once differentially expressed cDNAs are identified and isolated, the inserts can be amplified by PCR (12, 47) from phage by using the T3 and T7 sequences as primers. The PCR products are then run on agarose gels and the PCR products are isolated, purified by phenol–chloroform and chloroform extraction, and α-^{32}P random primer labeled for Northern analysis. cDNA clones displaying the appropriate specificity, that is, expressed in E11 but not E11-NMT (putative progression–suppression-related genes) or expressed in E11-NMT but not E11 (putative progression–induction-related genes), are sequenced directly using the dideoxynucleotide chain-termination technique of Sanger *et al.* (48) with T7 DNA polymerase (Pharmacia, Piscataway, NJ). Alternatively, sequencing can be conducted on exonuclease III-deleted cDNAs that are generated using the Promega (Madison, WI) Erase-a-Base kit. To avoid problems of compression artifacts during sequencing, dGTP analogs (Pharmacia) are also used for specific sequencing reactions.

Employing the subtractive hybridization scheme described in this article, we have identified a series of cDNA clones that display a differential pattern of hybridization in E11-NMT versus E11 cells (Fig. 10). These cDNA clones have been isolated from subtracted E11-NMT libraries (Fig. 9) and termed progression–induction-related (*Prog*$^+$) genes. Specific cDNAs have been identified that hybridize to a series of RNA transcripts present in E11-NMT cells, but not in CREF or E11 cells. In addition, the *Prog*$^+$ genes are not expressed in the E11 × E11-NMT hybrid (IIId), which does not express the progression phenotype (Table I). An increased expression of the *Prog*$^+$ genes is observed in both R1 and R2 hybrid cells, formed between CREF and E11-NMT, which display an enhanced expression of the progression phenotype (Table I). In addition, a differential pattern of expression of the various *Prog*$^+$ genes is observed in the F1 and F2 hybrids, formed between CREF and E11-NMT, and E11 × E11-NMT hybrids (3b and A6), which display a suppression in the progression phenotype (Table I; and Zao-Zhong Su and Paul B. Fischer, unpublished data, 1992). These genes are presently being characterized to determine their genetic identity and their potential role(s) in mediating the progression phenotype.

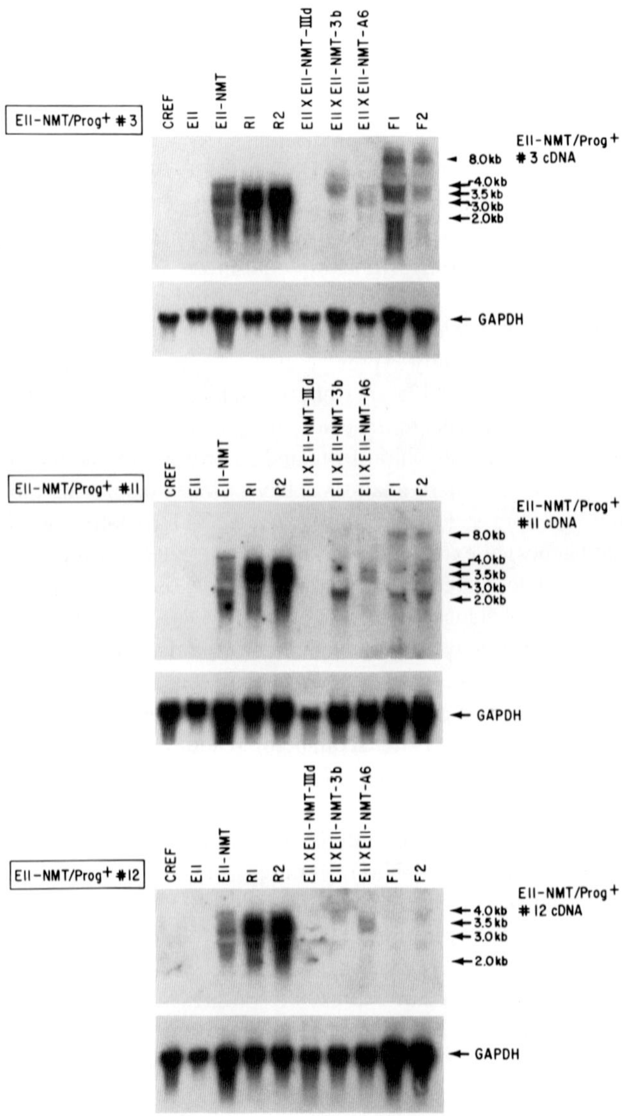

FIG. 10 Northern analysis of total RNA isolated from CREF, E11, E11-NMT, and various hybrids formed between CREF^{cap} and E11-NMT^{neo} and E11^{hyg} and E11-NMT^{neo}. Twenty micrograms of RNA was size fractionated on 0.8% agarose gels and transferred to nylon membranes (in 20× SSC). Gels were stained with ethidium bromide to determine relative quantities of RNA for each sample. The cDNA inserts (from E11-NMT/Prog$^+$ #3, #11, and #12) were isolated by digestion with *Xho*I and *Eco*RI. The inserts were then labeled with $[^{32}P]$dCTP (Amersham) with a multiprime labeling kit (Amersham). Both pre- and posthybridizations were performed at 42°C in 50% formamide in 5× SSC, 1× Denhardt's solution,

Additional Approaches Employing Subtractive Hybridization to Identify Differentially Expressed Genes

As with any molecular biological technique, improvements in the efficiency and sensitivity of subtractive hybridization protocols are inevitable. Many of the currently employed modifications in the subtractive hybridization scheme involve the use of PCR to amplify cDNA sequences that are located adjacent to oligonucleotides of known sequences (18, 23, 49, 50). By employing primers complementary to the universal primer sequences, it is possible to amplify subtracted cDNA libraries employing picogram quantities of first-strand cDNA as well as small quantities of starting mRNA. A novel approach described by Patanjali *et al.* (23) permits the construction of uniform-abundance (normalized) cDNA libraries. This approach should be applicable for the detection and analysis of cDNA corresponding to species of mRNA that are present in low abundance in a small proportion of cells in a complex tissue. In this procedure a kinetic approach is employed to produce cDNA libraries containing approximately equal representations of the totality of sequences in a specific preparation of poly(A)$^+$ RNA. A selected size range of cDNA fragments is then randomly primed and cloned into a phage vector. By using PCR, inserts are amplified followed by denaturation and self-annealing under optimized conditions. Following extensive but not complete reannealing, the single-stranded fraction is obtained, which is relatively depleted of the majority of abundant species of cDNA. Libraries constructed in this manner are suitable for cDNA subtraction, screening, or selection by hybridization for rare or low-abundance species of mRNA (23). This approach should find wide application for identifying differentially expressed genes expressed at low levels in a small subset of cells in a complex tissue, that is, tumor foci in a background of normal cells, cells displaying a specific developmental commitment lineage, and so on.

An interesting application of subtractive hybridization has been reported using

---

0.2% SDS, and containing 100 μg/ml of salmon sperm DNA. Blots were then washed and exposed for autoradiography. The same blots were stripped in 0.1 × SSC and 0.1% SDS for 2 hr at 80°C and reprobed with a multiprime [^{32}P]dCTP-labeled GAPDH probe. The cell types included in this assay were as follow: CREF; E11 (Prog$^-$); E11-NMT^{neo} (Prog$^+$); E11^{hyg} × E11-NMT^{neo} hybrids displaying an E11 phenotype (IIId, 3b, and A6) (Prog$^-$); CREF^{cap} × E11-NMT^{neo} hybrids with a round morphology (R1 and R2) (Prog$^+$); and CREF^{cap} × E11-NMT^{neo} hybrids with a flat morphology (F1 and F2) (Prog$^-$). The cDNAs isolated from subtracted E11-NMT libraries included E11-NMT/Prog$^+$ #3 and E11-NMT/Prog$^+$ #11 [which detect differentially expressed RNA species in the various cell types (except CREF, E11, and E11^{hyg} × E11-NMT^{neo} hybrid IIId) with approximate sizes of 8.0, 3.9, 3.5, 3.0, and 2.0 kb], and E11-NMT/Prog$^+$ #12 (which detects differentially expressed RNA species in E11-NMT and R1 and R2 hybrids, with approximate sizes of 3.9, 3.5, 3.0, and 2.0 kb).

two disparate cell lines, human melanoma and human lung carcinoma, resulting in the identification of novel melanoma-associated gene sequences (50). In this study 156 complementary DNA clones were isolated, using a subtracted melanoma-complementary DNA probe (melanoma minus lung carcinoma) after screening 2×10^4 independent recombinants of a melanoma expression library by *in situ* plaque hybridization. The cDNA clones were then amplified by PCR, screened for duplications, and grouped into 53 discrete genes. An interesting finding was that most of the genes identified represented rare messages, present at <1 copy per 200 molecules of mRNA in a typical cell. Partial sequence data obtained for 20 complementary DNA clones indicated 8 novel human genes. Employing Northern blotting, mRNA transcripts for five of the genes were detected in a variety of cell types. This approach may therefore be applicable for the identification of novel tumor-associated sequences and possibly novel tumor-associated antigens.

Summary and Future Perspectives

Technological advances in molecular biology have resulted in the development of highly efficient methodologies for the identification and cloning of gene(s) that display differential expression in both closely related and disparate cell types (8–24, 49, 50). These approaches will result in an improved understanding of the molecular basis of many biologically relevant processes in humans, including normal development, cellular differentiation, carcinogenesis, and aging. On a practical note, a clearer definition of the genetic elements regulating processes such as cancer may permit the future development of techniques, including those employing targeted antisense RNA molecules and genetically engineered immunological reagents (6, 51), to inhibit and even reverse the neoplastic process. Additionally, identification of specific genes that can directly promote (progression inducing) or inhibit (progression suppressing) the process of tumor progression will provide powerful tools for predicting cancer progression and developing strategies for inhibiting the process of tumor evolution to a more aggressive phenotype. In this respect, the applications of molecular biology will have direct medical applications for both the understanding and therapy of specific human diseases, including cancers, developmental abnormalities, and immunologically based diseases.

Acknowledgments

The contributions of our colleagues to the various studies described in this article are greatly appreciated. They include Drs. Lee E. Babiss, Barbara Driscoll, Gregory J. Duigou, Yuen-Kai T. Fung, Harold S. Ginsberg, Deborah S. Iman, Wen-Shing Liaw, Jerry W. Shay, I. Bernard Weinstein, and Stephen G. Zimmer. We also thank Dr. Anthony J. Raffo for critically reviewing this article. The present studies were supported by National Cancer Institute Grants CA35675 and CA43208. Dr. Paul B. Fisher is a Chermow Research Scientist in the Departments of Pathology and Urology, Columbia University, College of Physicians and Surgeons, New York, New York.

References

1. P. B. Fisher, *in* "Mechanisms of Tumor Promotion. III: Tumor Promotion and Carcinogenesis *In Vitro*" (T. J. Slaga, ed.), p. 57. CRC Press, Boca Raton, Florida, 1984.
2. I. B. Weinstein, *Cancer Res,* **48,** 4135 (1988).
3. J. M. Bishop, *Cell* **64,** 235 (1991).
4. R. Sager, *Science* **246,** 1406 (1989).
5. L. M. Liotta, P. S. Steeg, and W. G. Stetler-Stevenson, *Cell* **64,** 327 (1991).
6. P. B. Fisher and P. T. Rowley, *in* "Current Status of Differentiation Therapy of Cancer" (S. Waxman, G. Rossi, and F. Takaku, eds.), p. 201. Raven, New York, 1991.
7. G. J. Duigou, Z.-z. Su, L. E. Babiss, B. Driscoll, Y.-K. T. Fung, and P. B. Fisher, *Oncogene* **6,** 1813 (1991).
8. S. L. Berger and A. R. Kimmel (eds.), "Guide to Molecular Cloning Techniques." Academic Press, Orlando, Florida, 1987.
9. J. Sambrook, E. F. Fritsch, and T. Maniatis (eds.), "Molecular Cloning: A Laboratory Manual," 2nd Ed. Cold Spring Harbor Lab. Press, Cold Spring Harbor, New York, 1989.
10. M. D. Scott, H.-H. Westphal, and R. W. J. Rigby, *Cell* **34,** 557 (1983).
11. M. M. Davis, D. I. Cohen, E. A. Nielsen, M. Steinmetz, W. E. Paul, and L. Hood, *Proc. Natl. Acad. Sci. U.S.A.* **81,** 2194 (1984).
12. S. W. Lee, C. Tomasetto, and R. Sager, *Proc. Natl. Acad. Sci. U.S.A.* **88,** 2825 (1991).
13. S. M. Hedrick, D. I. Cohen, E. A. Nielsen, and M. M. Davis, *Nature (London)* **308,** 149 (1984).
14. C. Schneider, R. M. King, and L. Philipson, *Cell* **54,** 787 (1988).
15. A. J. Fornace, Jr., I. Alamo, Jr., and M. C. Hollander, *Proc. Natl. Acad. Sci. U.S.A.* **85,** 8800 (1988).
16. G. D. Yancopoulos, E. M. Oltz, G. Rathburn, J. E. Berman, R. K. Smith, R. D. Lansford, P. Rothman, A. Okada, G. Lee, M. Morrow, K. Kaplan, S. Prockop, and F. W. Alt, *Proc. Natl. Acad. Sci. U.S.A.* **87,** 5759 (1990).
17. L. M. Matrisian, G. T. Bowden, P. Krieg, G. Furstenberger, J.-P. Briand, P. Leroy, and R. Breathnach, *Proc. Natl. Acad. Sci. U.S.A.* **83,** 9413 (1986).
18. G. P. Owens, W. E. Hahn, and J. J. Cohen, *Mol. Cell. Biol.* **11,** 4177 (1991).
19. J. Fragnoli, N. J. Holbrook, and A. J. Fornace, Jr., *Anal. Biochem.* **187,** 364 (1990).
20. M. A. Frohman, M. K. Dush, and G. R. Martin, *Proc. Natl. Acad. Sci. U.S.A.* **85,** 8998 (1988).
21. C. Timblin, J. Battey, G. Asouline, and M. Wigler, *Nucleic Acids Res.* **18,** 1587 (1990).
22. I. Wieland, J. Battey, and W. M. Kuehl, *Proc. Natl. Acad. Sci. U.S.A.* **87,** 2720 (1990).
23. S. Patanjali, S. Parimoo, and S. M. Weissman, *Proc. Natl. Acad. Sci. U.S.A.* **88,** 1943 (1991).
24. E. Hara, T. Kato, S. Nakada, S. Sekiya, and K. Oda, *Nucleic Acids Res.* **19,** 7097 (1991).

25. H. L. Sive and T. St John, *Nucleic Acids Res.* **16,** 10937 (1988).
26. P. B. Fisher, I. B. Weinstein, D. Eisenberg, and H. S. Ginsberg, *Proc. Natl. Acad. Sci. U.S.A.* **75,** 2311 (1978).
27. P. B. Fisher, J. H. Bozzone, and I. B. Weinstein, *Cell* **18,** 695 (1979).
28. P. B. Fisher, N. I. Goldstein, and I. B. Weinstein, *Cancer Res.* **39,** 3051 (1979).
29. L. E. Babiss, D. L. Guernsey, and P. B. Fisher, *Cancer Res.* **45,** 6017 (1985).
30. P. B. Fisher, K. Dorsch-Hasler, I. B. Weinstein, and H. S. Ginsberg, *Nature (London)* **281,** 591 (1979).
31. L. E. Babiss, S. G. Zimmer, and P. B. Fisher, *Science* **228,** 1099 (1985).
32. G. J. Duigou, L. E. Babiss, D. S. Iman, J. W. Shay, and P. B. Fisher, *Mol. Cell. Biol.* **10,** 2027 (1990).
33. G. J. Duigou, L. E. Babiss, and P. B. Fisher, *Ann. N.Y. Acad. Sci.* **567,** 302 (1989).
34. Z. Z. Su, C. A. O'Brian, and P. B. Fisher, manuscript in preparation (1992).
35. P. B. Fisher, R. A. Mufson, I. B. Weinstein, and J. B. Little, *Carcinogenesis* **2,** 183 (1981).
36. P. B. Fisher, L. E. Babiss, I. B. Weinstein, and H. S. Ginsberg, *Proc. Natl. Acad. Sci. U.S.A.* **79,** 3527 (1982).
37. L. E. Babiss, W.-S. Liaw, S. G. Zimmer, G. C. Godman, H. S. Ginsberg, and P. B. Fisher, *Proc. Natl. Acad. Sci. U.S.A.* **83,** 2167 (1986).
38. H. Hermo, Jr., G. J. Duigou, L. E. Babiss, and P. B. Fisher, *Carcinogenesis* **8,** 967 (1987).
39. Z. Z. Su, D. Grunberger, and P. B. Fisher, *Mol. Carcinog.* **4,** 231 (1991).
40. F. M. Ausubel, R. Brent, R. E. Kingston, D. D. Moore, J. G. Seidman, J. A. Smith, and K. Struhl (eds.), "Current Protocols in Molecular Biology," Wiley, Media, Pennsylvania, 1992.
41. J. M. Chirgwin, A. E. Przybyla, R. J. MacDonald, and W. J. Rutter, *Biochemistry* **18,** 5294 (1979).
42. J. M. Short, J. M. Fernandez, J. A. Sorge, and W. D. Huse, *Nucleic Acids Res.* **16,** 7583 (1988).
43. A. R. Kimmel and S. L. Berger, *in* "Guide to Molecular Cloning Techniques" (S. Berger and A. Kimmel, eds.), Methods in Enzymology, Vol. 152, p. 307. Academic Press, Orlando, Florida, 1987.
44. U. Gubler, *in* "Guide to Molecular Cloning Techniques" (S. Berger and A. Kimmel, eds.), Methods in Enzymology, Vol. 152, p. 325. Academic Press, Orlando, Florida, 1987.
45. J. R. Duguid, R. G. Rohwer, and B. Seed, *Proc. Natl. Acad. Sci. U.S.A.* **85,** 5738 (1988).
46. J. L. R. Rubenstein, A. E. J. Brice, R. D. Ciaranello, D. Denney, M. H. Porteus, and T. B. Usdin, *Nucleic Acids Res.* **18,** 4833 (1990).
47. U. B. Rasmussen, P. Basset, and J.-Y. Daniel, *Nucleic Acids Res.* **17,** 3308 (1989).
48. F. Sanger, S. Nicklen, and A. R. Coulson, *Proc. Natl. Acad. Sci. U.S.A.* **74,** 5463 (1977).
49. C.-J. Kho, and H. Zarbl, *Technique* **3,** 58 (1991).
50. J. T. Hutchins, R. J. Deans, M. S. Mitchell, C. Uchiyama, and J. Kan-Mitchell, *Cancer Res.* **51,** 1418 (1991).
51. J. A. Leon, N. I. Goldstein, and P. B. Fisher, *Pharmacol. Ther.* in press (1993).

[5] λ Phage-Based Systems for Rescue of Chromosomal Markers

William C. Summers and Peter M. Glazer

Introduction

The recovery of DNA sequences or genes that have been inserted into chromosomal DNA becomes increasingly difficult as the size of the genome increases. Simple approaches such as plasmid rescue, whereby the introduced sequences adjacent to, or within, a plasmid genome are excised, ligated, and transfected into *Escherichia coli,* become inefficient for sequences inserted into mammalian genomes. In addition, if an efficient rescue is needed, such as in the study of mutagenesis, in which many individual genes must be recovered, plasmid rescue has proved inadequate. An alternative approach is afforded by the use of the λ bacteriophage packaging reaction, which gives a high efficiency of excision and packaging of inserted λ sequences into infectious phage particles. If the gene of interest is cloned into the λ genome prior to its introduction into the host cell genome, it will be in a form that is recoverable with high efficiency by this *in vitro* reaction.

Although this approach has obvious applications to genetic manipulations between diverse prokaryotic organisms, it is with mammalian cells that this system has been particularly useful. For selectable genes that can be cloned into the available space in the λ vectors, it has been possible to introduce such a gene [e.g., the herpes simplex virus (HSV) *TK* gene and the α-globin cluster] into mammalian cells and then recover the gene by excision and packaging with λ packaging extracts (1–3). This approach has obvious extensions to cloning genes that complement functions that can be selected for in mammalian cell cultures. The serious limitation to this method of gene cloning, however, is the size limitation on the DNA fragment to be cloned; it must be less than about 15–20 kbp, sufficient to fit in the available space in the λ phage particle, and large enough to contain the eukaryotic transcription signals as well as the coding sequences of the gene. As such, it is satisfactory for viral genes and cDNA clones, but is less likely to be of general use in isolating genomic clones. Good reviews of various λ vectors and their size restrictions are found in Williams and Blattner (4, 5). Insertion vectors have no minimum size for the cloned sequences but are limited to smaller inserts (usually less than 10 kbp) than replacement vectors, which can accept larger inserts but often have minimum size constraints.

Glazer *et al.* (6) applied this method to design a vector system that allows efficient recovery of integrated, chromosomal sequences to be used as target genes in

Methods in Molecular Genetics, Volume 1

studies of mutagenesis. Thus, a mammalian cell line or transgenic mouse is constructed that carries the integrated λ vector, which in turn harbors the target sequence (6, 7). In their case the target gene was the *supF* gene of *E. coli*. Mutations in this gene result in the loss of suppressor function and are easily scored by plating the λ bacteriophage on the appropriate *E. coli* indicator strain. The recovered mutants are, if they are from single-copy cell lines, all from different chromosomal DNA molecules, and they are easily sequenced to determine the specific lesion in each mutant. Even if the cell lines have multiple copies, the statistical likelihood of recovery of two phages from the same chromosome is very low. This method has been applied subsequently to the β-galactosidase gene target by Gossen *et al.* (8) and Kohler *et al.* (9) and to the *lacI* gene target by Kohler *et al.* (10).

Principle of Method

The first step involves the design and construction of the vector to be integrated and recovered. In a typical application the gene of interest, for example, the *E. coli* *supF* gene, is cloned into a plasmid vector along with a genetic marker that is selectable in mammalian cells, for example, the bacterial *neo* gene, which confers resistance to the neomycin analog G418. This plasmid is linearized at a convenient restriction endonuclease cleavage site and inserted into a λ vector such as λ gt2 (11) or more elaborate derivatives. Alternatively, a cDNA expression library can be inserted into the λ vector. This DNA is then self-ligated to form intact cohesive end sequences (*cos* sites) on each end of the complete λ vector. The two *cos* sites together with some flanking DNA are necessary for the excision and packaging of the integrated λ genome (12–14). This DNA is then introduced into the mammalian cell by the appropriate gene transfer method, for example, the calcium phosphate precipitation technique or microinjection, and the transformants selected in the presence of G418, a neomycin analog that is toxic to mammalian cells, or they are selected for the presence and expression of the gene to be isolated. To recover the integrated λ genome containing the gene of interest, the chromosomal DNA is isolated in high molecular weight form and then used as a substrate in the λ packaging reaction. This set of reactions results in the specific excision of the λ sequences by cleavage at the *cos* sites, and assembly of the excised DNA into viable phage particles. The particles thus formed are plated on an *E. coli* host to produce clonal bacteriophage stocks of individual rescuants.

Because the DNA substrate in the packaging reaction frequently is isolated from eukaryotic sources or possibly from prokaryotic organisms that lack the appropriate DNA modification systems, such substrates are at risk for restriction endonuclease attack both in the packaging reaction itself and in the subsequent growth of the packaged phage on the *E. coli* host. Various approaches have been used to minimize these problems. Rosenberg (15) prepared packaging reactions from an *E. coli* strain deficient in the *Eco*K restriction system and found that the yield of the packaging

TABLE I Rescue of Phage from Mouse DNA by λ *in Vitro* Packaging Reactions[a]

Packaging extract	Host cells for plating phage: SY204 (K12 *hsdR*$^-$ mcrAB$^+$ *mrrA*$^+$	Host cells for plating phage: C1a (no known restriction systems)
K12 R$^+$	5	100
K12 R$^-$	23	1700

[a] Comparison of different sources of packaging extracts and host strains for plating packaged phage. The number of phage rescued per microgram of mouse chromosomal DNA, divided by the number of copies of the λ sequences per cell, is given. Host cells for plating were either *E. coli* SY204 (K12, which lacks the *hsdR* restriction system, but contains the *mcrA*, *mcrB*, and *mmrA* restriction systems) or *E. coli* C1a, which lacks all known K12 restriction systems. Packaging extracts were prepared from either *E. coli* K12 strains containing the *hsdR*, *mcrAB*, and *mmrA* restriction systems (R$^+$) or from strains engineered to lack the K12 restriction systems (R$^-$).

reaction was increased about 10-fold. Gossen and Vijg (16) found that the use of *E. coli* strain C as the host for plating the packaging reactions markedly improved the recovery of packaged phage. Strain C is deficient in the *hsdR*, *mcrAB*, and *mrrA* restriction systems (16, 17). A similar high efficiency has been reported for a commercial packaging and recovery system that presumably is prepared from such restriction-deficient *E. coli* strains. We have also observed that restriction activity in both the packaging extracts and in the host bacteria used for plating the rescued phage can reduce the efficiency of the rescue process (Table I). The use of *E. coli* K12-derived restriction-deficient strains to produce the packaging extract and the use of an *E. coli* C host bacterium yielded an approximately 300-fold increase in the yield of rescued phage. The use of *E. coli* C1a by itself as the plating bacteria improves rescue efficiency by up to 75-fold. Suitable λ packaging extracts can be produced by a straightforward procedure from several liters of bacterial culture or can be purchased commercially [Promega (Madison, WI) uses the strains from Rosenberg (15) and Stratagene (La Jolla, CA) uses unidentified and currently unavailable strains]. For our purposes the Stratagene extract is superior to the Promega extract, but the former can be equaled by extracts made as described below.

Materials and Reagents

Because many of the procedures, material, and reagents are commonly used in molecular genetic laboratory work and are well described in other readily available sources, we will describe in detail only the special features that are germane to the λ rescue system.

λ *Vectors*

A wide variety of vectors are available that offer multiple cloning sites and accept inserts of varying sizes. Williams and Blattner (4, 5) reviewed the early vectors; other listings can be found in Sambrook *et al.* (11). It should be noted that many λ vectors contain amber mutations in the *S* gene (lysis function) and require hosts with appropriate amber suppressors for growth. Because *E. coli* C is the most efficient host for phage rescued from mammalian cells, it might be necessary to introduce the appropriate suppressor into this strain for use with a specific λ vector.

Host Bacteria

Escherichia coli C is available from the American Type Culture Collection (ATCC, Rockville, MD) (No. 13706). *Escherichia coli* K12 strains carrying *mcrA, mcrB, mrrA,* and *hsdR* are widely available. An analysis of a variety of restriction-deficient strains has been published by Raleigh *et al.* (17), and many of these strains are available from the ATCC or are commercially available (e.g., from New England BioLabs, Beverly, MA).

Packaging Extracts

Commercial extracts lacking restriction activity are available (Promega, Stratagene); however, they are extremely expensive. Extracts can be prepared as described by Hohn (12) and Enquist and Sternberg (18). *Escherichia coli* K12 strains constructed by Hohn (BHB 2688 and BHB 2690) give efficient packaging extracts but retain restriction activity. We have obtained from Dr. N. Murray (University of Edinburgh) a restriction-deficient strain of *E. coli* K12 (NM759) that produces high-activity extracts. We are also constructing new packaging strains by lysogenization of restriction-deficient *E. coli* K12 strains with the appropriate derivatives of λ phage with complementary amber mutations in the *E* and *D* genes. Because these phage also carry the *b2* deletion to prevent excision of the prophage during extract production, lysogenization is infrequent and complicates such strain constructions. The method of Enquist and Weisberg (19), however, can be used to select such rare lysogens. The use of *E. coli* C as a source of extracts, surprisingly, is not optimal. Extracts made from *E. coli* K12-derived strains appear to be superior in terms of packaging efficiency.

The preparation of the packaging extracts follows the method of Hohn (12) and is described in detail in Sambrook *et al.* (11). Bacteria lysogenic for complementary λ mutants (*Eam* and *Dam*) are used to produce λ packaging enzymes and proteins by thermal induction of the lysogens. Either freeze-thaw or sonic disruption of the cells is used to extract the appropriate proteins. Cultures (1.5 liter) of BHB 2688 and BHB 2690 (or NM 759) will yield approximately 5–10 ml of packaging extract, which can be stored as 100 pairs of complementary packaging reactions.

Methods

Introduction of λ DNA into Cells

Self-Ligation of λ DNA

λ DNA is prepared from the shuttle vector or from the appropriate mammalian expression library. The DNA is self-ligated at a concentration of 300 μg/ml. This ligation enhances preservation of the intact *cos* sites and the flanking packaging signals on transfection into mammalian cells. Prior to ligation the DNA is heated at 50°C for 5 min to melt out circular monomers. The reactions are made up as follows: 200 μl λ DNA at 300 μg/ml, 25 μl 10× ligase buffer (supplied by ligase manufacturer, New England Biolabs), and 1500 units T4 DNA ligase. The ligation reaction is carried out at 37°C to enhance intermolecular ligations. Alternatively, an excess of λ phage "arms" (i.e., fragments from the two ends of the λ genome) can be ligated to the λ vector. This procedure will also protect the *cos* sites and reduce the frequency of concatamers.

Construction of Substrates in Cells

The calcium phosphate method has given good results because integration of multiple copies is essential in the case of the self-ligated vectors. The tandem integration facilitates efficient rescue. Single copies of λ vectors can be rescued only if the *cos* sites are protected and flanking packaging signals are provided, for example by ligation of λ "arms." The λ *supFneo* vector (6, 7) contains the *neo* gene for resistance to G418. We have also constructed a vector derived from λ L47.1 (20) that contains the herpes simplex virus thymidine kinase (HSV *TK*) gene in addition to the *supF* gene (P. M. Glazer and W. C. Summers, unpublished observations). For vectors that lack selectable markers, cotransformation with plasmid DNA containing a selectable marker is often successful (21).

Preparation of Cellular DNA

Once the cells from which the λ DNA is to be rescued are identified and grown up, DNA is prepared from the cells for use as a substrate in the λ packaging reactions. High molecular weight cellular DNA is necessary for efficient packaging. Gentle handling of the viscous DNA is essential; avoidance of shear-producing steps such as rapid pipetting and vortexing is important.

Monolayer cells are best lysed *in situ* after washing with phosphate-buffered saline (PBS). Four milliliters of lysis buffer per 10^7 cells is added to the monolayer [lysis buffer: 10 m*M* Tris-HCl (pH 8.0), 100 m*M* NaCl, 25 m*M* ethylenediaminetetraacetic acid (EDTA), 0.5% sodium dodecyl sulfate (SDS), 200 μg/ml proteinase K). The lysed cell suspension becomes viscous and is gently poured into a 15-ml tube. This lysed cell suspension is incubated for 3 hr at 37°C and RNase (heated at 100°C for 5 min) is added to a final concentration of 10 μg/ml. The RNA is digested by incubation for 1 hr at 37°C. One volume of water-saturated phenol is added and the stoppered tube is gently rocked to mix the aqueous and phenol phases until a homogeneous emulsion is formed. The tube is centrifuged at room temperature for 5 min at 3000 *g* in a standard benchtop centrifuge such as the Sorvall (Norwalk, CT) GLC. The upper aqueous layer, containing the DNA, is removed with a wide-bore pipette and extracted with equal volumes of a 1:1 mixture of phenol and chloroform and then with chloroform alone. The DNA is transferred to a 50-ml tube and 2 vol of 95% ethanol is added. After mixing gently, the precipitated DNA is wound up on a clean glass rod, washed once in 70% ethanol, and dried in air. This DNA is resuspended in 200 $\mu l/10^7$ cells of 10 m*M* Tris-HCl (pH 7.4). Because the subsequent packaging reactions are Mg^{2+} dependent, EDTA is omitted from the buffer at this step. Optimum DNA concentrations are in the range of 0.5–1.0 $\mu g/\mu l$. A completely dissolved and homogeneous solution is critical. Such a solution will be viscous and homogeneous, without "lumps."

Packaging Reactions

The high molecular weight cell DNA is used as a substrate in the λ packaging reactions. The reaction volume depends on the amount of DNA from which the phage DNA is to be rescued. The best results are obtained when the DNA solution, at 1 $\mu g/\mu l$, constitutes no more than 15% of the total volume of the packaging reaction (i.e., the sum of the volumes of the freeze-thaw extract, the sonicated extract, and the DNA solution). In our hands the optimum ratio of sonicated extract to freeze-thaw extract is 2:1, and the optimum DNA concentration is 0.15 $\mu g/\mu l$. The components of the packaging reaction are thawed on ice from storage at −80°C and combined quickly as soon as they have thawed. Mixing of the DNA into the reaction is accomplished by repeated pipetting with a wide-bore pipette tip (e.g., Bio-Rad, Richmond, CA). The reaction is incubated at 37°C for 2–3 hr and then it is diluted with SM buffer (see below) to at least 1 ml, but more if necessary, depending on the expected phage titer. The resulting phage suspension is assayed and analyzed as described below.

Assay of Phage

The basic methods for plating λ phage are described in the classic text by Adams (22) as updated by Murray (23) and Arber *et al.* (24). Host bacteria are prepared by growth of the desired strain, for example, *E. coli* C, in minimal medium with 0.5% (w/v) maltose, to induce the synthesis of the *malA* protein that is part of the λ receptor. Dilutions of packaged λ are made in λ suspension medium [SM buffer: 6 m*M* Tris-HCl (pH 7.2), 10 m*M* $MgCl_2$, 50 mg/liter gelatin] and 0.05-ml aliquots are added to 0.1 ml of host bacteria (2–5 × 10^8/ml) in 12-ml sterile glass tubes. After 15 min of adsorption at room temperature, 3 ml of melted top agar (45°C) is added and the contents mixed by swirling and then poured onto room temperature LB agar plates. After the top agar has solidified, the plates are incubated at 37°C for 12–18 hr and the plaques observed. Plaques of interest are picked with a sterile loop into SM buffer and phage stocks prepared by either the liquid lysis method or by making plate lysates (11).

The DNA of the rescued phage can be analyzed by methods appropriate to the purpose of the experiment: genetic tests, sequence analysis, or restriction fragment length polymorphism (RFLP) studies are common examples.

Comments

Because 100% efficient packaging of pure λ DNA should yield about 2 × 10^{10} plaque-forming units (pfu) per microgram DNA, the best available packaging systems, which yield 1–2 × 10^9 pfu/μg, are still only 5–10% efficient. If the substrate is not pure phage DNA, but instead chromosomal DNA with a few integrated λ genomes, the efficiency falls even further. For example, under the best conditions given in Table I, we obtained 1700 pfu/μg total cell DNA; this value corresponds to about 4 × 10^7 pfu/μg of phage DNA, only about 5% of the efficiency of the extract when it packaged pure λ DNA. The reasons for this loss of efficiency are not yet clear. In reconstruction experiments in which mammalian DNA was added to λ packaging reactions, the presence of the mammalian DNA competed with and inhibited the packaging of the pure λ DNA. Low-affinity, nonspecific binding of one or more essential components of the packaging reaction to the mammalian DNA could explain such interference. The presence of so much mammalian DNA might titrate one or more of the DNA-binding proteins involved in the packaging reactions. If so, it might be possible to improve the efficiency by addition of the limiting protein(s). One obvious candidate for such a DNA-binding component is the terminase protein complex, which must bind to the *ter* site and cut it after the DNA is encapsidated in the λ prohead structure (12). The availability of purified terminase preparations now makes it possible to test such notions. A recent report that heterologous proteins, such as p16 of phage φ29, which binds to the ends of λ phage genomes with short φ29 terminal sequences to make hybrid proheads, suggests yet another approach to possible improvement in packaging efficiencies and novel vector construction (25).

Sequences from both ends of the λ phage DNA are needed for the recognition by the terminase and the start of the packaging reaction, at least *in vivo* (13). Although linear, single λ genomes are packaged well, circular monomers are poor substrates for packaging. Concatameric DNA, which is the physiological substrate in phage growth, is also a good substrate. For DNA that is integrated, the presence of two intact flanking *cos* sites is required (14). Indeed, the reported high efficiencies for commercial λ packaging reactions are achieved with self-ligated λ DNA as substrate for packaging. With self-ligation at relatively high DNA concentrations, it is probable that one will obtain multiple tandem copies of the vector rather than circular monomers. If this is undesirable, the ligation of the left and right terminal fragments of the λ genome to the right and left termini of the complete DNA will result in a λ genome with sufficient additional sequences to allow efficient packaging from integrated vectors, but without the deliberate bias in favor of multiple, tandem copies.

As with any mammalian cell containing inserted foreign genes, cell lines and transgenic mice are at risk for rearrangement and loss of such sequences. This is especially so if the inserted sequences are tandem arrays of multiple copies. Care must be exercised in maintaining and checking the cell lines and transgenic animals. Not all inserted sequences may be recoverable by the λ rescue system. Because the rescue requires two intact, flanking *cos* sites, their loss or rearrangement will render the intervening λ sequences unrescuable.

The *E. coli* K12 lysogenic strains employed by Hohn (12) each carry a *recA* mutation in the bacterium and a *red* mutation in the prophage to minimize recombination between the rescued phage and the unexcised prophage from the lysogenic strain used to produce the packaging extract. The use of other strains without *recA* mutations may allow sufficient recombination to yield a small fraction of phages that are artifacts of the packaging system. Occasional preparations of packaging extracts have low backgrounds of phage even in the absence of added DNA substrate. This background may represent excised and packaged prophage as well as recombinant phage, both from the lysogenic host bacteria used to make the extracts.

Although the emphasis in the use of the λ rescue system has been on the recovery of DNA sequences from mammalian cells, it is clear that such a system is of use in cloning genes in prokaryotic organisms, especially those for which no genetic system has been developed or for which no temperate phages or compatible plasmids are yet known.

Acknowledgments

We gratefully acknowledge help and advice from K. B. Low, M. Gottesman, N. Murray, B. Hohn, W. P. Summers, L. Enquist, and N. Sternberg. E. Gunther generously communicated unpublished results. This work was supported by NIH Grant CA 39238.

References

1. W. Lindenmeier, H. Hauser, I. G. de Wilke, and G. Shutz, *Nucleic Acids Res.* **10,** 1243 (1982).
2. Y. Lau and Y. W. Kan, *Proc. Natl. Acad. Sci. U.S.A.* **80,** 5225 (1983).

3. Y. Lau and Y. W. Kan, *Proc. Natl. Acad. Sci. U.S.A.* **81,** 414 (1984).
4. B. G. Williams and F. R. Blattner, *J. Virol.* **29,** 555 (1979).
5. B. G. Williams and F. R. Blattner, *in* "Genetic Engineering" (J. K. Setlow and A. Hollaender, eds.), Vol. 2, p. 201. Plenum, New York, 1980.
6. P. M. Glazer, S. N. Sarkar, and W. C. Summers, *Proc. Natl. Acad. Sci. U.S.A.* **83,** 1041 (1986).
7. W. C. Summers, P. M. Glazer, and D. Malkevich, *Mutat. Res.* **220,** 263 (1989).
8. J. A. Gossen, W. J. de Leeuw, C. H. Tan, E. C. Zwarthoff, F. Berends, P. H. Lohman, D. L. Knook, and J. Vijg, *Proc. Natl. Acad. Sci. U.S.A.* **86,** 7971 (1989).
9. S. W. Kohler, G. S. Provost, P. L. Kretz, M. J. Dycaico, J. S. Sorge, and J. M. Short, *Nucleic Acids Res.* **18,** 3007 (1990).
10. S. W. Kohler, G. S. Provost, A. Fieck, P. L. Kretz, W. O. Bullock, J. S. Sorge, D. L. Putman, and J. M. Short, *Proc. Natl. Acad. Sci. U.S.A.* **88,** 7958 (1991).
11. J. Sambrook, E. F. Frisch, and T. Maniatis, "Molecular Cloning: A Laboratory Manual," 2nd Ed. Cold Spring Harbor Lab. Press, Cold Spring Harbor, New York, 1989.
12. B. Hohn, *in* "Recombinant DNA," Part F (R. Wu, ed.), Methods in Enzymology, Vol. 68, p. 299. Academic Press, New York, 1979.
13. B. Hohn, *Proc. Natl. Acad. Sci. U.S.A.* **80,** 7456 (1983).
14. B. Hohn, *J. Mol. Biol.* **98,** 93 (1975).
15. S. M. Rosenberg, *Gene* **39,** 313 (1985).
16. J. A. Gossen and J. Vijg, *Nucleic Acids Res.* **16,** 9343 (1989).
17. E. A. Raleigh, N. E. Murray, H. Revel, R. M. Blumenthal, D. Westaway, A. D. Reith, P. W. J. Rigby, J. Elhai, and D. Hanahan, *Nucleic Acids Res.* **16,** 1563 (1988).
18. L. Enquist and N. Sternberg, *in* "Recombinant DNA," Part F (R. Wu, ed.), Methods in Enzymology, Vol. 68, p. 281. Academic Press, New York, 1979.
19. L. W. Enquist and R. A. Weisberg, *Virology* **72,** 147 (1976).
20. W. A. Loenen and W. J. Brammar, *Gene* **10,** 249 (1980).
21. M. Perucho, D. Hanahan, and M. Wigler, *Cell* **22,** 309 (1980).
22. M. H. Adams, "Bacteriophage." Wiley (Interscience), New York, 1959.
23. N. E. Murray, *in* "Lambda II" (R. W. Hendrix, J. W. Roberts, F. W. Stahl, and R. A. Weisberg, eds.), p. 395. Cold Spring Harbor Lab. Press, Cold Spring Harbor, New York, 1983.
24. W. Arber, L. Enquist, B. Hohn, N. E. Murray, and K. Murray, *in* "Lambda II" (R. W. Hendrix, J. W. Roberts, F. W. Stahl, and R. A. Weisberg, eds.), p. 433. Cold Spring Harbor Lab. Press, Cold Spring Harbor, New York, 1983.
25. L. E. Donate and J. L. Carrascosa, *Virology* **182,** 534 (1991).

Section II

Cell Hybrids

[6] Preparation and Properties of Microcell Hybrids

Amelia A. Langston and R. E. K. Fournier

Introduction

Spontaneous fusion of mammalian cells to form somatic cell hybrids was first reported over 30 years ago by Barski *et al.* (1). The two technical developments that established cell fusion as a feasible and general genetic technique were the discovery that fusion could be induced by exposing cells to ultraviolet (UV)-inactivated Sendai virus (2) or polyethylene glycol (3), and the development of biochemical selection systems based on intergenic complementation (4). Since that time, cell hybrids have been powerful tools for gene mapping and analysis of genes involved in development (5, 6), disease (7), and malignancy (8).

The earliest cell fusion experiments involved combining two intact cells to form whole-cell hybrids; these generally contain complete chromosome sets from both parents. When parental cells of different species were used, the hybrids tended to segregate chromosomes during passage in culture (9). Segregating hybrid populations have been useful for gene mapping, but their heterogeneous and unstable karyotypes have limited their utility for more complex genetic tests.

The subject of this article is microcell fusion, a technique for generating hybrid cell lines with relatively simple karyotypes (10). Microcell hybrids generally contain only one or a few donor chromosomes or chromosome fragments. As cloned dominant selectable markers have been developed, it has become possible to transfer virtually any chromosome or chromosome region from one cell to another. The genotypic properties of microcell hybrids make them useful tools for gene mapping and genetic analysis of complex cellular phenotypes. Microcell hybrids can also be used as starting materials for cloning DNA markers (11) or expressed genes (12) from defined chromosomal regions.

The steps involved in microcell fusion are shown in Fig. 1. The procedure can be divided into several parts: micronucleation of donor cells, enucleation in the presence of cytochalasin B, purification of microcells, fusion with intact recipients, and biochemical selection of microcell hybrids. This article presents an overview of considerations that are important for each of these steps. The major methodological alternatives are discussed, and detailed protocols are presented for procedures that we find useful in the laboratory.

Methods in Molecular Genetics, Volume 1

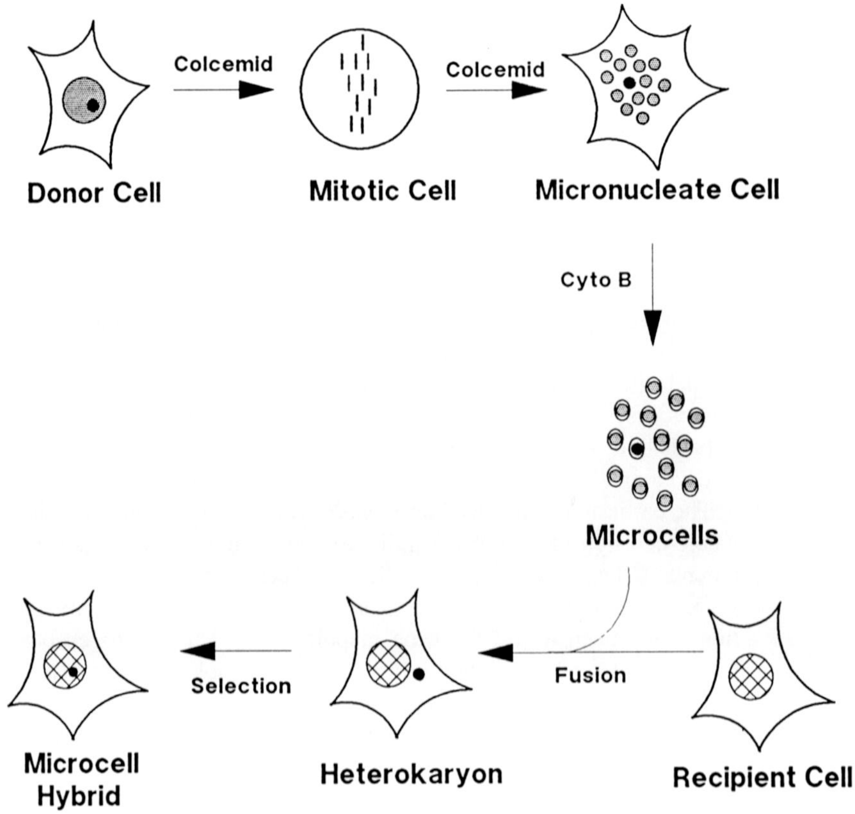

FIG. 1 Diagrammatic representation of the steps involved in microcell fusion. Cyto B, Cytochalasin B.

Preparation for Microcell Fusion

Reagents and Materials

Reagents

Phosphate-buffered saline (PBS): 137 m*M* NaCl, 2.7 m*M* KCl, 4.3 m*M* Na_2HPO_4, 1.47 m*M* KH_2PO_4 (pH 7.4)

Colcemid: Demecolcine (Cat. No. D7385; Sigma, St. Louis, MO) powder is stored desiccated and protected from light at −20°C. Stock solutions of 1 mg/ml in 0.9% (w/v) NaCl are stable for at least 6 months at −20°C if protected from light

Concanavalin A (ConA) (Cat. No. C2631; Sigma): Powder is stored desiccated at 4°C

WSC cross-linker: 1-Cyclohexyl-3-(2 morpholinoethyl)carbodiimide metho-*p*-toluene sulfonate (Cat. No. C1011; Sigma) powder is stored desiccated at −20°C

Cytochalasin B (Cat. No. C6762; Sigma): Powder is stored desiccated and protected from light at −20°C. Stock solutions of 2 mg/ml in dimethyl sulfoxide (DMSO) are stable indefinitely at 4°C if protected from light

Phytohemagglutinin P (PHA-P) (Cat. No. 3110-56-4; Difco, Detroit, MI): Powder is a lyophilized stock consisting of equal parts by weight of PHA-P and salt. PHA-P solutions are always prepared fresh; the powder is stored at 4°C

Polyethylene glycol (PEG) 1500 (Cat. No. 011-14806/C; Koch-Light, Suffolk, England): Crystals are stored at room temperature

Hoechst dye No. 33258: Bisbenzimide (Cat. No. B2883; Sigma) powder is stored at room temperature. Stock solutions of 50 mg/ml in PBS are stable indefinitely at 4°C if protected from light

Acetoorcein: 0.5% (w/v) orcein (Cat. No. O7380; Sigma) in 50% (v/v) acetic acid is prepared by refluxing for 2–4 hr. The solution can be stored indefinitely at room temperature

Percoll (Cat. No. 17-0891-01; Pharmacia, Piscataway, NJ): Supplied as a sterile suspension in aqueous solution (for suspension enucleation protocol)

Special Materials

High-speed refrigerated centrifuge: For example, Sorvall (Norwalk, CT) RC-5B (SA600 rotor) or Beckman (Fullerton, CA) J2-21 (JA20 rotor)

Polycarbonate 50-ml centrifuge tubes (Nalgene; Nalge, Rochester, NY) with lids

Hot-wire cutter connected to a variable-voltage transformer

Swinnex 25-mm filter holders (Millipore, Bedford, MA); 8- and 5-μm polycarbonate filters (Nuclepore, Pleasanton, CA)

Gradient mixing device (LKB, Gaithersburg, MD) and peristaltic pump (only for unit gravity sedimentation)

Selection Schemes

An essential element of any microcell fusion experiment is biochemical selection for the transfer of a particular donor-encoded marker. It is also generally necessary to eliminate intact donor cells that have escaped micronucleation and enucleation, either by counterselection or particle purification. Finally, in some crosses it is desirable to eliminate selectively whole-cell hybrids.

Prior to the development of cloned selectable markers, endogenous markers, including thymidine kinase (TK), hypoxanthine-guanine phosphoribosyltransferase (HPRT), and adenine phosphoribosyltransferase (APRT), were employed extensively. Selective retention of these genes requires mutant recipient cell lines deficient in the relevant enzyme. Codominant markers, such as resistance to the cardiac glycoside ouabain, are also useful for biochemical selections; this particular agent is used extensively to eliminate human donor cells, which are innately sensitive, in microcell fusions with rodent recipients (13).

Cloned selectable genes, primarily of prokaryotic or viral origin, are now widely used in somatic cell genetics (reviewed in Ref. 14). This allows any donor chromosome or chromosomal region to be selectively marked prior to microcell fusion, and wild-type recipient cells can be employed. Several of these genes permit both positive and negative selection, including *gpt* (15), herpes simplex virus (HSV) *tk-neo* (16), and hygromycin phosphotransferase–thymidine kinase (*HyTK*; 17) fusion transgenes. Introducing selectable transgenes into donor and/or recipient cells provides great flexibility in devising an appropriate selective strategy.

Mycoplasma Testing

Mycoplasma infection causes changes in cell genotype and phenotype and results in poor hybrid yields. Prior to microcell fusion, both donor and recipient cell lines should be tested for mycoplasma contamination by fluorescent Hoechst 33258 staining (18) or culture methods (19). Contaminated cultures should be discarded.

Preparation of Plastic Bullets and Concanavalin A Cross-Linking

Enucleation of micronucleate cell populations is generally accomplished by centrifugation of cell monolayers in the presence of cytochalasin B. We find it convenient to attach cells to bullet-shaped pieces of plastic that fit round end down into 50-ml centrifuge tubes. Plastic "bullets" are cut from tissue culture plates using a hot wire cutter, and they can be reused many times until the plastic begins to fatigue and crack.

Cells must be firmly attached to the surface for effective enucleation to occur. Even poorly adherent cells or cells that grow in suspension can be enucleated by attaching them to plastic surfaces to which concanavalin A (ConA) has been covalently linked (20). We use ConA-coated bullets routinely, even for adherent cells, because it minimizes contamination of the crude microcell preparation with intact cells, which can dislodge from untreated surfaces during enucleation. Concanavalin A-cross-linked bullets can be prepared ahead of time and stored in PBS at 4°C.

Preparation of Plastic Bullets

1. Prepare a cardboard bullet template measuring 24 × 86 mm with one rounded end. It should fit loosely, rounded end down, into a 50-ml polycarbonate centrifuge tube.

2. Trace the template outline onto the bottom surface of 150-mm tissue culture dishes. Four bullets can be prepared from each dish.

3. Bullets are cut from the tissue culture plates using a hot nickel-chromium resistant wire attached to a variable-voltage transformer. The edges of the bullets are then polished with sandpaper.

4. Bullets are stored in 95% (v/v) ethanol.

Concanavalin A Cross-Linking of Bullets

5. Remove the bullets from ethanol and place them in sterile 150-mm plates, five bullets per plate, with edges not touching. Allow the ethanol to evaporate by placing the plates, tops ajar, in a laminar flow hood. Typically, 12–24 bullets will be processed in each fusion experiment.

6. Prepare ConA solution (15 mg/ml in 0.9% NaCl) and incubate at 37°C for at least 30 min. Remove undissolved solids by gravity filtration through Whatman (Clifton, NJ) No. 1 paper, and sterilize the filtrate by passing it through a 0.2-μm filter (e.g., Gelman Acrodisc, Ann Arbor, MI).

7. Prepare WSC cross-linker (75 mg/ml in 0.9% NaCl) and sterilize by membrane filtration (0.2-μm filter).

8. When the bullets are dry, pipette 0.6 ml of WSC solution onto the surface of each bullet. Overlay with 0.6 ml of ConA and spread the mixture over the entire surface of each bullet. Incubate for 1–2 hr at room temperature.

9. Remove the ConA–WSC solution by aspiration, and rinse the bullets twice with sterile PBS. Bullets may be stored in 50-ml culture tubes in sterile PBS at 4°C.

10. After use, the bullets can be soaked in 1% SDS, wiped with a clean cloth, rinsed with water, stored in ethanol, and reused.

Micronucleation of Donor Cells

The initial step in microcell fusion is to induce donor cells to become micronucleate. This can be accomplished by either of two general approaches. The first method involves prolonged mitotic arrest (21, 22), whereas the second approach employs a short mitotic arrest followed by exposure of mitotic cells to conditions that enhance micronucleation.

Prolonged mitotic arrest is a simple and widely used method that effectively micronucleates most cell types. In this procedure, exponentially growing cultures are exposed to Colcemid, an inhibitor of microtubule polymerization. Cells initially accumulate in metaphase, but they eventually proceed through the block and reenter interphase, becoming micronucleate in the process. The major limitation of prolonged mitotic arrest is that some slowly growing cells, particularly diploid human cells, may micronucleate at low frequency. In these cases, micronucleation frequency can often be enhanced by quasisynchronization of the cells, for example by low serum arrest, prior to Colcemid treatment.

Sequential treatment of transformed cells with Colcemid and cytochalasin B yields micronucleate cells at high frequency (23). The procedure involves exposure of cells to Colcemid for 8–16 hr, selective mitotic detachment, and treating mitotic cells with cytochalasin B for 4–6 hr. This method has the disadvantage that large numbers of mitotic cells are required. Furthermore, it works poorly for diploid cells, which tend to become binucleate when treated in this manner. For these reasons, this method is not used widely.

An alternate method is mitotic arrest followed by cold shock (24) or brief exposure to hypotonic medium (25). This method works well for micronucleation of human lymphoblasts (26), which are reportedly difficult to micronucleate by prolonged mitotic arrest. In this case, purification of crude microcell preparations is essential because large numbers of contaminating particles are produced.

Prolonged Mitotic Arrest with Colcemid

Efficient micronucleation by prolonged mitotic arrest depends on careful determination of the optimal Colcemid concentration and time of exposure. Colcemid concentrations in the range 0.01–0.1 μg/ml effectively induce micronucleation of most rodent cells; however, within that range the effective concentration for a given cell line is usually narrow. Micronucleation of human cells, particularly diploid cells, may require 10–20 μg/ml Colcemid (27, 28), and such concentrations generally produce cytotoxicity. Table I summarizes conditions that effectively micronucleate

TABLE I Micronucleation of Selected Cell Lines

Cell line	Colcemid (μg/ml)	Time (hr)	Approximate percentage micronucleation
Mouse L cells	0.02	40–48	90
Mouse embryo fibroblasts	0.04	36	70
Mouse 3T6	0.04	48	60
Mouse 10T 1/2	0.01	48	99
Mouse C2C12 (myoblast)	0.01	48	50
Mouse B78 (melanoma)	0.06	48	90
Chinese hamster ovary (CHO)	0.04	48	80
Rat-2 fibroblast	0.025	48	80
Human foreskin fibroblast	10–20	48	40
Human Rh18 (rhabdomyosarcoma)	10	24	50

a variety of commonly used cell lines. These conditions are merely guidelines, and each cell line to be employed as a microcell donor should be tested individually.

The kinetics of mitotic arrest and subsequent micronucleation in Colcemid depend on the generation times of the donor populations. In general, rapidly growing cells micronucleate faster and more efficiently than slowly growing lines. Most cells require Colcemid exposure for times corresponding to one to two cell generations. The hybrid yield of microcells generated by Colcemid exposure for more than two cell generations is low (10), so that optimal exposure is generally the shortest time compatible with good micronucleation.

Optimal conditions for micronucleating any particular cell line can be determined in simple calibration experiments. Donor cells are seeded at 20–40% confluence, incubated overnight, and Colcemid is added to various concentrations spanning the expected effective range. Plates are assessed for micronucleation and cell toxicity periodically over the course of 48 hr. For most adherent cells, visualization under phase contrast is generally sufficient to assess the micronucleation index semiquantitatively (see Fig. 2). For cells growing in suspension or for adherent cells that remain rounded, it may be useful to harvest cells at intervals and to stain them with acetoorcein or Hoechst 33258 to visualize micronucleate, mononucleate, and mitotic cells. For rodent cells, micronucleation indexes of 60–90% are generally at-

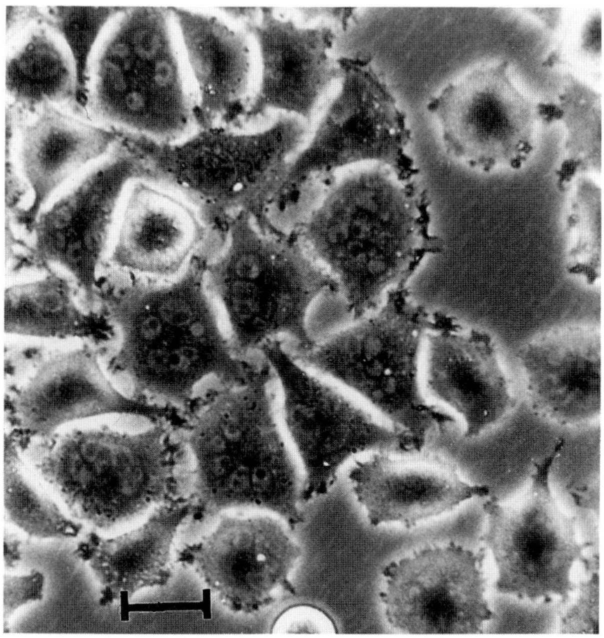

FIG. 2 Micronucleation of mouse L cells, visualized under phase contrast after exposure to Colcemid (0.02 μg/ml) for 48 hr. Bar: 5 μm.

tainable. Micronucleation of human cells is more variable, and the percentage micronucleation may be as low as 20–30% even under optimal conditions.

Procedure

1. Cultures should be growing exponentially before the addition of Colcemid. On the day prior to arrest, plate the donor cells such that they will be 30–40% confluent at the time of Colcemid addition. Colcemid arrests are generally performed in nonselective medium.
2. Sixteen to 24 hr later, add Colcemid to the appropriate concentration and incubate for 24–48 hr, as appropriate.
3. Harvest the micronucleate cells by trypsinization and pool the suspensions. Count an aliquot of the pool and save an aliquot for quantitation of micronucleate cells.
4. Centrifuge the donor cells and resuspend in PBS to a cell concentration of approximately 1×10^6/ml. Clumps of cells should be broken by trituration prior to plating.

Micronucleation may be quantitated by either acetoorcein or Hoechst 33258 staining, both of which allow visualization of nuclear structures. In general, Hoechst-stained specimens are easier to visualize and interpret.

a. *Acetoorcein staining:* Mix a drop of the cell suspension with a drop of acetoorcein solution on a slide, overlay a coverslip, and view under transmitted light after 1–2 min. Nuclear material stains red, cytoplasm pink, so that partitioning is evident in micronucleate cells.
b. *Hoechst staining:* Pellet the cells and fix them in 3:1 (v/v) methanol–acetic acid. Place a drop of the suspension on a microscope slide and air dry. Overlay with Hoechst 33258 solution (0.5 μg/ml) and stain for 1 min. Rinse with water, and mount a coverslip with mounting buffer (50% glycerol in PBS). Under UV illumination (excitation 365 nm; emission 480 nm), micronucleate cells are clearly distinguishable from their mononucleate counterparts.

Enucleation of Micronucleate Populations

Enucleation of micronucleate donor cells is accomplished by centrifugation in the presence of cytochalasin B, a fungal metabolite that interferes with cytokinesis. These conditions result in extrusion of the nucleus or micronuclei, each surrounded by a thin rim of cytoplasm and plasma membrane (29).

A convenient method of enucleation involves attachment of the donor cells to a solid surface that can be centrifuged in cytochalasin B-containing medium (29). In this case, cytoplasts remain attached to the surface while extruded nuclear material is sheared from the cells by centrifugal force and pelleted to the bottom of the tube. The use of ConA-cross-linked surfaces facilitates cell attachment and increases the efficiency of enucleation for both adherent and nonadherent cells (20). Cell suspensions can also be enucleated using cytochalasin-containing Ficoll (30) or Percoll gradients (27). In these cases, enucleation results from shear forces arising from density differences of nucleoplasm vs cytoplasm.

Enucleation from Plastic Bullets

Enucleation from ConA-cross-linked surfaces is a useful method for adherent cell types. Many cells that grow in suspension can also be enucleated in this way. Although a number of different enucleation geometries have been devised, we find it convenient to attach cells to bullet-shaped pieces of plastic that fit into 50-ml centrifuge tubes, as discussed above. This allows for efficient processing of large numbers of cells ($2-3 \times 10^7$) in a single 30-min centrifugation. Cells can also be enucleated directly from T-flasks or other surfaces, but these procedures are cumbersome and offer no real advantages over bullet enucleation.

The efficiency of this method depends on firm attachment of the cells to the substrate and on generation of sufficient centrifugal force to shear the nuclei from the cells. The optimal speed and time of centrifugation are variables that should be determined empirically for each donor line. In general, centrifugation at 25,000–30,000 *g* for 30–60 min at 30°C is appropriate.

Following each spin, remove one bullet and examine the surface under phase-contrast illumination to assess the degree of enucleation. Completely enucleated monolayers will contain only cytoplasmic remnants attached to the surface. Incompletely enucleated monolayers will have threads of material, resembling beads on a string, attached to the cytoplasmic remnants. If significant numbers of micronuclei remain attached, longer centrifugation is required, and the bullets and tubes may be simply returned to the centrifuge. If contamination of the crude microcell pellet with intact cells occurs, a prespin in the absence of cytochalasin B may be helpful in removing loosely attached cells prior to enucleation (10), but this is rarely necessary when processing ConA-coated surfaces.

Procedure

1. Under sterile conditions, remove bullets from PBS and place into 150-mm plates, cross-linked side up.
2. Overlay approximately 1×10^6 cells (in 1.0–1.5 ml PBS) onto the surface

of each bullet, and allow the cells to settle onto the bullet. If undisturbed, the cells will settle in 10–15 min. Cells that settle onto ConA-coated surfaces, unlike ordinary tissue culture plastic, seem to be fixed in place within minutes.

3. When the cells are attached, flood each plate with 40 ml complete medium and incubate the plates at 37°C to allow the cells to spread. Complete spreading does not seem required for effective enucleation from these surfaces, but most adherent cell lines spread completely within 1 to 2 hr. Suspension cells will not appear to spread; they will nonetheless be firmly tethered.

4. Cytochalasin-induced enucleation does not occur if the temperature of the medium is less than about 25°C (28). Therefore the medium in the tubes must be warm during enucleation. In the Sorvall RC-5B, temperature settings of 28–34°C generally give internal temperatures of about 30°C during centrifugation. The appropriate temperature settings for each individual centrifuge should be determined empirically. Use those settings to prewarm the empty rotor and chamber by spinning at enucleation speed for 10–15 min.

5. Prepare enucleation medium: serum-free medium containing 10 μg/ml cytochalasin B (diluted from a 2-mg/ml stock in DMSO). Pour approximately 40 ml into each tube and keep warm.

6. When the donor cells have flattened on the bullets, transfer one or two (back-to-back) bullets in each tube and cap securely. Centrifuge at 25,000–30,000 *g* for 30 min at about 30°C.

7. After centrifugation, a small microcell pellet should be apparent in the bottom of each tube. Remove one bullet and examine under phase-contrast illumination to monitor the degree of enucleation; if incomplete, respin the tubes. Once enucleation is complete, remove the bullets from the tubes (place in water; they should not be allowed to dry before cleaning) and decant the cytochalasin-containing medium. Enucleation medium can be reused to process a second set of bullets or filter sterilized for future use.

8. Resuspend each microcell pellet in about 1 ml serum-free medium and pool the suspensions. Record the suspension volume and reserve an aliquot for particle quantitations.

Suspension Enucleation

Suspension enucleation should be considered if attachment of cells to ConA-treated surfaces is ineffective. Suspension enucleation protocols generally involve centrifugation through Ficoll (30) or Percoll (27) gradients in the presence of cytochalasin B. Large numbers of cells (1×10^8 cells/20-ml tube) can be processed in this way, but the overall efficiency of enucleation is lower than monolayer enucleation methods. The crude microcell preparations generated by suspension enucleation are heterogeneous and must be purified to remove contaminating whole cells, as described in the following section.

The simplest suspension enucleation method involves centrifugation of cells through a cytochalasin B-containing Percoll density gradient that is formed during centrifugation (27). Following centrifugation, two or more bands will be visible within the gradient, each containing a mixture of whole cells, karyoplasts, and microcells. Cytoplasts and membrane fragments equilibrate in the gradient above these nuclei-containing bands.

Procedure

1. Prewarm the centrifuge and rotor to about 30°C.
2. Prepare a 1:1 (v/v) mixture of Percoll and culture medium with serum, and add cytochalasin B to a final concentration of 10–20 μg/ml.
3. Suspend the donor cells in the Percoll–medium solution to a cell concentration of $<5 \times 10^6$/ml.
4. Divide the cell suspension into 20-ml polycarbonate gradient tubes, filled to capacity, and centrifuge the tubes at 44,000 *g* for 60 min.
5. After centrifugation, two bands should be visible near the middle of the density gradient. Carefully remove both bands from each tube, taking care to avoid aspirating material from the top of the gradient.
6. Pool the crude preparation and centrifuge at 500–800 *g* for 5–10 min to remove the Percoll. Resuspend the pellet in 40–50 ml of serum-free medium, and reserve an aliquot for particle quantitations.

Particle Quantitations

The crude microcell pellet is a heterogeneous mixture of particles including microcells, cytoplasmic vesicles, whole nuclei (karyoplasts), and intact cells. The relative proportions of these constituents will depend on the characteristics of the donor cells and the efficiency of enucleation. The composition of the crude pellet is determined by staining with either acetoorcein or Hoechst 33258, as described above. Both dyes preferentially stain nuclear material, making it possible to distinguish microcells, cytoplasmic vesicles, and karyoplasts or intact cells. The total number of particles is determined by counting an aliquot of the pellet using a hemacytometer; the number of microcells in the preparation can then be calculated. These quantitations are particularly helpful in defining optimal conditions for enucleation. If particle purification (discussed in the next section) is performed, an aliquot of the final microcell preparation should also be stained and counted to assess the efficiency of purification and recovery of microcells.

Purification of Microcell Preparations

If the crude preparation contains more than 10% intact cells and/or karyoplasts, significant numbers of whole-cell hybrids and/or donor cell clones will survive unless they can be eliminated by biochemical selection. In situations in which direct counterselection is not possible, purification of contaminated microcell preparations is usually necessary. Two alternatives for particle purification are presented: membrane filtration and unit gravity sedimentation.

Membrane Filtration

Purification through polycarbonate filters is the simplest and most widely used method of microcell purification (23, 27). We find sequential filtration through membranes of 8- and 5-μm pore size to be adequate for excluding whole cells and karyoplasts, although an additional 3-μm filtration is performed in some laboratories. Filtration will result in some loss of microcells due to clogging and nonspecific adherence of particles to plastic, but this can be minimized by diluting the crude enucleation preparation and use of an adequate number of filters.

Filters are mounted in Swinnex adaptors (Millipore, Bedford, MA), wrapped in foil, and autoclaved prior to fusion. For filtration, adapters are mounted on sterile 10-ml syringes from which plungers have been removed (in processing the microcell pellet from 24 bullets, a minimum of 3 or 4 pairs of filters should be used). The pellet is resuspended in serum-free medium, 10 ml/filter set. The suspension will usually flow through the 8-μm membrane by gravity; application of gentle pressure is necessary for the 5-μm filtration. Filters may fail if set improperly in the adapter or if excessive pressure is used to force the suspension through, so each aliquot should be checked prior to pooling the filtrate. The final microcell suspension is centrifuged at 750–1000 *g* for 10 min at room temperature to pellet microcells for fusion. An aliquot of the pooled filtrate should be reserved for particle quantitations.

Unit Gravity Sedimentation

Purification by unit gravity sedimentation is an efficient technique for separating karyoplasts and whole cells from microcells; however, it is time consuming and requires special equipment. It is most applicable to situations in which the enucleation pellet is heavily contaminated with larger particles, particularly if donor cells cannot be eliminated by counter selection.

The principle of the method is separation of whole cells and karyoplasts from microcells on the basis of differences in size. A linear gradient of 1–3% bovine serum albumin (BSA) is generated in a 50-ml syringe, using a gradient mixing device and peristaltic pump. The sample is allowed to sediment through the gradient for 3–4 hr. Whole cells and karyoplasts fall to the bottom of the gradient, while purified microcells can be recovered from the top half of the gradient. Preparation of the density gradient apparatus has been described by Hecht *et al.* (31).

Suspension/Monolayer Fusion

Use of the chemical fusogen polyethylene glycol (PEG) has largely replaced inactivated Sendai virus as the method of choice for most types of somatic cell fusion (3). Polyethylene glycol is widely available and produces rapid, efficient, and reproducible fusion.

Microcells are first allowed to agglutinate with the monolayer of recipient cells in the presence of phytohemagglutinin P (PHA-P). The microcells are then fused with recipient cells by brief exposure to PEG. A fraction of the microcell heterokaryons will subsequently divide to form microcell hybrid clones. We observe efficient microcell fusion with microcell–recipient cell ratios between 1:1 and 5:1. This generally requires $0.5–2 \times 10^7$ microcells/25-cm^2 recipient flask.

The most important variable at this stage of the procedure is the concentration of PEG used for fusion, which is generally 44–50% (w/w) (23). After PHA-P-induced agglutination, it is usually possible to achieve efficient fusion of most cell types with 44–46% PEG, but some relatively nonfusogenic recipients require higher concentrations. For an initial fusion involving a given donor–recipient pair, it is informative to fuse several different recipient flasks, each at a different PEG concentration.

Procedure

1. Recipient cells are plated into 25-cm^2 flasks on the day prior to fusion such that the monolayer will be 70–80% confluent at the time of fusion. An extra flask of recipient cells should be prepared to be processed as a self-fused control, without addition of microcells.
2. Prepare PHA-P (200 μg/ml in serum-free medium) and filter sterilize (the stock powder is 50% PHA-P by weight).
3. Prepare 1 ml/recipient flask of 42–50% (w/w) PEG in serum-free medium, dissolve at 37°C, and filter sterilize.

4. Resuspend the microcell pellet in 1 ml serum-free medium per recipient flask. The suspension should be vigorously pipetted to break clumps.

5. Rinse each recipient monolayer with 10 ml serum-free medium. Add 1 ml of the microcell suspension to each flask, followed by 1 ml of PHA-P. Tilt the flasks to spread microcells over the monolayer and incubate at 37°C for 10 min. Prior to fusion, check the flasks under the microscope to ensure that the microcells have agglutinated to the recipient monolayer.

6. Fuse one flask at a time, as follows: aspirate PHA-P, invert the flask, and add 1 ml of PEG. Turn the flask over, begin timing, and slowly rock the flask back and forth to spread the PEG over the monolayer. After 60 sec, aspirate the PEG and rinse rapidly with three changes of serum-free medium.

7. Add complete nonselective medium containing antibiotics and incubate for 24 hr.

8. The following day, split each flask 1:10 or 1:20 into selective medium. Microcell hybrid clones should be visible within 2–4 weeks.

It is not possible to predict the number of microcell clones that will be generated in a microcell fusion experiment, as each donor–recipient pair will behave differently. Other variables, such as the length of Colcemid exposure during micronucleation and the selection scheme, may also affect hybrid yields (10, 23). In general, the efficiency of microcell fusion should be on the order of $0.5–5 \times 10^{-5}$ relative to the number of recipient cells fused. If a given fusion does not produce microcell hybrid clones, it is important to critically examine the data collected during the experiment regarding micronucleation, enucleation, and composition of the final microcell preparation. Other seemingly minor events such as inappropriate temperature during enucleation, especially high temperature, may also compromise hybrid yield. Several fusion experiments may be required to optimize all of the variables that contribute to efficient microcell transfer.

Genotypic Analysis of Microcell Hybrid Clones

Genotypic characterization of the hybrid collection is an essential part of any microcell fusion experiment. It is important to establish whether chromosome fragmentation and/or cotransfer of unwanted material has occurred during fusion. Cytogenetic techniques and DNA marker analysis provide complementary information with respect to chromosomal content and homogeneity of each clonal population.

Fluorescence *in situ* hybridization to metaphase chromosomes is a convenient cytogenetic method for initial characterization of a collection of interspecific hybrids (32). The species specificity of genomic repetitive sequences makes it possible to visualize donor chromosomes by using a probe consisting of genomic DNA from the relevant species. In analyzing human–rat hybrids, we have found it useful to use an *Alu*-consensus sequence as the probe, which produces a reverse G-banded pattern of hybridization. This allows identification of the transferred chromosome(s) as well as visualization of rearrangements, deletions, and translocations. Chromosome-specific centromeric probes and chromosome painting probes are also commercially available. Conventional G banding can be performed on spreads prior to *in situ* hybridization, providing high-resolution confirmation of the hybrid karyotype.

DNA marker analysis, either by Southern blotting (33) or polymerase chain reaction (PCR) analysis (34), is useful for confirming the hybrid genotypes and in characterizing the chromosomal contents of fragment-containing hybrids. A complete collection of human chromosome-specific PCR primers is now available (35), and should allow rapid characterization of large hybrid collections.

Applications

Hybrid cell lines containing a particular chromosome or chromosome fragment are useful for a number of different genetic studies. A comprehensive collection of karyotypically simple hybrid cells is available from the NIGMS Human Genetic Mutant Cell Repository (Camden, NJ). Panels of hybrids with defined genotypes are useful for chromosome assignment and mapping of cloned genes. Alternatively, such cell lines can be used as microcell donors for transfer of a chromosome or chromosomal region into a specific cellular background to test for a particular phenotypic change. Finally, monochromosomal hybrids can be used to generate collections of secondary hybrid lines containing overlapping chromosome fragments.

Genetic Analysis of Hybrid Phenotypes

Somatic cell genetic methods have been used for many years to study mechanisms involved in complex cellular phenotypes. Suppression of tumorigenicity (8), complementation of DNA repair defects (36), extinction of tissue-specific gene expression (6), and induction of cellular senescence (37) are among the phenotypic changes that have been observed in cell hybrids. Refinements in cell fusion techniques and recombinant DNA techniques have facilitated detailed genetic studies of these effects.

The earliest studies suggesting a discrete genetic basis for such hybrid phenotypes employed karyotypic analysis of segregating whole-cell hybrids (38, 39). Changes in a specific hybrid phenotype were correlated with loss of one or several chromosomes in segregating hybrid populations. Segregating populations are generally heterogeneous, both karyotypically and phenotypically, making conclusive genetic studies difficult. Nevertheless, segregation analysis remains an important initial experimental approach for identifying a subset of chromosomes that can be independently tested for the ability to confer a given phenotype.

Nonrandom chromosomal abnormalities associated with certain types of cancer have also provided important clues regarding the genetic basis of malignancy (40). Cell fusion techniques offer experimental tools with which to test the significance of such cytogenetic changes.

Microcell fusion is the most useful method for direct testing of individual "candidate" chromosomes implicated in hybrid phenotypes. An ideal donor for such experiments is a karyotypically simple interspecific hybrid in which the chromosome of interest is selectively marked and retained. Such donor cell lines are generally available. Generation of karyotypically simple human–rodent hybrid cell lines has generally been accomplished by random integration of a dominant selectable marker into primary fibroblast (41) or lymphoblast cultures (26), followed by microcell fusion into rodent recipient cells. Careful genotypic characterization of hybrid clones is essential, because many hybrid clones generated by these methods will not be monochromosomal.

In experiments involving hybrid phenotypes, it is useful to have the ability either to retain or eliminate the introduced chromosome, depending on the selective conditions employed. In this regard, endogenous selectable genes such as *HPRT, APRT,* or *TK,* and the cloned selectable markers *gpt, HyTK,* and *TK-neo,* offer attractive systems for designing genetic tests that include segregation analysis of backselectants.

Deletion Hybrids

Spontaneous chromosome fragmentation occurs in hybrids generated by microcell fusion, making this a useful method for producing hybrid lines containing subchromosomal fragments. The frequency of fragment-containing hybrids generated by conventional microcell fusion varies from one cross to another (42–44). The precise determinants of chromosome fragmentation have not been studied systematically, although experience suggests that the frequency increases with increases in the duration of mitotic arrest. The genotypes of fragment-containing hybrids produced by conventional microcell fusion techniques are generally simple, with individual hybrids retaining one or a few subchromosomal fragments (44). Preferential retention of centromeric fragments has been observed (44), as with both chromosome-mediated gene transfer and irradiation–fusion methods (45). Extensive sequence rearrangement within fragments does not seem to occur following microcell fusion.

Irradiation of donor cells prior to fusion with a nonirradiated recipient is another method by which fragment-containing hybrids can be produced (46). Exposure to ionizing radiation induces chromosome breaks, resulting in transfer of chromosome fragments, many of which will be stably retained in the hybrid cells by translocation onto recipient chromosomes. The utility of radiation-reduced hybrids has been established as a result of the development of molecular methods for assessing the genotypes of such lines, coupled with statistical methods of data analysis (45). The major limitation of radiation-reduced hybrids is that a large amount of extraneous donor DNA is cotransferred with the chromosome fragments of interest, making such hybrids inappropriate materials for genetic studies of complex phenotypes.

It is possible to irradiate isolated microcells prior to fusion, thus generating more extensive chromosome fragmentation than with conventional microcell fusion (47). The resulting hybrids differ from those obtained with whole-cell irradiation–fusion protocols in that there is minimal cotransfer of extraneous donor chromosomal material. This consideration is pertinent to experiments that involve scoring a phenotypic change in hybrid cells. Although experience with this approach is limited, it appears to combine the attractive features of both microcell fusion and irradiation fusion, and may become a useful method for generating genotypically simple hybrid cell lines containing chromosome fragments.

Fragment-containing microcell hybrids provide genetic information that is complementary to data obtained by linkage analysis. This is largely because their genotypes can now be defined with precision by combining techniques such as conventional marker analysis, fluorescence *in situ* hybridization, and pulsed-field gel electrophoresis. Fragment-containing hybrids have proved to be extremely useful materials for high-resolution mapping (7, 44), isolation of genomic markers (11), and cloning of expressed genes from defined chromosomal regions (12).

Acknowledgments

We thank our colleagues in the Division of Basic Sciences, especially M. J. Thayer, for contributing information on micronucleation parameters; M. J. Thayer, G. Bulla, and M. Shapero for critically reading this article; and J. Torgerson for secretarial help in preparing the manuscript. R.E.K.F. was supported by Grant GM26449 from the National Institute of General Medical Sciences. A.L. was supported by Physician Scientist Award HD00936 from the National Institute of Child Health and Human Development.

References

1. G. Barski, S. Sorieul, and F. Cornefert, *J. Natl. Cancer Inst.* **26,** 1269 (1961).
2. Y. Okada and F. Murayama, *Exp. Cell Res.* **40,** 154 (1965).
3. G. Pontecorvo, *Somat. Cell Genet.* **1,** 397 (1975).
4. J. W. Littlefield, *Science* **145,** 709 (1964).
5. R. L. Davidson, *Annu. Rev. Genet.* **8,** 195 (1974).
6. H. Gourdeau and R. E. K. Fournier, *Annu. Rev. Cell Biol.* **6,** 69 (1990).
7. J. W. Fountain *et al.*, *Am. J. Hum. Genet.* **44,** 58 (1989).

8. E. J. Stanbridge, *Cancer Surv.* **7,** 317 (1988).
9. M. C. Weiss and H. Green, *Proc. Natl. Acad. Sci. U.S.A.* **58,** 1104 (1967).
10. R. E. K. Fournier and F. H. Ruddle, *Proc. Natl. Acad. Sci. U.S.A.* **74,** 319 (1977).
11. S. A. Ledbetter, D. L. Nelson, S. T. Warren, and D. H. Ledbetter, *Genomics* **6,** 475 (1990).
12. K. W. Jones, M. Chevrette, M. H. Shapero, and R. E. K. Fournier, *Nature Genet.* in press (1992).
13. R. S. Kucherlapati, R. M. Baker, and F. H. Ruddle, *Cytogenet. Cell Genet.* **14,** 362 (1975).
14. M. A. Eglitis, *Hum. Gene Ther.* **2,** 195 (1991).
15. R. C. Mulligan and P. Berg, *Proc. Natl. Acad. Sci. U.S.A.* **78,** 2072 (1981).
16. F. Schwartz, N. Maeda, O. Smithies, *et al., Proc. Natl. Acad. Sci. U.S.A.* **88,** 10416 (1991).
17. S. D. Lupton, L. L. Brunton, V. A. Kalberg, and R. W. Overell, *Mol. Cell. Biol.* **11,** 3374 (1991).
18. T. R. Chen, *Exp. Cell Res.* **104,** 255 (1977).
19. M. F. Barile, G. P. Bodey, J. Snyder, D. B. Riggs, and M. W. Grabowski, *J. Natl. Cancer Inst.* **36,** 155 (1966).
20. T. V. Gopalakrishnan and E. B. Thompson, *Exp. Cell Res.* **96,** 435 (1975).
21. A. Levan, *Hereditas* **40,** 1 (1954).
22. S. G. Phillips and D. M. Phillips, *J. Cell Biol.* **40,** 248 (1969).
23. R. E. K. Fournier, *Proc. Natl. Acad. Sci. U.S.A.* **78,** 6349 (1981).
24. S. L. Schor, R. T. Johnson, and A. M. Mullinger, *J. Cell Sci.* **19,** 281 (1975).
25. O. V. Zatsepina, V. I. Poliakov, and S. Chentsov, *Eur. J. Cell Biol.* **26,** 277 (1982).
26. D. Warburton, S. Gersen, M.-T. Yu, C. Jackson, B. Handelin, and D. Housman, *Genomics* **6,** 358 (1990).
27. J. A. Sanford and E. Stubblefield, *Somat. Cell Mol. Genet.* **13,** 279 (1987).
28. C. A. McNeill and R. L. Brown, *Proc. Natl. Acad. Sci. U.S.A.* **77,** 5394 (1980).
29. T. Ege and N. R. Ringertz, *Exp. Cell Res.* **87,** 378 (1974).
30. W. H. Wigler and I. B. Weinstein, *Biochem. Biophys. Res. Commun.* **63,** 669 (1975).
31. T. T. Hecht, N. H. Ruddle, and F. H. Ruddle, *Fed. Proc.* **34,** 995 (1975).
32. P. Lichter, S. Ledbetter, D. Ledbetter, and D. Ward, *Proc. Natl. Acad. Sci. U.S.A.* **87,** 6634 (1990).
33. E. Southern, *J. Mol. Biol.* **98,** 503 (1975).
34. D. L. Nelson, *Curr. Opinion Genet. Devel.* **1,** 62 (1991).
35. C. Abbott and S. Povey, *Genomics* **9,** 73 (1991).
36. E. C. Friedberg, K. Henning, C. Lambert, P. J. Saxon, R. A. Schultz, G. S. Sekhon and E. J. Stanbridge, *Basic Life Sci.* **52,** 257 (1990).
37. O. M. Pereira-Smith and J. R. Smith, *Proc. Natl. Acad. Sci. U.S.A.* **85,** 6042 (1988).
38. A. M. Killary and R. E. K. Fournier, *Cell* **38,** 523 (1984).
39. E. J. Stanbridge, R. R. Flandemeyer, D. W. Daniels, and W. Nelson-Rees, *Somat. Cell Genet.* **7,** 699 (1981).
40. S. Heim and F. Mittelman, "Cancer Cytogenetics." Alan R. Liss, New York, 1987.
41. T. G. Lugo, B. Handelin, A. M. Killary, D. E. Housman, and R. E. K. Fournier, *Mol. Cell Biol.* **7,** 2814 (1987).
42. R. E. K. Fournier and J. A. Frelinger, *Mol. Cell. Biol.* **2,** 539 (1982).

43. R. E. K. Fournier and R. G. Moran, *Somat. Cell Genet.* **9,** 69 (1983).
44. R. J. Leach, M. J. Thayer, A. J. Schafer, and R. E. K. Fournier, *Genomics* **5,** 167 (1989).
45. D. R. Cox, M. Burmeister, E. R. Price, S. Kim, and R. M. Myers, *Science* **250,** 245 (1990).
46. S. J. Goss and H. Harris, *Nature (London)* **255,** 1445 (1975).
47. S. F. Dowdy, D. J. Scanlon, C. L. Fusching, G. Casey, and E. J. Stanbridge, *Genes Chromosomes Cancer* **2,** 318 (1990).

[7] Techniques and Applications of Microcell-Mediated Chromosome Transfer to Mammalian Cells

Karla A. Henning and Errol C. Friedberg

Introduction

The experimental techniques of DNA-mediated gene transfer (DMGT), chromosome-mediated gene transfer (CMGT), and microcell-mediated chromosome transfer (MMCT) are historically related. Eighteen years after Avery *et al.* (1) reported their celebrated experiments on the transformation of *Pneumococcus* with purified DNA, Szybalska and Szybalski (2) reported the genetic transformation of $HPRT^-$ human cells using hypoxanthine–aminopterin–thymidine (HAT) selection, following the introduction of human DNA from $HPRT^+$ cells. The vagaries of the technical requirements for the uptake of DNA in calcium phosphate precipitates by mammalian cells were not fully appreciated for another decade (3). In the interim the genetic transformation of mammalian cells by the introduction of partially purified metaphase chromosomes emerged as a less capricious technique, credited to McBride and Ozer (4) for its first convincing demonstration. The ability to transfer genes carried in isolated metaphase chromosomes is now well documented in the cells of both mammalian and nonmammalian species, and this topic has been extensively reviewed (5–9).

Chromosome-mediated gene transfer has proved to be a useful technique for the identification and mapping of genes. However, its utility is limited by the relatively small size of the chromosomal fragments (transgenomes) that are usually incorporated into recipient cells. Additionally, transformation by the transgenome is associated with the stable integration of pieces of DNA into the genome of recipient cells. Hence, genes cannot be easily retrieved and those identified by CMGT cannot be readily shuttled back and forth from one cell to another for further analysis.

Microcell-Mediated Chromosome Transfer

General Considerations

The essential feature of MMCT is the ability to transfer entire intact chromosomes from one cell to another and hence to isolate and maintain individual chromosomes from a particular species in cells from a different species, so-called interspecific somatic cell hybrids. The ability to isolate, propagate, and transfer entire chromo-

Methods in Molecular Genetics, Volume 1

somes offers expanded applications for somatic cell genetics. Microcell-mediated chromosome transfer essentially derives from the technique of somatic cell hybridization, except that instead of fusing donor cells carrying their entire chromosomal content to recipient cells, the donor cells contain only a portion of the chromosomal content (in the extreme, a *single* chromosome of interest).

In the mid-1970s Ege, Ringertz, and colleagues (10, 11) observed that enucleation of normal mononucleate cells by treatment with cytochalasin B allows the harvesting of *minicells* by low-speed centrifugation. These minicells were found to have an intact nucleus surrounded by a thin rim of cytoplasm between the nuclear membrane and the plasma membrane. If prior to enucleation the cells were induced to undergo micronucleation by treatment with mitotic inhibitors such as Colcemid (demecolcine), the same procedure yielded a pellet consisting of even smaller structures, each of which contains a micronucleus. These structures were termed *microcells* and could be fused to recipient cells in much the same way that whole cells can. Ege and Ringertz (11) prophetically noted that "the demonstration that microcells can be fused with normal cells indicates that they offer a new possibility of introducing a few chromosomes of one cell into another cell." Independent studies by Schor *et al.* (12) also yielded microcells (which they termed *minisegregants*) by storing growing cells in the cold and then returning them to normal growth conditions.

Technique of Microcell-Mediated Chromosome Transfer

Both fibroblasts and lymphoblasts have successfully been used to produce microcells for chromosome transfer experiments. However, different cell types and species may micronucleate with different efficiency and specific experimental conditions such as the concentration of Colcemid should be determined empirically (13).

Microcell Formation from Fibroblasts

1. Donor cells are split into six 25-cm^2 straight-neck flasks (Nunc, Roskilde, Denmark) with 5 ml of growth medium and incubated at 37°C until the cells reach 10–25% of confluency. Colcemid (GIBCO, Grand Island, NY) is added to a final concentration of 0.02 μg/ml for human–mouse hybrids (using mouse A9 cells) and the flasks are further incubated for 48 hr at 37°C.

2. For enucleation, the flasks are filled to the neck with 10 μg/ml cytochalasin B (Sigma, St. Louis, MO) in serum-free medium and preincubated for 30 min at 37°C. The flasks are then placed with the bottoms facing out, in a Beckman (Fullerton, CA) JA-14 rotor containing 100 ml of water in each well as a cushion and are centrifuged at 24,000 *g* (12,500 rpm) for 70 min at 34°C in a Beckman J2-21M centrifuge with slow acceleration and deceleration.

3. The supernatant is poured off and the microcell pellets are resuspended in

serum-free medium. The suspension is filtered through 5- and 3-μm polycarbonate filters (Nuclepore, Pleasanton, CA) to remove large microcells and any whole cells that may have become dislodged during the centrifugation.

Microcell Formation from Suspension Cultures

The production of microcells from cells growing in suspension can be achieved using a Percoll density gradient procedure (14).

1. To promote micronuclei formation, donor cells in the log phase of growth are exposed to 0.5 μg/ml Colcemid for 24 hr, then pelleted and resuspended in a hypotonic solution consisting of equal volumes of conditioned medium and sterile water, for 10 min at 37°C. The cells are then centrifuged and resuspended in fresh growth medium and incubated for another 24 hr at 37°C.
2. To achieve enucleation, the cells are pelleted and 10^8 cells are resuspended in 20 ml of a mixture of equal volumes of Percoll (Pharmacia, Piscataway, NJ) and growth medium containing 20 μg/ml cytochalasin B. The cells are then centrifuged at 19,000 rpm in polycarbonate gradient tubes (Nalgene; Nalge, Rochester, NY) in a Sorvall (Norwalk, CT) SS-34 rotor at 34°C for 70 min. Usually six to eight tubes are prepared for each fusion experiment.
3. The two visible bands in each tube are removed and added to 25 ml of serum-free medium and centrifuged at 1500 rpm for 5 min to remove the Percoll. The pellets are resuspended in serum-free medium and filtered through a series of polycarbonate filters as described above to obtain a relatively pure microcell population.

Fusion and Hybrid Selection

1. The microcell suspension produced by either of the above procedures is pelleted in a tabletop centrifuge for 15 min at 2200 rpm at room temperature, and carefully resuspended in 2 ml of phytohemagglutinin P (Sigma) (50 μg/ml) in serum-free medium. The microcell suspension is then added to the recipient monolayer in a 25-cm^2 flask previously rinsed with serum-free medium.
2. The microcells are allowed to settle by gravity and to attach for 20–30 min at room temperature. The supernatant is removed and the attached microcells are fused to recipient cells with 2 ml of 50% (w/v) polyethylene glycol (PEG) 1500 in 75 m*M* *N*-2-hydroxyethylpiperazine-*N*′-2-ethanesulfonic acid (HEPES; Boehringer Mannheim, Indianapolis, IN) for 60 sec.
3. The monolayer is quickly rinsed three times with phosphate-buffered saline (PBS) to remove the PEG and then incubated overnight at 37°C in growth medium. The fused cells are then trypsinized and split into eight 100-mm tissue culture dishes containing growth medium.
4. After 2 or 3 days of incubation at 37°C to allow expression, the cells are fed with growth medium containing the appropriate selective agent (see *Genetic Selection for Transferred Chromosomes*). After the clones reach a sufficient size they are

trypsinized using cloning cylinders, and transferred to 48-well plates in which individual clones can be expanded for further analysis.

Genetic Selection for Transferred Chromosomes

In the absence of selection chromosomes donated from microcells or minisegregants are readily lost following their fusion to recipient cells. The X chromosome, chromosome 16, and chromosome 17 can be selected by growth of cells in selective media based on the expression of the endogenous biochemical markers hypoxanthine phosphoribosyltransferase (HPRT), adenine phosphoribosyltransferase (APRT), and thymidine kinase (TK), respectively (13). However, relatively few mammalian chromosomes contain known constitutive genes that allow biochemical selection. Additionally, this strategy limits the range of recipient cells to those that are mutant at the loci that encode the markers mentioned above.

The induction of mutations in certain genes can confer a phenotype that provides for selection of a particular chromosome carrying that gene. For example, following the treatment of mouse fibroblasts with chemical mutagens, Landolph and Fournier (15) isolated ouabain-resistant (Oua^r) mutant variants. They then used these 8-azaguanine-resistant human cell lines as recipients to select a mouse chromosome carrying the *Oua*^{r} gene. However, this strategy also has obvious limitations for the more general selection of chromosomes.

In the mid-1980s several investigators addressed this problem by devising strategies for the integration of cloned foreign genes, which express dominant selectable markers, into the donor chromosomes prior to their transfer (16, 17). This experimental manipulation offers the potential of transferring and selecting any chromosome in the karyotype of a mammalian cell, and hence for generating panels of monochromosomal interspecific hybrids, each of which carries a single karyotypically normal chromosme from the species of interest. The range of genetic markers that are typically used for the selection of cells carrying donated chromosomes is relatively modest. However, a larger number of recessive and dominant selectable systems could in theory be employed for this purpose. The following section describes these selectable markers briefly, beginning with those used most extensively in practice (also see Table I).

Dominant Selectable Markers

The most widely applicable methods for selection of transferred chromosomes take advantage of dominant selectable markers that code for resistance to various drugs. The exact level of drug necessary for selection can vary from cell line to cell line, and should be determined empirically. For example, the use of the antibiotic Geneticin (G418) for selection of *neo*-tagged human chromosomes (see below) in mouse A9 cells requires 400 μg/ml G418 (effective concentration). However,

TABLE I Selectable Markers for Mammalian Cells

Marker gene	Product
Dominant selectable	
Tn*5 neo*	Aminoglycoside phosphotransferase
Escherichia coli hph	Hygromycin B phosphotransferase
E. coli gpt	Xanthine–guanine phosphoribosyltransferase
Salmonella typhimurium hisD	Histidinol dehydrogenase
E. coli trpB	Tryptophan synthase β subunit
Bacillus cereus bsr	Blasticidin S deaminase
E. coli asnA	Asparagine synthase
Recessive	
HPRT	Hypoxanthine–guanine phosphoribosyltransferase
TK	Thymidine kinase
APRT	Adenine phosphoribosyltransferase
Amplifiable	
MDR1	P-glycoprotein
DHFR	Dihydrofolate reductase
ADA	Adenosine deaminase
ODC	Ornithine decarboxylase
CAD	CAD (carbamoyl-phosphate synthase, aspartate carbamoyltransferase, dihydroorotase)

maintenance of the same chromosomes in human simian virus 40 (SV40)-transformed fibroblasts may require only 250 μg/ml of G418.

1. One of the simplest and most frequently used dominant selectable markers is the bacterial neomycin resistance gene (*neo*), which encodes aminoglycoside 3′-phosphotransferase (APH). Two different APH enzymes have been cloned from the bacterial transposons Tn*5* and Tn*601*, both of which confer resistance to antibiotics such as kanamycin and neomycin in bacterial cells. When expressed in mammalian cells, the Tn*5 neo* gene confers resistance to the neomycin analog, G418 (18).

2. The *Escherichia coli hph* gene codes for hygromycin B phosophotransferase, and confers resistance to the aminocyclitol antibiotic hygromycin (19). As with *neo*, the *hph* gene can be used as a dominant selectable marker in mammalian cells when expressed from a suitable eukaryotic promoter (20, 21). Although not used in chromosome-transfer studies published to date, the ease of the selection and the ability to select for both *neo* and *hph* simultaneously should make it a useful marker in future studies.

3. The *E. coli gpt* gene encodes xanthine–guanine phosphoribosyltransferase (XGPRT), which can convert xanthine to xanthine monophosphate (XMP) much more efficiently than the mammalian HPRT enzyme. In an SV40 expression vector,

gpt can be used as a dominant selectable marker (22–24), allowing mammalian cells to grow in medium containing hypoxanthine, aminopterin, and thymidine (HAT), along with xanthine and mycophenolic acid, an inhibitor of inosine monophosphate (IMP) dehydrogenase.

4. The *Salmonella typhimurium hisD* gene encodes histidinol dehydrogenase, which catalyzes the two-step NAD^+-dependent oxidation of L-histidinol to L-histidine. When expressed in mammalian cells the *hisD* gene allows cells to grow in medium containing histidinol in place of histidine (25, 26). Histidinol is also toxic to mammalian cells through inhibition of histidyl-tRNA and selection for *hisD* can alternatively be carried out in the presence of histidine if high levels of histidinol are employed.

5. The *E. coli trpB* gene encodes the β subunit of tryptophan synthase. Expression of this gene in mammalian cells allows growth in medium containing indole in place of tryptophan (25).

6. Two additional dominant selectable markers have been described that could potentially be useful in chromosome transfer studies. When expressed from the SV40 promoter, the blasticidin S-resistance (*bsr*) gene from *Bicillus cereus* confers resistance to the microbial antibiotic blasticidin S, an inhibitor of protein synthesis (27). Cells that express the *E. coli* gene coding for asparagine synthase (*asnA*) can grow in asparagine-free medium containing the glutamine analog albizziin, because the bacterial enzyme can utilize ammonia as an amide donor, whereas the mammalian enzyme requires glutamine (28).

Recessive Markers

Most early chromosome transfer studies were performed using recessive selection systems. The use of these genes as selectable markers requires the use of appropriate mutant cell lines and is therefore not as widely applicable as the previously described dominant selection systems.

1. A popular selection involves the use of HAT medium, which can be used to select for the presence of the *HPRT* gene. Translocations between the X chromosome and untagged autosomes can be maintained by selection for the *HPRT* gene. Hypoxanthine–guanine phosphoribosyltransferase is involved in the purine salvage pathway, and cells that are deficient in HPRT are unable to grow in the presence of HAT medium. However, cells that are deficient in HPRT can grow in the presence of 6-thioguanine, allowing selection of HPRT-deficient cell lines.

2. Another popular recessive marker is thymidine kinase (TK), which is involved in the salvage pathway for thymidine nucleotides. Cells that are deficient in TK are unable to grow in HAT medium. The *TK* gene has been cloned from the herpes simplex virus (29) and from chicken cells (30) and can be used as a selectable marker in TK-deficient cell lines. Cells that are deficient in thymidine kinase can be selected in the presence of the thymidine analog 5-bromodeoxyuridine.

3. A third recessive marker used historically is the *APRT* gene. Adenine phosphoribosyltransferase catalyzes the conversion of adenine into AMP. Cells deficient in APRT will not grow in the presence of azaserine and adenine. Selection against APRT is possible with 2,6-diaminopurine (reviewed in Ref. 31).

Amplifiable Markers

A third class of selectable markers consists of mammalian genes that have been shown to be amplified by growth of cells in increasing levels of the appropriate drug. The corresponding cDNAs, when overexpressed, can be used as dominant selectable markers. Although not often used for chromosome transfer experiments to date, they are potential candidates for future studies.

1. The human multidrug resistance gene (*MDR1*) codes for P-glycoprotein, an energy-dependent multidrug efflux pump. When overexpressed, the *MDR1* cDNA confers resistance to colchicine (32).
2. Dihydrofolate reductase (DHFR) catalyzes the reduction of folate to tetrahydrofolate, and can be inhibited in the presence of methotrexate, a folate analog. Two methods have been developed that allow selection in cells with wild-type levels of DHFR. The *DHFR* cDNA is either placed under control of a strong promoter, or a mutant *DHFR* cDNA is used that allows transfected cells to grow in a level of methotrexate that is lethal to wild-type cells (33, 34).
3. Adenosine deaminase (ADA) is a nonessential protein, expressed at low levels in most cells. Adenosine deaminase catalyzes the irreversible conversion of adenine nucleosides to their inosine analogs and can be inhibited in the presence of 2′-deoxycoformycin. If cells are grown in the presence of high concentrations of adenosine or its analog 9-β-D-xylofuranosyladenine (Xyl-A), ADA is required to detoxify these compounds, and can therefore be used as a selectable marker (31).
4. Cells can be made resistant to higher-than-normal levels of the ornithine decarboxylase (ODC) suicide-substrate inhibitor, difluoromethylornithine, by transfer of an ODC-expressing plasmid (35).
5. The multifunctional protein CAD is involved in *de novo* pyrimidine biosynthesis. It includes carbamoyl-phosphate synthase, aspartate carbamoyltransferase, and dihydroorotase. *N*-Phosphonoacetyl-L-aspartate (PALA) is a transition state analog inhibitor of aspartate carbamoyltransferase that causes UMP depletion. *CAD* may be able to function as a dominant selectable marker if high levels of PALA are employed (36, 37).

Tagging Chromosomes with Selectable Genes

The tagging of donor chromosomes with selectable genes can be performed in several ways. A commonly used procedure is to transfect cells with plasmids carrying the genes using calcium phosphate coprecipitation (38, 39). Calcium phosphate-mediated gene transfer is relatively inefficient in most diploid cell lines, thus prompting the more extensive use of immortalized cell lines for this purpose. Additionally, it has been suggested that calcium phosphate-mediated gene transfer induces chromosome rearrangements in transfected cells (26). An alternative procedure is to use a defective amphotropic retrovirus to tag chromosomes. Such retroviral vectors infect cells efficiently, thus facilitating the use of diploid cell lines with normal karyotypes. Additionally, these vectors integrate into human chromosomes in single copy, with physically defined ends, thereby facilitating subsequent physical mapping of the integration sites (26, 40).

It should be pointed out that the rearrangement of donor chromosomes is not necessarily always disadvantageous for studies in somatic cell genetics. Extensive rearrangements can certainly complicate the assignment of genes to particular chromosomes and are counterproductive to the aim of generating comprehensive panels of interspecies monochromosomal hybrids carrying karyotypically normal, genetically tagged chromosomes. However, as indicated later, in some cases the presence of chromosomal rearrangements can facilitate the refined subchromosomal mapping and molecular cloning of genes.

An aspect of chromosome selection that has not been adequately addressed is the tagging of single chromosomes already present in an interspecific hybrid, but which cannot be identified or maintained in recipient cells following chromosome transfer because they do not carry selectable markers. Transfection of hybrids carrying untagged chromosomes with plasmids, or infection with retroviral vectors, will obviously not result in the selective tagging of these chromosomes, although it is certainly possible, albeit laborious, to consummate a second round of MMCT, to select for new monochromosomal hybrids and to screen these for hybrids carrying just the tagged chromosome of interest by DNA hybridization against the integrated plasmid. Alternatively, the advent of techniques for homologous recombination should facilitate selective tagging of individual chromosomes by incorporating DNA sequences for homologous regions of the chromosome of interest.

Characterization of Single Human Chromosomes in Interspecific Hybrids

The ability to isolate interspecific hybrids that carry single human chromosomes allows one to perform many genetic manipulations that would be otherwise difficult. The human chromosomal content of somatic cell hybrids is usually determined by cytogenetic analysis of banded chromosomes in metaphase spreads. In rodent–human hybrids, G-11 staining can be used to differentiate the human chromosomes from the rodent chromosomes on the basis of color (41), but the identity of the human chromosomes must still be determined by chromosome-banding techniques (42). This type of analysis is laborious and requires skilled cytogenetic training, particularly in light of the fact that the human material must be analyzed in a rodent background. These limitations can lead to problems when trying to determine the precise human chromosomal content of a particular hybrid. Isozyme analysis, or Southern blot analysis with a panel of human chromosome-specific probes, can provide verification of the presence or absence of specific chromosomes. However, the prevalence of chromosomal rearrangements, deletions, and insertions of human material in somatic cell hybrids, often at a level undetectable by conventional cytogenetics, is well established. Several new technologies have been developed that have increased the ease and sensitivity of the analysis of human chromosomes in interspecific hybrids.

Fluorescence in Situ Hybridization

Fluorescence *in situ* hybridization (FISH) of metaphase spreads with labeled human DNA is a powerful molecular cytogenetic tool for the identification of human DNA in interspecific hybrids (43). Translocations of human material to rodent chromosomes, which may not be readily detected by conventional cytogenetics, are easily visible by this method (44). "Chromosome painting" with chromosome-specific libraries allows one to identify particular human chromosomes, and can help visualize and identify translocations involving two or more human chromosomes (45, 46). Fluorescence *in situ* hybridization with chromosome-specific DNA probes can also verify the presence or absence of specific chromosomal regions, and can help identify chromosomal rearrangements, including deletions (47). Under conditions that suppress the hybridization signal from ubiquitous repetitive sequences, FISH with cosmid clones can be used to produce physical maps of human chromosomal regions (48).

Chromosome-Specific Polymerase Chain Reaction

Primers for the polymerase chain reaction (PCR) have been developed for genes on the X chromosome and each of the 22 autosomes that are capable of amplifying the human genes but not their rodent equivalents, or that give a different product for the human and rodent genes (49, 50). This application of PCR allows rapid analysis of the chromosomal content of somatic cell hybrids, uses small amounts of material, and can greatly facilitate the rapid screening of large numbers of cell clones.

IRS-PCR

Polymerase chain reaction with primers directed to interspersed repetitive sequences (IRS-PCR) can also be used to facilitate the identification of the chromosomal content of somatic cell hybrids. The specific amplification of human DNA in somatic cell hybrids can be achieved with primers directed to the human-specific portion of the short interspersed element *Alu* (51) and/or to the human long interspersed element, L1Hs (52). The DNA probe sets derived in this way from a somatic cell hybrid can be used for reverse *in situ* hybridization onto normal human metaphase spreads to determine the human DNA content of the hybrid (48, 53). IRS-PCR can also be used to produce a "PCR karyotype" of a somatic cell hybrid (54), in which a chromosome-specific pattern of amplification products can be visualized by agarose gel electrophoresis in the presence of ethidium bromide. Identification of hybrids containing multiple human chromosomes can be achieved by hybridization of IRS-PCR products against a reference panel of monochromosomal or highly reduced polychromosomal hybrids representing each human chromosome. This method is also useful in the identification of small or translocated fragments that cannot be identified by standard cytogenetics (54, 55).

Applications of Microcell-Mediated Chromosome Transfer in Somatic Cell Genetics

Identification of Genes and Their Mapping to Particular Chromosomes

The utility of MMCT in mammalian genetics is highly dependent on the ability to detect phenotypic changes associated with the introduction and expression of new genes in recipient cells. Such complementable phenotypes have been largely restricted to two general areas of cellular physiology: defective cellular responses to DNA damage and neoplastic transformation.

Genes Involved in Cellular Responses to DNA Damage

Defects in cellular responses to DNA damage are typically associated with increased sensitivity to killing by DNA-damaging agents. Notable examples are cells from individuals with the hereditary excision repair-defective diseases xeroderma pigmentosum (XP) and Cockayne syndrome (CS), both of which are characterized by profound sensitivity to killing by ultraviolet (UV) radiation and a variety of so-called UV-mimetic chemicals (56–59). Additionally, cells from patients with ataxia telangiectasia (AT) are hypersensitive to ionizing radiation and to so-called X-ray-mimetic chemicals (57), and cells from patients with Fanconi's anemia (FA) are abnormally sensitive to agents that cross-link DNA (57). The identification of genes that correct these phenotypes following MMCT has thus far been restricted to XP and AT.

Xeroderma pigmentosum is an autosomal recessive disorder characterized at the cellular level by defective nucleotide excision repair. At least 10 genetic complementation groups have been defined, based on the correction of defective repair synthesis in heterodikaryons generated by fusion of cells from different individuals (56–59). The simplest interpretation of the multiple genetic complementation groups is that XP is a single-gene disease with extensive locus heterogeneity (60).

However, direct confirmation of this interpretation obviously requires the identification of distinct genes. In this respect MMCT has been most helpful. Microcell-mediated chromosome transfer of human chromosomes tagged with the *E. coli neo* gene has resulted in the correction of sensitivity to killing by UV radiation in XP cells from genetic complementation groups A (61–63), F (64), and D (65). Cytogenetic and molecular characterization of the complementing chromosomes mapped the putative *XP-A, XP-F,* and *XP-D* genes to chromosomes 9, 15, and 19, respectively. Deciphering rearrangements in the chromosomes that corrected XP group A and D cells facilitated subchromosomal mapping of the *XP-A* (66) and *XP-D* (67) genes, and identified the cloned human *ERCC2* gene as a candidate gene for the *XP-D* locus (65) (see *Subchromosomal Mapping of Genes by Microcell-Mediated Chromosome Transfer*).

Ataxia telangiectasia is also an autosomal recessive disease (57). In addition to increased sensitivity to killing by agents that promote strand breakage in DNA, AT cells in culture are characterized by a persistent DNA synthesis when exposed to ionizing radiation, referred to as radioresistant DNA synthesis (57). Analysis of the correction of radioresistant DNA synthesis in heterodikaryons indicates the existence of at least four genetic complementation groups designated A/B, C, D, and E (68). Genetic linkage analysis has mapped the *AT-A* and *AT-C* loci to the region 11q22-23 (69–71). Microcell-mediated chromosome transfer has identified the additional presence of the *AT-D* locus on chromosome 11 (72–74) and refined chromosomal mapping has localized this gene to the region 11q22-23 as well (74).

Phenotypic correction of CS and FA cells by MMCT has not been reported, although the phenotypes of cells from these diseases should readily lend themselves to such an approach. However, correction of the increased frequency of site chromatid exchangers (SCEs) in Bloom's syndrome has been reported following the transfer of a single human chromosome cytogenetically characterized as a normal chromosome 15 (75).

Genes Associated with Tumorigenicity

Microcell-mediated chromosome transfer of single genetically tagged human chromosomes has been associated with suppression of the tumorigenic phenotype of several human tumor cell lines, thereby identifying a number of putative tumor suppressor genes. The introduction of chromosome 11 suppressed the tumorigenicity of HeLa cells (76), Wilms' tumor cells (77), SiHa cells derived from a human cervical carcinoma (78), and A204 cells derived from a human rhabdomyosarcoma (79). This chromosome had no effect on the tumorigenicity of cell lines established from a human renal cell carcinoma (79). However, the tumorigenicity of this cell line was suppressed by the introduction of chromosome 3 (80), and the transfer of human chromosome 6 resulted in the suppression of tumorigenicity of various human melanoma cell lines (81). The introduction of chromosome 1, 6, or 9 suppressed tumorigenicity of a human endometrial carcinoma cell line (82). Finally, chromosome 17 suppresses the tumorigenicity of a human neuroblastoma cell line (83).

Complementation of Miscellaneous Phenotypes

A variety of different studies are illustrative of the utility of MMCT for defining diverse gene functions. Human cells are normally sensitive to diphtheria toxin, whereas mouse cells are resistant to this toxin. This difference is believed to result in part from the expression of a toxin-specific receptor in human cells, the gene for which has been mapped to chromosome 5 (84). Microcell-mediated chromosome transfer of an intact human chromosome 5 to a mouse hepatoma cell line was associated with the acquisition of sensitivity to diphtheria toxin.

The phenotype of immortality in a Syrian hamster cell line can be suppressed by fusion to human diploid cells with a limited life span in culture (85). Consistent with the observation that hybrid clones that did not senesce had lost both copies of chromosome 1, MMCT of human chromosome 1 to the Syrian hamster cells suppressed their immortalized phenotype. This result was not observed following the introduction of human chromosome 11, suggesting that suppression of tumorigenicity by chromosome 11 is not related to the acquisition of senescence (85).

Brief exposure of mouse fibroblast cell lines such as C3H 10T1/2 to 5-azacytidine results in their differentiation to muscle cells (86). This process requires the activation of a gene designated *myoD*, which in human cells is located on chromosome 11 (87). The introduction of a single human chromosome 11 derived from human primary skin fibroblasts into 10T1/2 cells is accompanied by the activation of the human *myoD* gene and the conversion of the mouse fibroblasts to muscle cells. However, this result was not observed when primary human fibroblasts that served as the donor source of chromosome 11 were fused to the 10T1/2 cells (87). These experiments suggest that activation of the human *myoD* gene is repressed in human cells by one or more loci located on chromosomes other than chromosome 11.

Subchromosomal Mapping of Genes by Microcell-Mediated Chromosome Transfer

The ability to correlate the presence of only a portion of a transferred chromosome with a complementable phenotype provides the obvious potential for mapping genes to defined regions of chromosomes. The loss of chromosomal material can occur by a variety of mechanisms either spontaneously, or as the result of perturbations designed to promote breakage and/or rearrangement. The cytogenetic characterization of interspecific hybrids carrying single human chromosomes frequently reveals the presence of chromosomal abnormalities such as rearrangements and/or breakage. The events that lead to such karyotypic abnormalities have not been extensively studied. During an earlier discussion (see *Tagging Chromosomes with Selectable Genes*) we alluded to the suggestion that the integration of plasmids carrying dominant selectable markers into the chromosomes of transfected cells, in particular integration of multiple plasmids, may promote chromosomal instability. Additionally, the presence of rearranged chromosomes is to be expected if the cell types from which the donor chromosomes are derived is aneuploid to begin with. Finally, little is known about the dynamics of chromosome segregation in interspecies hybrids. But it seems reasonable to presume that, in addition to segregating entire chromosomes, rodent cells may also promote rearrangements of human chromosomes that are most consistent with their genetic stability.

Deciphering complex chromosomal rearrangements associated with phenotypic complementation can be problematic. The increasing availability of new molecular strategies for chromosome analysis, particularly the use of chromosomal "fingerprinting" by PCR, as described earlier, has simplified this problem enormously and has provided for immediate subchromosomal mapping of several genes carried on single human chromosomes. A particularly dramatic example is provided by the analysis of a chromosome that corrects a number of different phenotypes in AT cells from genetic complementation group D (74). Extensive cytogenetic characterization of the correcting chromosome revealed the presence of an abnormal chromosome 18. However, molecular hybridization with probes from the region 11q22-23, to which the *AT-A* and *AT-C* loci had been mapped, revealed the presence of translocated sequences from this region of the genome (74). These results were consistent with and extended the independent observation that MMCT of a normal chromosome 11 results in correction of the X-ray sensitivity of AT-D cells (72).

Irradiation Microcell-Mediated Chromosome Transfer

Chromosomal reduction can occur spontaneously, or can be provoked by exposing cells to agents such as X rays or γ rays (88–90). Reduction of tagged human chromosomes can be effected by irradiation of monochromosomal hybrid cell lines carrying the chromosome of interest. An alternative approach is to irradiate the microcells prepared from the hybrids and then generate a second generation of hybrids by fusing the irradiated microcells to rodent cells. This technique is referred to as X-irradiation microcell-mediated chromosome transfer (XMMCT) (91). This innovation affords the opportunity of selecting a panel of hybrids that carry many different deletions of the tagged chromosome. Irradiation of the microcells instead of the hybrid cells reduces the potential for translocation of rodent chromosomal material to the human chromosome. Additionally, in generating a secondary panel of hybrids, only those hybrids that carry mitotically stable reduced chromosomes survive the selection. The utilization of selection conferred by the presence of a dominant selectable marker on the reduced chromosome is necessarily limited by deletions that retain both the gene of interest and the selectable marker. If these are widely separated on the chromosome, it may be difficult to generate informative deletions.

Acknowledgments

This work in the authors' laboratory was supported by research Grants CA 44247 from the United States Public Health Service and DE-FG03-88ER60676 from the United States Department of Energy. We thank our laboratory colleagues for critical review of the manuscript.

References

1. T. Avery, C. M. McCleod, and M. McCarty, *J. Exp. Med.* **79,** 137 (1944).
2. E. H. Szybalska and W. Szybalski, *Proc. Natl. Acad. Sci. U.S.A.* **48,** 2026 (1962).
3. G. Scangos and F. H. Ruddle, *Gene* **14,** 1 (1981).
4. O. W. McBride and H. L. Ozer, *Proc. Natl. Acad. Sci. U.S.A.* **70,** 1258 (1973).
5. K. Willecke, *Theor. Appl. Genet.* **52,** 97 (1978).
6. O. W. McBride and J. L. Peterson, *Annu. Rev. Genet.* **14,** 321 (1980).
7. L. A. Klobutcher and F. H. Ruddle, *Annu. Rev. Biochem.* **50,** 533 (1981).
8. A. J. R. de Jonge and D. Bootsma, *Int. J. Cytol.* **92,** 133 (1984).
9. D. J. Porteus, *Trends Genet.* **3,** 177 (1987).
10. T. Ege, H. Hamberg, U. Krondahl, J. Ericsson, and N. R. Ringertz, *Nature (London)* **228,** 869 (1970).
11. T. Ege and N. R. Ringertz, *Exp. Cell. Res.* **87,** 378 (1974).
12. S. L. Schor, R. T. Johnson, and A. M. Mullinger, *J. Cell Sci.* **19,** 281 (1975).
13. P. J. Saxon and E. J. Stanbridge, *in* "Molecular Genetics of Mammalian Cells" (M. Gottesman, ed.), Methods in Enzymology, Vol. 151, p. 313. Academic Press, Orlando, Florida, 1987.
14. J. A. Sanford and E. Stubblefield, *Somat. Cell Mol. Genet.* **13,** 279 (1987).
15. J. R. Landolph and R. E. K. Fournier, *Mutat. Res.* **107,** 447 (1983).
16. D. L. Nelson, J. H. Weis, M. J. Pryzborski, R. C. Mulligan, J. G. Seidman, and D. E. Housman, *J. Mol. Appl. Genet.* **2,** 563 (1984).
17. E. S. Srivatsan, E. J. Stanbridge, P. J. Saxon, P. J. Sambrook, J. J. Trill, and J. A. Tischfield, *Cytogenet. Cell Genet.* **38,** 227 (1984).
18. P. J. Southern and P. Berg, *J. Mol. Appl. Genet.* **1,** 327 (1982).
19. L. Gritz and J. Davies, *Gene* **25,** 179 (1983).
20. K. Blochlinger and H. Diggelmann, *Mol. Cell. Biol.* **4,** 2929 (1984).
21. R. F. Santerre, N. E. Allen, J. N. Hobbs, Jr., R. N. Rao, and R. J. Schmidt, *Gene* **30,** 147 (1984).
22. R. C. Mulligan and P. Berg, *Science* **209,** 1422 (1980).
23. R. C. Mulligan and P. Berg, *Proc. Natl. Acad. Sci. U.S.A.* **78,** 2072 (1981).
24. R. C. Mulligan and P. Berg, *Mol. Cell. Biol.* **1,** 449 (1981).
25. S. C. Hartman and R. C. Mulligan, *Proc. Natl. Acad. Sci. U.S.A.* **85,** 8047 (1988).
26. D. Warburton, S. Gersen, M.-T. Yu, C. Jackson, B. Handelin, and D. Housman, *Genomics* **6,** 358 (1990).
27. M. Izumi, H. Miyazawa, T. Kamakura, I. Yamaguchi, T. Endo, and F. Hanaoka, *Exp. Cell. Res.* **197,** 229 (1991).

28. M. Cartier and C. P. Stanners, *Gene* **95,** 223 (1990).
29. F. Colbere-Garapin, S. Chousterman, F. Horodniceanu, P. Kourilsky, and A.-C. Garapin, *Proc. Natl. Acad. Sci. U.S.A.* **76,** 3755 (1979).
30. M. Perucho, D. Hanahan, L. Lipsich, and M. Wigler, *Nature (London)* **285,** 207 (1980).
31. R. J. Kaufman, *Genet. Eng.* **9,** 155 (1987).
32. S. E. Kane, D. H. Reinhard, C. M. Fordis, I. Pastan, and M. M. Gottesman, *Gene* **84,** 439 (1989).
33. M. J. Murray, R. J. Kaufman, S. A. Latt, and R. A. Weinberg, *Mol. Cell. Biol.* **3,** 32 (1983).
34. C. C. Simonsen and A. D. Levinson, *Proc. Natl. Acad. Sci. U.S.A.* **80,** 2495 (1983).
35. T.-R. Chiang and L. McConlogue, *Mol. Cell. Biol.* **8,** 764 (1988).
36. B. Robert de Saint Vincent, S. Delbrück, W. Eckhart, J. Meinkoth, L. Vitto, and G. Wahl, *Cell* **27,** 267 (1981).
37. G. M. Wahl, B. Robert de Saint Vincent, and M. L. DeRose, *Nature (London)* **307,** 516 (1984).
38. R. S. Athwal, M. Smarsh, B. M. Searle, and S. S. Deo, *Somat. Cell Mol. Genet.* **11,** 177 (1985).
39. P. J. Saxon, E. S. Srivatsan, G. V. Leipzig, J. H. Samashima, and E. J. Stanbridge, *Mol. Cell. Biol.* **5,** 140 (1985).
40. D. Housman, *Somat. Cell Mol. Genet.* **13,** 441 (1987).
41. M. Bobrow and J. Cross, *Nature (London)* **251,** 77 (1974).
42. R. S. Verma and A. Babu, "Human Chromosomes, Manual of Basic Techniques." Pergamon, New York, 1989.
43. D. Pinkel, T. Straume, and J. W. Gray, *Proc. Natl. Acad. Sci. U.S.A.* **83,** 2934 (1986).
44. A. J. Giaccia, J. W. Evans, and J. M. Brown, *Genes Chromosomes Cancer* **2,** 248 (1990).
45. P. Lichter, T. Cremer, J. Borden, L. Manuelidis, and D. C. Ward, *Hum. Genet.* **80,** 224 (1988).
46. D. Pinkel, J. Landegent, C. Collins, J. Fuscoe, R. Seagraves, J. Lucas, and J. Gray, *Proc. Natl. Acad. Sci. U.S.A.* **85,** 9138 (1988).
47. P. Lichter, A. L. Boyle, T. Cremer, and D. C. Ward, *GATA* **8,** 24 (1991).
48. P. Lichter, C.-J. Tang, K. Call, G. Hermanson, G. A. Evans, D. Housman, and D. C. Ward, *Science* **247,** 64 (1990).
49. C. Abbott and S. Povey, *Genomics* **9,** 73 (1991).
50. S. Theune, J. Fung, S. Todd, A. Y. Sakaguchi, and S. L. Naylor, *Genomics* **8,** 511 (1991).
51. D. L. Nelson, S. A. Ledbetter, L. Corbo, M. F. Victoria, R. Ramirez-Solis, T. D. Webster, D. H. Ledbetter, and C. T. Caskey, *Proc. Natl. Acad. Sci. U.S.A.* **86,** 6686 (1989).
52. S. A. Ledbetter, D. L. Nelson, S. T. Warren, and D. H. Ledbetter, *Genomics* **6,** 475 (1990).
53. C. Lengaur, H. Riethman, and T. Cremer, *Hum. Genet.* **86,** 1 (1990).
54. S. A. Ledbetter, J. Garcia-Heras, and D. H. Ledbetter, *Genomics* **8,** 614 (1990).
55. D. C. Bicknell, D. Markie, N. K. Spurr, and W. F. Bodmer, *Genomics* **10,** 186 (1991).

56. J. E. Cleaver and K. H. Kraemer, *in* "The Metabolic Basis of Human Disease" (C. R. Scriver, A. L. Beaudet, W. S. Sly, and D. Valle, eds.), p. 2949. McGraw-Hill, New York, 1989.
57. E. C. Friedberg, "DNA Repair." Freeman, San Francisco, 1985.
58. J. H. J. Hoeijmakers and D. Bootsma, *Cancer Cells* **2,** 311 (1990).
59. V. A. Bohr, M. K. Evans, and A. J. Fornace, *Lab. Invest.* **61,** 143 (1989).
60. E. C. Friedberg and K. H. Henning, *Mutat. Res.* in press (1993).
61. R. A. Schultz, P. J. Saxon, T. W. Glover, and E. C. Friedberg, *Proc. Natl. Acad. Sci. U.S.A.* **84,** 4176 (1987).
62. G. P. Kaur and R. S. Athwal, *Proc. Natl. Acad. Sci. U.S.A.* **86,** 8872 (1989).
63. K. Ishikazi, M. Oshimura, M. S. Sasaki, Y. Nakamura, and M. Ikenaga, *Mutat. Res.* **235,** 209 (1990).
64. P. J. Saxon, R. A. Schultz, E. J. Stanbridge, and E. C. Friedberg, *Am. J. Hum. Genet.* **44,** 474 (1989).
65. W. L. Flejter, L. D. McDaniel, D. Johns, E. C. Friedberg, and R. A. Schultz, *Proc. Natl. Acad. Sci. U.S.A.* **89,** 261 (1992).
66. K. A. Henning, R. A. Schultz, G. S. Sekhon, and E. C. Friedberg, *Somat. Cell Mol. Genet.* **16,** 395 (1990).
67. W. L. Flejter, L. D. McDaniel, M. Askari, E. C. Friedberg, and R. A. Schultz, *Genes Chromosomes Cancer* **5,** 335 (1992).
68. N. G. J. Jaspers, R. A. Gatti, C. Baan, P. C. Linssen, and D. Bootsma, *Cytogenet. Cell Genet.* **49,** 259 (1988).
69. R. A. Gatti *et al., Nature (London)* **336,** 577 (1988).
70. O. Sanal, S. Wei, T. Foroud, U. Malhotra, P. Concannon, P. Charmley, W. Salser, K. Lange, and R. A. Gatti, *Am. J. Hum. Genet.* **47,** 860 (1990).
71. Y. Ziv, G. Rotman, M. Frydman, J. Dagan, T. Cohen, T. Foroud, R. A. Gatti, and Y. Shiloh, *Genomics* **9,** 373 (1991).
72. K. Komatsu, S. Kodama, Y. Okumura, M. Koi, and M. Oshimura, *Mutat. Res.* **235,** 59 (1990).
73. Y. Ejima, M. Oshimura, and M. S. Sasaki, *Int. J. Radiat. Biol.* **58,** 989 (1990).
74. C. Lambert, R. A. Schultz, M. Smith, C. Wagner-McPherson, L. D. McDaniel, T. Donlon, E. J. Stanbridge, and E. C. Friedberg, *Proc. Natl. Acad. Sci. U.S.A.* **88,** 5907 (1991).
75. L.D. McDaniel and R. A. Schultz, *Proc. Natl. Acad. Sci. U.S.A.,* **89,** 7968 (1992).
76. P. Saxon, E. S. Srivastan, and E. J. Stanbridge, *EMBO J.* **5,** 3461 (1986).
77. B. E. Weismann, P. J. Saxon, S. R. Pasquale, G. R. Jones, A. G. Geiser, and E. J. Stanbridge, *Science* **236,** 175 (1987).
78. M. Koi, H. Morita, H. Yamada, H. Satoh, J. C. Barrett, and M. Oshimura, *Mol. Carcinog.* **2,** 12 (1989).
79. M. Oshimura, H. Kugoh, M. Koi, M. Shimizu, H. Yamada, H. Satoh, and J. C. Barrett, *J. Cell. Biochem.* **42,** 135 (1990).
80. M. Shimizu, J. Yokota, N. Mori, T. Shuin, M. Shinoda, M. Terada, and M. Oshimura, *Oncogene* **5,** 185 (1990).
81. J. M. Trent, E. J. Stanbridge, H. L. McBride, E. U. Meese, G. Casey, D. E. Araujo, C. M. Witkowski, and R. B. Nagle, *Science* **247,** 568 (1990).

82. H. Yamada, N. Wake, S.-I. Fujimoto, J. C. Barrett, and M. Oshimura, *Oncogene* **5,** 1141 (1990).
83. S. A. Bader, C. Fasching, G. M. Brodeur, and E. J. Stanbridge, *Cell Growth Differ.* **2,** 245 (1991).
84. H. Hayes, Y. Kaneda, T. Uchida, and Y. Okada, *Chromosoma* **96,** 26 (1987).
85. O. Sugawara, M. Oshimura, M. Koi, L. A. Annab, and J. C. Barrett, *Science* **247,** 707 (1990).
86. S. M. Taylor and P. A. Jones, *Cell* **17,** 771 (1979).
87. M. J. Thayer and H. Weintraub, *Cell* **63,** 23 (1990).
88. S. J. Goss and H. Harris, *Nature (London)* **255,** 680 (1975).
89. S. J. Goss and H. Harris, *J. Cell Sci.* **25,** 17 (1977).
90. T. Glaser, D. Housman, W. H. Lewis, D. Gerhard, and C. Jones, *Somat. Cell Mol. Genet.* **15,** 477 (1989).
91. S. F. Dowdy, D. J. Scanlon, C. L. Fasching, G. Casey, and E. J. Stanbridge, *Genes Chromosomes Cancer* **2,** 318 (1990).

[8] Functional Analysis of Tumor Suppressor Genes

Steven F. Dowdy and Eric J. Stanbridge

Introduction

The introduction of genetic material from a normal cell into a malignant one via somatic cell fusion first demonstrated the presence of genetic material capable of suppressing the tumorigenic phenotype of a malignant cell (1, 2). Although the process generally involves the introduction of an entire normal genome into the malignant cell, information regarding the chromosomal location of the putative tumor suppressor gene(s) and cellular attributes of the resultant nontumorigenic somatic cell hybrids was obtained (3). However, much of this information was limited due to the nature of somatic cell hybridization. Molecular cloning of a few candidate tumor suppressor genes has been accomplished. However, the procedures used (positional cloning) are labor intensive and functional studies of tumor suppression have been reported only for *p53* and *Rb* cDNAs.

The refined process of transferring a single selectable chromosome via microcell-mediated chromosome transfer (MMCT) into malignant cells allows investigation into the functional role of tumor suppressor genes following transfer of limited genetic information, and retains cis and trans regulation of physiological levels of expression (4, 5). The use of the MMCT technique has documented the suppression of tumorigenicity of several malignant cell lines following the introduction of normal chromosomes (6, 7). However, the ability to reduce further the chromosomal region that contains the tumor suppressor activity is limited by the availability of selectable chromosomes of interest. In general, the available selectable chromosomes include naturally occurring balanced chromosomal translocations involving a selectable gene such as the hypoxanthine phosphoribosyltransferase (*HPRT*) locus on the Xq chromosomal arm, or chromosomes containing integrated vectors with a dominant biochemical selectable marker, for example, the bacterial *neo* or *gpt* genes (5, 8, 9).

Several investigators have developed methods that reduce the amount of chromosomal DNA that enters the cell either by chromosome-mediated gene transfer (CMGT) (10) or radiation-reduced hybrids (11). Goss and Harris (12) described a radiation-induced method to generate deletions of human chromosomes. The method involves the lethal irradiation of human fibroblasts that are rescued by subsequent fusion to hamster cells. These radiation-induced hybrids resulted in the isolation of fragments of human DNA integrated into the hamster chromosome. Cox *et al.* (13) have used a similar approach to isolate the telomeric region of human chromosome 4p. Hybrids generated by these methods are exceptionally useful for

Methods in Molecular Genetics, Volume 1

the mapping of genes and the isolation of probes in these regions but, due to the fragmented nature of the human chromatin, offer little in the way of usable reagents for the functional analysis of tumor suppression.

All of the above procedures have been proven to generate deletions of human chromosomes; however, all of these procedures lack the ability to enrich for specific arm deletions of a biochemically selectable chromosome while avoiding its fragmentation. In addition, these methods generate a large percentage of chromosomal translocations to other human or rodent chromosomes, which are unacceptable chromosomal reagents for functionally analyzing the suppression of tumorigenicity. Therefore, we devised the irradiation microcell-mediated chromosome transfer (XMMCT) procedure to enrich for specific arm deletions of chromosome 11, while maintaining the intactness of the chromosome and decreasing the frequency of recombination with other host chromosomes (14).

In this article we describe a modification of the microcell-mediated chromosome transfer procedure for generating deletions of selectable chromosomes in hybrids containing a single selectable human chromosome in a rodent background. The XMMCT procedure greatly enriches for deleted/rearranged selectable chromosomes that are intact and not translocated to a rodent chromosome. The benefits of XMMCT include (1) a significant decrease in translocation of the desired chromosome to rodent chromosomes, (2) only hybrids containing mitotically stable selectable chromosomes that are able to survive the selection protocol, thus reducing screening time, (3) enrichment for deletions contained on a specific chromosomal arm, and (4) the ability to transfer the deleted chromosome to another mammalian host, irrespective of species. The microcell hybrids generated from an XMMCT fusion can be readily analyzed by cytogenetic and restriction fragment length polymorphism (RFLP) analysis. Moreover, when the reduced human chromosome is sequestered in a rodent cell, the resulting microcell hybrid contains a haploid equivalent of the selectable human chromosome, therefore both polymorphic and nonpolymorphic DNA markers can be utilized to determine the extent of the deletions. In addition, the ability to generate an apparent unlimited assortment of deletions/rearrangements greatly facilitates the process of mapping DNA markers.

General Strategy

The procedure, designated irradiation microcell-mediated chromosome transfer (XMMCT), involves the radiation of microcells generated from monochromosome hybrids containing a single selectable human chromosome in a rodent background (14). The microcells are filtered to isolate a population of microcells primarily containing single chromosomes, thus precluding the probability of a translocation occurring to a mouse chromosome. The microcells are exposed to a dose of γ radiation in the range of 400 to 500 rad to ensure single or limited chromosomal "hits," while avoiding the extreme fragmentation of the chromosome seen in the Goss–Harris procedure (11, 12). The irradiated microcells are then fused to a new recipient rodent cell line and selected for the mitotic stability of the biochemical marker, which enriches for deletions of the opposite chromosomal arm in relationship to the selectable biochemical marker. The XMMCT hybrids are then selected in the appropriate growth medium and colonies are isolated and expanded. The XMMCT hybrids are first analyzed cytogenetically for the presence of a single intact human chromosome by using Giemsa-11 staining. Translocated human–mouse chromosomes, although useful for mapping studies, are not useful for subsequent functional analysis of tumor suppressor genes. The hybrids containing single human chromosomes are then analyzed cytogenetically by G banding and on the molecular level by Southern blotting analysis. The characterized human chromosomes present in the XMMCT hybrids are then transferred into recipient tumorigenic human cell lines. These human microcell hybrids are analyzed for alteration of neoplastic properties such as tumorigenic potential in nude mice, growth conditions *in vitro,* and so on.

Materials and Methods

Cell Lines

A9, an HPRT$^-$ mouse fibrosarcoma cell line derived from L cells, has been described by Littlefield (15). MCH 701.8 is derived by transferring a naturally occurring balanced X;11p translocated chromosome [der(11)t(X;11p)(q21;q13)] from the human fibroblast cell line GM3322 (NIGMS Human Genetic Mutant Cell Repository) into A9 cells by MMCT (14). All of the cell lines are routinely assayed for the presence of *Mycoplasma* by cultural methods (16) and/or by the 4,6-diamidino-2-phenylindole (DAPI) method (17), and have been routinely found to be negative by both methods.

Media and Culture Conditions

Parental rodent and human cell lines, and microcell hybrids, are maintained at 37°C and 5% CO_2 on Dulbecco's modified Eagle's medium (DMEM) supplemented with 10% (v/v) fetal bovine serum (FBS; Hazleton, Lenexa, KS) plus 2m*M* L-glutamine, 100 IU/ml penicillin, 0.1 mg/ml streptomycin, and 0.05 μg/ml Fungizone (amphotericin B). Biochemical selection for the retention of the t(X;11) chromosomes is performed by the addition of 100 μM hypoxanthine, 40 μM methotrexate, and 1.6 m*M* thymidine (HAT) to the medium. Each translocated chromosome retains the q portion of the X chromosome containing the *HPRT* gene. The cells are routinely passaged by washing in 1× phosphate-buffered saline (PBS; pH 7.2) and then adding a solution of 1× PBS containing 0.05% (w/v) trypsin and 0.02% (v/v) ethylenediaminetetraacetic acid (EDTA).

Fusion Reagents

Polyethylene glycol 1000 (PEG; Baker Chemical, Phillipsburg, NJ) is diluted to 48% (w/v) with DMEM (no serum). The diluted PEG solution is detoxified by filtration through a Bio-Rex RG mixed-bed resin (20–50 mesh; Bio-Rad Laboratories, Richmond, CA). Phytohemagglutinin P (PHA-P; Difco, Detroit, MI) is diluted to 50 μg/ml in DMEM (no serum). Cytochalasin B (Sigma Chemical, St. Louis, MO) is diluted to 10 μg/ml in DMEM (no serum). Colcemid (Calbiochem-Behring, San Diego, CA) is diluted to 1 mg/ml in double-distilled H_2O. All solutions are filter sterilized through a 0.2-μm filter. Eight- and 5-μm polycarbonate filters are obtained from Nuclepore (Pleasanton, CA).

Irradiation Source

The microcells in 15-ml polypropylene tubes are irradiated with 400–500 rad of γ radiation from a constant two-point ^{137}Cs source.

Procedures

Microcell-Mediated Chromosome Transfer

Microcell-mediated chromosome transfer is performed by treating donor A9 microcell hybrid cell lines with 0.02 μg/ml of Colcemid for 48 hr in six T-25 Nunc tissue culture flasks (Nunclon, Roskilde, Denmark; this is an extremely important point). Micronuclei are harvested by filling the flasks with DMEM containing 10 μg/ml cytochalasin B and then centrifuging the flask in a fixed-angle rotor (JA-14; Beckman, Fullerton, CA) with a 100-ml H_2O cushion at 24,000 *g* (14,000 rpm) for 65 min at 34°C. The medium in each flask is carefully aspirated to leave the microcell pellets intact, with approximately 3 ml of medium remaining in the flask to resuspend the pellets.

The microcells are then successively filtered through 8- and 5-μm polycarbonate filters in series to remove large microcells containing multiple chromosomes, contaminating whole cells, and karyoplasts. The filtrate is centrifuged at 1400 rpm in a clinical centrifuge containing a swinging bucket rotor at room temperature for 5 min, and the microcell pellet is then resuspended in 2 ml of DMEM containing 50 μg/ml phytohemagglutinin P. The filtered microcells are overlaid onto recipient A9 cells, allowed to attach for 15 min, then fused with 2 ml 48% detoxified PEG 1000 for 60 sec. The microcell hybrids are then extensively washed three times with PBS and allowed to recover overnight in DMEM plus 10% FBS. The following day the cells are split 1:8 and placed on DMEM plus 10% FBS and HAT selection. Independent colonies are isolated and expanded for cytogenetic and DNA analyses.

Irradiation Microcell-Mediated Chromosome Transfer

Microcells are produced from the monochromosome hybrid donor containing a single human chromosome in a mouse A9 cell as described above, and used for irradiation studies. Prior to the fusion step with the recipient, 2 ml of filtered microcells in DMEM plus PHA-P are placed in a 15-ml polypropylene sterile tube and irradiated with a dose of 480 rad of γ radiation from a constant two-point ^{137}Cs source. This dose was chosen because it is approximately triple the LD_{50} for mammalian fibroblast cell lines (18) and thus would ensure a relatively high number of chromosomal "hits" while avoiding fragmentation. The irradiated microcells are then fused to A9 recipient cells and treated as described above. Twenty to 50 or more independent colonies should be isolated from multiple dishes. The XMMCT procedure is outlined in Fig. 1.

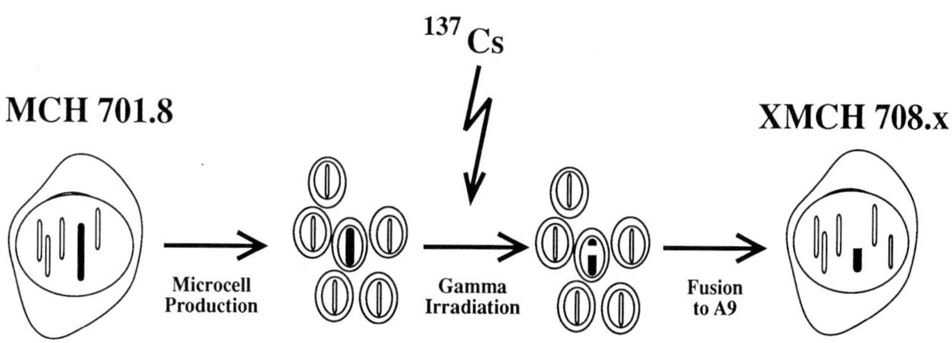

FIG. 1 Schematic diagram representing the irradiation, microcell-mediated chromosome transfer technique. After microcell production, the microcells are exposed to γ radiation and fused to an A9 cell. The filled-in chromosome represents the human t(X;11p) chromosome present in MCH 701.8 and the missing region of this chromosome represents an interstitial deletion.

Chromosomal Analysis of Irradiated Microcell Hybrids

Cytogenetic analysis is performed essentially as described by Rooney and Czepulkowski (19). Semiconfluent T-25 tissue culture flasks from each XMMCT cell line are arrested in metaphase by incubation in DMEM containing 0.05 μg/ml of vinblastin sulfate for 1.5–3 hr at 37°C. The mitotic cells are separated from nonmitotic cells by a metaphase shake. The cells are then centrifuged at 1400 rpm for 5 min at room temperature. Two milliliters of 0.75 *M* KCl, preheated to 37°C, is then added to the cells and incubated for 10 min at 37°C. The cells are fixed by the addition of Carnoy's fixative for 20 min, then centrifuged. The fixative is aspirated off, and the procedure repeated twice more. After the last centrifugation step, the cells are resuspended in Carnoy's fixative and dropped onto prewetted glass slides. The slides are dried in an oven overnight at 60°C.

XMMCT hybrids are analyzed by Giemsa-11 (G-11) staining (20) and Giemsa banding (G banding). Giemsa stock solution is made by dissolving 1 g of Giemsa powder in 66 ml of glycerol and then adding 66 ml of methanol this solution. G-11 staining is performed by staining the slides in a Giemsa–phosphate buffer (PB) solution containing 3 ml Giemsa stock solution and 37.5 ml PB, pH 11, for 1–4 min. The slides are then rinsed twice for 4 min in water and dried. A minimum of 25 metaphase spreads is analyzed for the presence of an intact human chromosome. Human chromosomes are identified by a pale blue staining compared to a deep magenta stain incorporated by mouse chromosomes. G banding is performed as follows: the dried slides are digested with 1% trypsin in PB, pH 7.5, for 25 sec. Following trypsinization, the slides are rinsed in buffer and then soaked in the same buffer solution for 2 min. The slides are stained in a Giemsa–PB solution containing 3 ml Giemsa stock solution and 50 ml PB, pH 6.8, for 1.5 min, and then allowed to dry. A minimum of 25 metaphase spreads is karyotyped for each cell line.

Southern Blot Analysis of Irradiated Microcell Hybrids

Genomic DNA is isolated from near-confluent T-25 tissue culture flasks; 10–20 μg of genomic DNA is digested with 8–12 units of restriction endonuclease in the manufacturer's buffer at 37°C overnight, and separated by electrophoresis through a 0.8% (w/v) agarose slab gel 1× TAE buffer [40 m*M* Tris (pH 7.8), 5 m*M* sodium acetate, and 1 m*M* EDTA]. The DNA is transferred to a nylon filter as per the instructions of the manufacturer. The filter is then prehybridized and hybridized to heat-denatured ^{32}P-labeled DNA at $>1 \times 10^6$ counts per minute (cpm)/ml of hybrization solution {5× SET [150 m*M* Tris (pH 8.0), 0.75 *M* NaCl, and 5 m*M* EDTA], 1× Denhardt's, 0.5% (w/v) sodium dodecyl sulfate (SDS), and 10% (w/v) dextran sulfate} and 100 μg/ml of heat-denatured salmon sperm DNA, at 65°C overnight. The filters are rinsed twice in 2× SET, 0.2% (w/v) SDS and washed in a large volume of 2× SET, 0.2% (w/v) SDS at 65°C for 20 min, then in 0.1× SET, 0.1% (w/v) SDS at 65°C for 1 hr. The washed filters are left damp and then exposed to Kodak (Rochester, NY) XAR5 film overnight at −70°C with an intensifying screen.

Tumorigenicity Analysis

The characterized deleted human chromosome from a XMMCT hybrid is transferred into a recipient tumorigenic human cell line via MMCT and colonies are selected and expanded. In general it is good practice to isolate 6 to 10 hybrids for characterization. The human XMMCT hybrids are then analyzed for the presence of the transferred deleted human chromosome by cytogenetic analysis and for the absence of any rodent chromatin. The presence of rodent chromatin is easily identified by Southern blot analysis using total ^{32}P-labeled rodent DNA as the probe. Any human XMMCT hybrid containing rodent DNA will hybridize to the highly repetitive ^{32}P-labeled rodent DNA.

The tumorigenic potential of each human XMMCT cell line is then determined by subcutaneous injection into the ventral surface of athymic *nu*/*nu* (nude) mice. In general, two near-confluent T-150 flasks contain >4 × 10^7 cells and are sufficient for four independent injections. The cells are trypsinized, counted in a hemacytometer chamber, centrifuged, and resuspended in DMEM plus 10% (v/v) fetal calf serum (FCS) at a concentration of 1 × 10^7 cells/0.2 ml. Each cell line is injected subcutaneously into the ventral surface of at least four individual nude mice. The injection sites should be examined on a regular basis until the mice are sacrificed.

Comments

There are several hypothetical fates of a chromosome selected by the XMMCT process. First, a single hit to the selectable chromosome occurring between the centromere and the biochemical marker, with no other chromosomes present in the microcell, would result in an unstable chromosomal fragment containing the selectable marker. Therefore, due to the mitotic instability of the marker, the recipient cell would not survive the selection process. Second, a single hit occurring between the centromere and the opposite chromosomal arm, in relation to the biochemical marker, would result in a mitotically stable deleted chromosome. Therefore, the recipient cell would survive the selection process. Third, two hits occurring to the selectable chromosome would have several possible results, including (1) interstitial deletion between the marker and the centromere with religation, (2) interstitial deletion of the opposite arm of the chromosome followed by religation with the terminal region, (3) a mitotically unstable chromosomal fragment or an inverted chro-

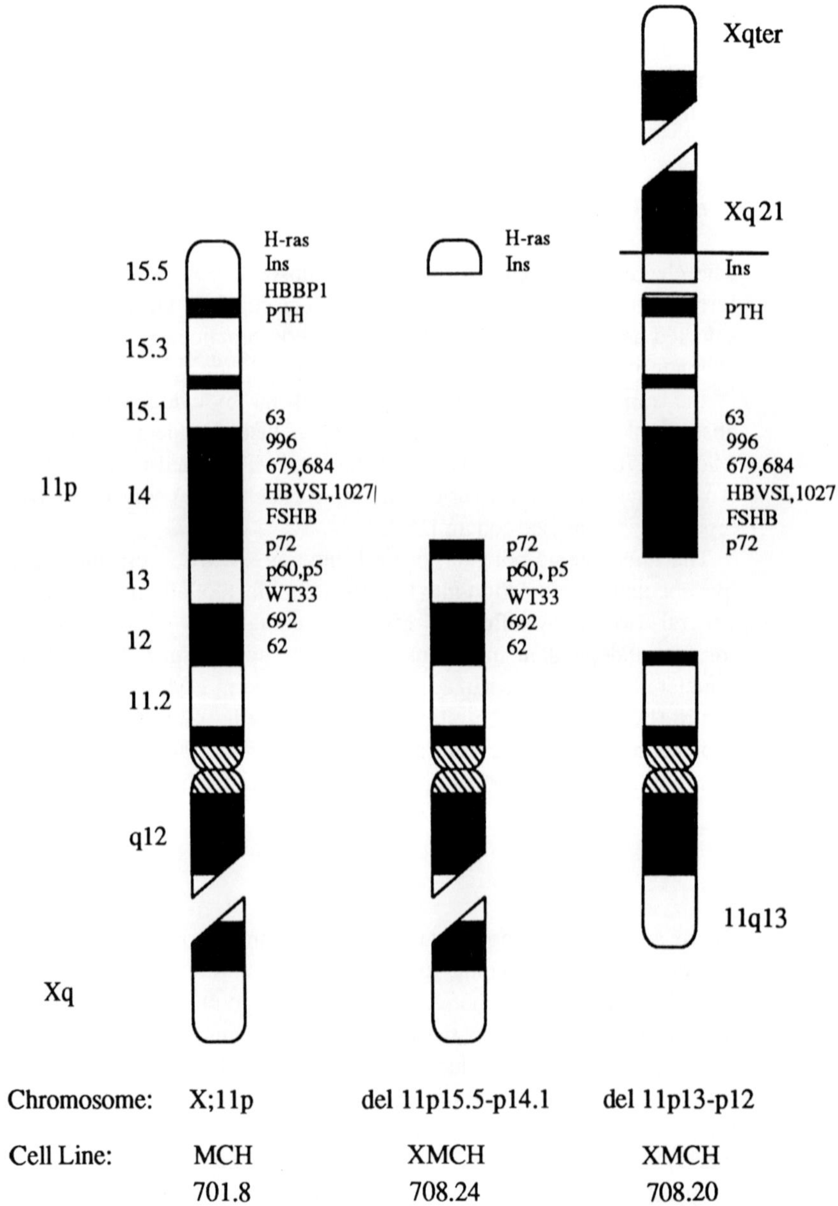

FIG. 2 Summary of Southern blot and cytogenetic results of the X;11p human chromosome present in MCH 701.8, and the resulting deleted 11p13-p12 chromosome in XMCH 708.20 and the deleted 11p15.5-p4.1 chromosome in XMCH 708.24 after XMMCT treatment (14, 20). The blank regions of the ideograms denote deletions detected by cytogenetic analysis,

mosome (generated by two hits flanking the centromere), and (4) indiscernible complex deletions and/or rearrangements of the selectable chromosome. The terminal deletion and interstitial deletions of the opposite arm would result in deletions contained on a specific chromosomal arm of a mitotically stable, selectable chromosome, making it an ideal reagent for functional studies into the suppression of tumorigenicity.

Using the XMMCT procedure described above, we have been able to generate successfully deletions on the human 11p arm present in MCH 701.8 encompassing both the 11p13 Wilms' tumor suppressor gene (*WT1*) and the 11p15 Wilms' tumor suppressor gene (*WT2*) (21). Of 25 chromosome-11 XMMCT clones initially isolated, 11 (45%) contained no apparent human chromatin by G-11, 9 (35%) contained a single untranslocated human chromosome, and 5 (20%) contained a translocated human–mouse chromosome. Two such characterized deleted t(X;11p) chromosomes are schematically represented in Fig. 2. The deleted 11p chromosomes appear to be stable in the A9 microcell hybrid background and are capable of being successfully transferred from the rodent XMMCT hybrids into other cell lines. The deleted 11p chromosomes have proved useful for both functional analysis of the 11p15 *WT2* gene and for mapping the 11p chromosomal arm. In theory, there is no limit to the extent of possible deletions that can be generated by XMMCT.

Acknowledgments

The studies that led to the development of this experimental procedure were supported by Grant CA19104 from the National Cancer Institute and the Council for Tobacco Research-

---

Southern blot analysis, or both. The DNA markers are positioned in their mapped position. The absence of a DNA marker denotes its loss on the chromosome based on Southern blot analysis. The blank diagonal truncation denotes a continuance of the Xq arm for diagrammatic simplification. The deleted X;11p chromosome in XMCH 708.24 contains a simple interstitial deletion encompassing the 11p15.5-p14.1 region. The deleted X;11p chromosome in XMCH 708.20 contains a complex rearrangement involving the inversion of the 11pter-q13 region with religation of the 11p15.5 subband to the Xq21 band. The horizontal line denotes the breakpoint and religation. In addition, the chromosome has undergone a deletion of the 11p13-p12 region and a submicroscopic deletion surrounding the *HBBP1* locus as detected by Southern blot analysis. Cytogenetic analysis was performed as described (19) on a minimum of 25 metaphase chromosome spreads. The DNA markers used are as described (14, 20). Ins, Insulin; HBVSI, hepatitis B viral insertion site I; FSHB, follicle-stimulating hormone B chain; HBBP1, β-globin pseudogene 1; PTH, parathyroid hormone; 62, 63, 679, 684 692, 996, 1027, and p5, p72, p60 are random genomic DNA fragments mapped to chromosome 11 as indicated.

USA, Inc. S.F.D. was supported by a Damon Runyon–Walter Winchell Cancer Research Fund Fellowship (DRG 1072).

References

1. H. Harris, O. J. Miller, G. Klein, P. Worst, and T. Tachibana, *Nature (London)* **223,** 363 (1969).
2. E. J. Stanbridge, *Nature (London)* **260,** 17 (1976).
3. E. J. Stanbridge and J. Wilkinson, *Proc. Natl. Acad. Sci. U.S.A.* **75,** 1466 (1978).
4. R. E. K. Fournier and F. H. Ruddle, *Proc. Natl. Acad. Sci. U.S.A.* **74,** 319 (1977).
5. P. J. Saxon, E. S. Srivatsan, G. V. Leipzig, J. H. Sameshima, and E. J. Stanbridge, *Mol. Cell. Biol.* **5,** 140 (1985).
6. B. E. Weissman, P. J. Saxon, S. R. Pasquale, G. R. Jones, A. G. Geiser, and E. J. Stanbridge, *Science* **236,** 175 (1987).
7. J. M. Trent, E. J. Stanbridge, H. L. McBride, E. U. Meese, G. Casey, D. E. Araujo, C. M. Witkowski, and R. B. Nagle, *Science* **247,** 568 (1990).
8. A. Tunnacliffe, L. V. Crawford, and P. Goodfellow, *Mol. Cell. Biol.* **4,** 1657 (1984).
9. T. G. Lugo, B. Handelin, A. M. Killary, D. E. Housman, and R. E. K. Fournier, *Mol. Cell. Biol.* **7,** 2814 (1987).
10. D. J. Porteous, J. E. Morten, G. Cranston, J. M. Fletcher, A. Mitchell, V. van Heyningen, J. A. Fantes, P. A. Boyd, and N. D. Hastie, *Mol. Cell Biol.* **6,** 2223 (1986).
11. S. J. Goss and H. Harris, *Nature (London)* **255,** 680 (1975).
12. S. J. Goss and H. Harris, *J. Cell Sci.* **25,** 17 (1977).
13. D. R. Cox, C. A. Pritchard, E. Uglum, D. Casher, J. Kobori, and R. M. Myers, *Genomics* **4,** 397 (1989).
14. S. F. Dowdy, D. J. Scanlon, C. L. Fasching, G. Casey, and E. J. Stanbridge, *Genes Chromosomes Cancer* **2,** 318 (1990).
15. L. W. Littlefield, *Exp. Cell Res.* **41,** 190 (1966).
16. E. J. Stanbridge, *Isr. J. Med. Sci.* **17,** 563 (1981).
17. W. C. Russell, C. Newman, and D. H. Williamson, *Nature (London)* **253,** 461 (1975).
18. J. B. Little, G. M. Hahn, E. Frindel, and M. Tubiana, *Radiology (Easton, Pa.)* **106,** 689 (1973).
19. C. W. Rooney and B. H. Czepulkowski, "Human Cytogenetics: A Practical Approach." IRL Press, Washington, D.C., 1986.
20. K. K. Friend, S. Chen, and F. H. Ruddle, *Somat. Cell Genet.* **2,** 183 (1976).
21. S. F. Dowdy, C. L. Fasching, D. Araujo, E. Livanos, K.-M. Lai, B. E. Weissman, and E. J. Stanbridge, *Science* **254,** 293 (1991).

[9] Chromosomal Mapping of cDNAs with Somatic Cell Hybrids Utilizing Polymerase Chain Reaction

Mihael H. Polymeropoulos and Carl R. Merril

Introduction

Efforts are underway to sequence and map expressed human genes as part of the Human Genome Program (1–3). One of the main strategies involves the identification of small fragments of cDNAs, followed by partial sequencing. These partially sequenced cDNAs serve as unique expression tag sites (ESTs) for mapping. In some cases these partial cDNA sequences identify sequences that share homology with previously known human or other mammalian genes. However, in the majority of cases they display no homology, and in these cases the sequences become the identifier for the expressed genes. They can be used to determine the chromosomal location of the cDNAs or as mapping identifiers (MIDs). A MID can refer to localization on a specific chromosome, or to subchromosomal localization such as a specific chromosomal band or a specific cloned chromosomal segment. Chromosomal segments may be cloned in yeast artificial chromosomes (YACs), cosmids, bacteriophage, or plasmid vectors.

Chromosomal assignment as well as chromosomal sublocalization can be achieved by the study of the segregation of the MIDs in panels of human–rodent somatic cell hybrids (3). Subchromosomal localization via direct visualization can be achieved by the technique of fluorescent *in situ* hybridization (FISH) (4). In addition, libraries of genomic clones, such as yeast artificial chromosomes, cosmids, and phages, can be screened with MID probes to achieve localization on specific DNA clones.

This article will concentrate on chromosomal mapping with somatic cell hybrids, utilizing the polymerase chain reaction (PCR). The somatic cell hybrids commonly employed were generated from human and rodent cell lines. Such hybrids characteristically lose human chromosomes while retaining the rodent chromosomes. Hybrid cell lines that have retained specific human chromosomes have been selected and used to assemble comprehensive chromosomal localization panels. These cell lines have been characterized by karyotyping and by hybridization with probes from known chromosomal localization. Most of the hybrid cell lines in use today have been extensively characterized by these hybridization and karyotyping methods.

Somatic Cell Hybrid Panels

Somatic cell hybrid panels may be of two types: first, there are panels in which more than one human chromosome is present in each of the cell lines. Chromosomal localization is possible, by studying discordance and concordance of a DNA marker in the different cell lines in the panel. Second, there are panels in which a single human chromosome is present in each cell line. These monochromosomal hybrid panels are relatively easy to interpret provided that the mapping technique utilizing the somatic cell hybrid panels is 100% reproducible, quantitatively and qualitatively. However, minute differences in technique can easily lead to a failure of assignment, because localization is based on the positive result from a single hybrid line. In contrast, somatic cell hybrid panels containing multiple human chromosomes have a redundancy that permits assignment even when a cell line fails to provide results because of an experimental artifact.

Characterization of the chromosomal content of each hybrid line is performed by the study of the segregation of DNA probes with known mapping assignments. Karyotyping of the somatic cell hybrid lines has shown that the human chromosomal content in each cell line can vary. For example, a cell line may contain a specific chromosome in only a small percentage of the metaphase spreads, while containing a different chromosome in up to 100% of the metaphases. This will mean that in the same cell line, single-copy genes will be represented in different numbers. This uneven representation of chromosomal numbers within a cell line accounts for one of the complications in the study of somatic cell hybrids: quantitative difficulties in interpreting the hybridization patterns for the presence or absence of a specific DNA segment within that cell line. Another problem that adds to the complexity in the characterization of somatic cell hybrid panels arises when human chromosomes are present in a fragmented and/or rearranged form within the rodent genome. The above comments illustrate that usefulness of the somatic cell hybrid panels as mapping tools depends on the accurate and detailed characterization of the human chromosomal content of each of the cell lines.

Many researchers have begun to employ the PCR in the study of these human–rodent somatic cell hybrid lines (5, 6). Polymerase chain reaction techniques are substantially more sensitive and rapid than the hybridization procedures. For these reasons, somatic cell hybrid panels previously characterized by karyotyping or hybridization often yield evidence of additional human chromosomal regions with the PCR methods.

cDNAs

cDNAs are created through the reverse transcription of the gene transcripts, the messenger RNAs. Messenger RNA can be divided into three topographical sequence elements: a 5′ untranslated region, a coding sequence, and a 3′ untranslated region. The 5′ untranslated region contains the transcriptional and translational control elements for the transcribed gene. The coding sequence contains the triplet codons that will be translated into the ultimate protein gene product, and the 3′ untranslated sequence contains additional regulatory elements necessary for the translation and stability of the messenger RNA.

At the genomic DNA level, genes are frequently interrupted by noncoding sequences, the introns. The size of these introns can vary greatly, between different genes as well as between introns within the same gene. The average size of an intron is approximately 600 base pairs (bp). In terms of location, introns are primarily found to interrupt coding sequences in contrast to noncoding regions of the gene or premessenger RNA. It has been estimated that 80% of the introns are found in the coding sequence, with the remaining 20% divided among the 5′ and 3′ untranslated sequences (7, 8). The introns are excised during the processing of the premessenger RNAs to form the mature form of the transcript, the messenger RNA.

It has been found that coding regions of messenger RNAs are more conserved than are the sequences in noncoding regions. These coding regions also appear to contain a high degree of homology within a family of genes as well as between genes of similar function in different species. In contrast, the highest degree of diversity has been found in the 3′ untranslated regions of the messenger RNAs (9). This characteristic of the 3′ untranslated region (UTR) often provides a region that can serve to generate probes that can act as specific identifiers of a specific single-copy gene. Conversely, the use of a whole cDNA clone to probe somatic cell hybrid lines for the purpose of identifying a chromosomal assignment for a cDNA can pose some problems when a portion of the whole cDNA sequence is highly conserved between human and rodent and/or when cDNA sequences belong to multigene families. In such cases multiple hybridization signals may arise on probing the somatic cell hybrids.

Polymerase Chain Reaction

The PCR offers a rapid method for the amplification of specific chromosomal regions with sequence tag sites (STSs), using primers derived from sequenced cDNAs. The reaction conditions to be discussed have been optimized for performance in a microtiter plate for the TECHNE (Princeton, NJ) thermocycler, although similar methods can be applied with any other thermocycler.

Polymerase Chain Reaction

Reagents

PCR buffer (10×): 500 m*M* KCl, 100 m*M* Tris-HCl (pH 8.3), 15 m*M* $MgCl_2$, 0.01% (w/v) gelatin

Taq DNA polymerase: 5 units/μl

dNTP mixture:

dATP (100 m*M*)	12.5 μl
dTTP (100 m*M*)	12.5 μl
dGTP (100 m*M*)	12.5 μl
dCTP (10 m*M*)	1.5 μl
deionized (dI) H_2O	961.0 μl

Oligonucleotide primer concentrations should be adjusted to 100 ng/μl in a final reaction volume of 15 μl.

Final reaction solution:

PCR buffer (10×)	1.50 μl
dNTP mixture	2.40 μl
dI H_2O	8.21 μl
Taq polymerase	0.09 μl
[α-^{32}P]dCTP	0.10 μl
primer 1	0.60 μl
primer 2	0.60 μl
DNA (50 ng/μl)	1.50 μl

Overlay the above 15-μl reaction mixture with 15 μl of light mineral oil to prevent evaporation.

Polymerase Chain Reaction Conditions

Each cycle consists of the following:

Denaturation	94°C for 1.4 min
Annealing	55°C for 1.0 min
Extension	72°C for 1.0 min

Perform the assay for 30 cycles, followed by a final extension at 72°C for 10 min.

Design of Oligonucleotide Primers

The 3′ untranslated regions offer the highest probability of a unique MID sequence that will not overlap species and gene family sequences, as previously discussed.

A

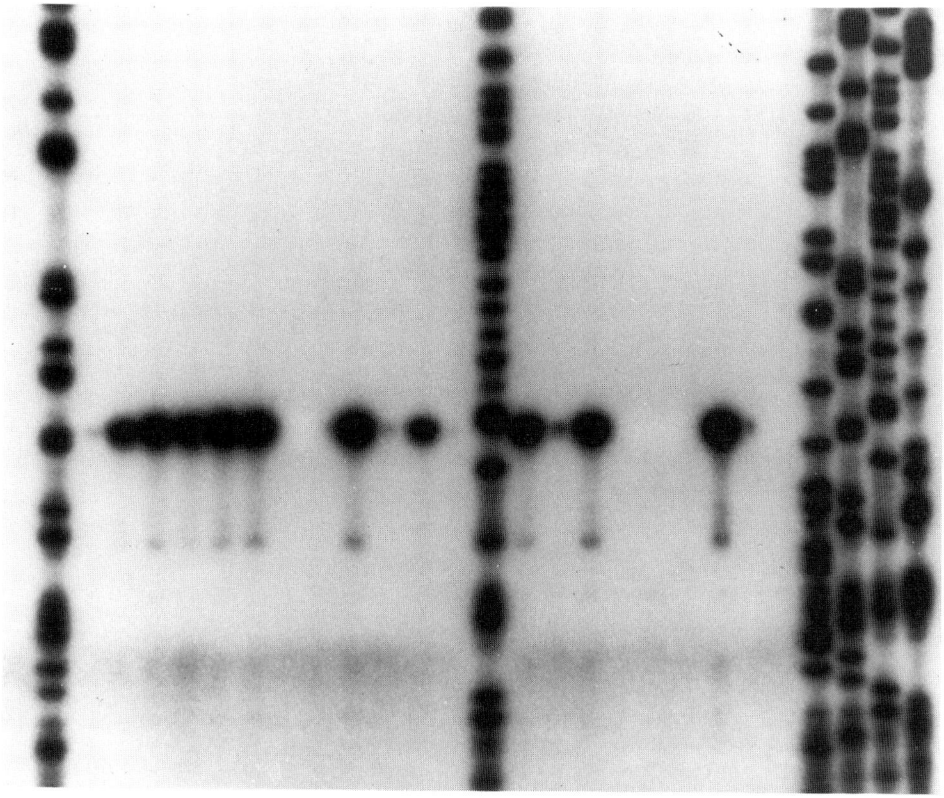

A 1 2 3 4 5 6 7 8 9 10 11 12 B 13 14 15 16 17 18 19 20 21 C D E F

FIG. 1 cDNA mapping studies utilizing PCR and human–rodent somatic cell hybrids. In (A)–(D) segregation of the P^{32}-labeled PCR product was analyzed by electrophoretic separation on 6% PAGE sequencing gels. Lanes A–F represent M13mp18 sequence size markers. Lanes 19–21 represent human, hamster, and mouse controls, respectively. (A) A chromosomal assignment of a cDNA in the absence of a rodent background. Lanes 2–6, 9, 11, and 13–15 contain only chromosome 3 as a common human chromosome. (B) A chromosomal assignment of a cDNA based on a quantitative difference when compared with the rodent background. Lanes 1–3 and 9 contain only chromosome 1 as a common human chromosome. (C) A chromosomal assignment of a cDNA based on an electrophoretic mobility difference of 5 bp between the human and mouse background. Lanes 1–3 and 9 contain only chromosome 1 as a common human chromosome. (D) Inability to establish a chromosomal assignment of a cDNA. Lanes 1, 7, 13–15, and 18 are not concordant with a single human chromosome.

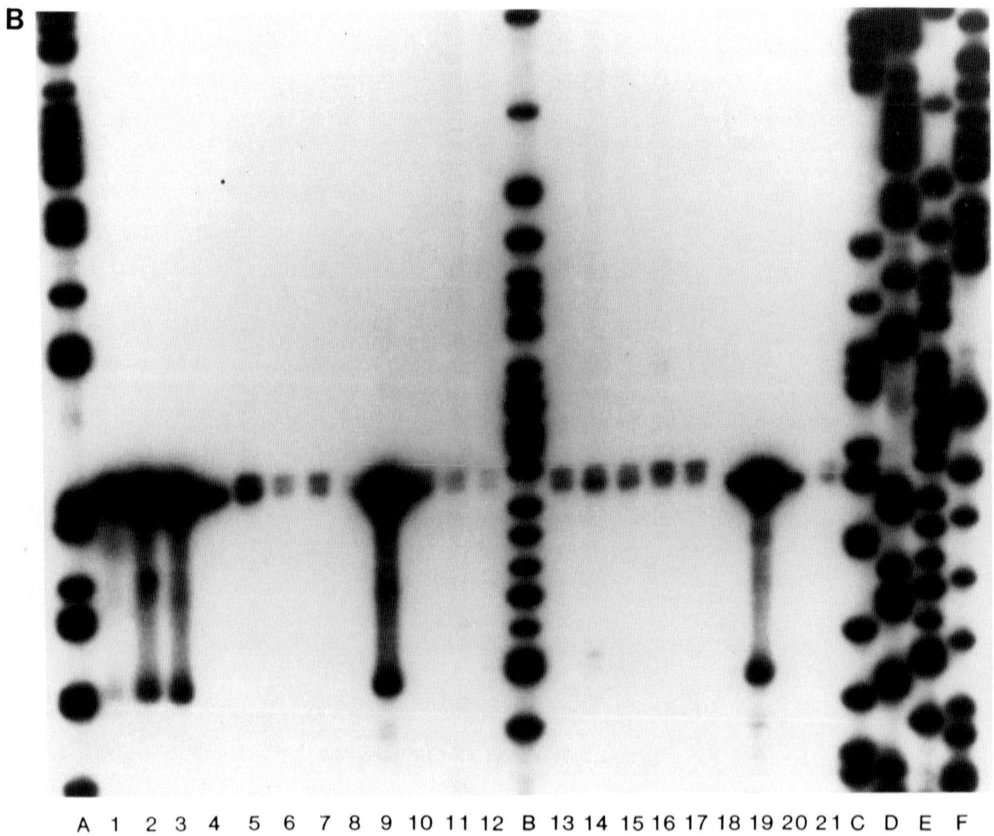

FIG. 1 (*continued*)

For the PCR, two oligonucleotide primers are designed to flank the target sequence of the 3′ untranslated region that would have the least probability of being interrupted by introns at the genomic level. This is a requirement because identification of the mapping location of the specific gene will be established by a study of the amplification signal from the chromosomal region as well as by the exact size of the amplified product. The size of the amplified product should match the predicted sequence from the known cDNA sequence.

We have developed the computer program INTRON to aid us in the identification of regions of the 3′ untranslated cDNA that are least likely to contain introns at the genomic DNA level (3). This computer program allows us to use the following observations: on excision of the introns from the pre-mRNA, most of the consensus sequence of the splice junction is lost. However, the dinucleotide GG, consisting of

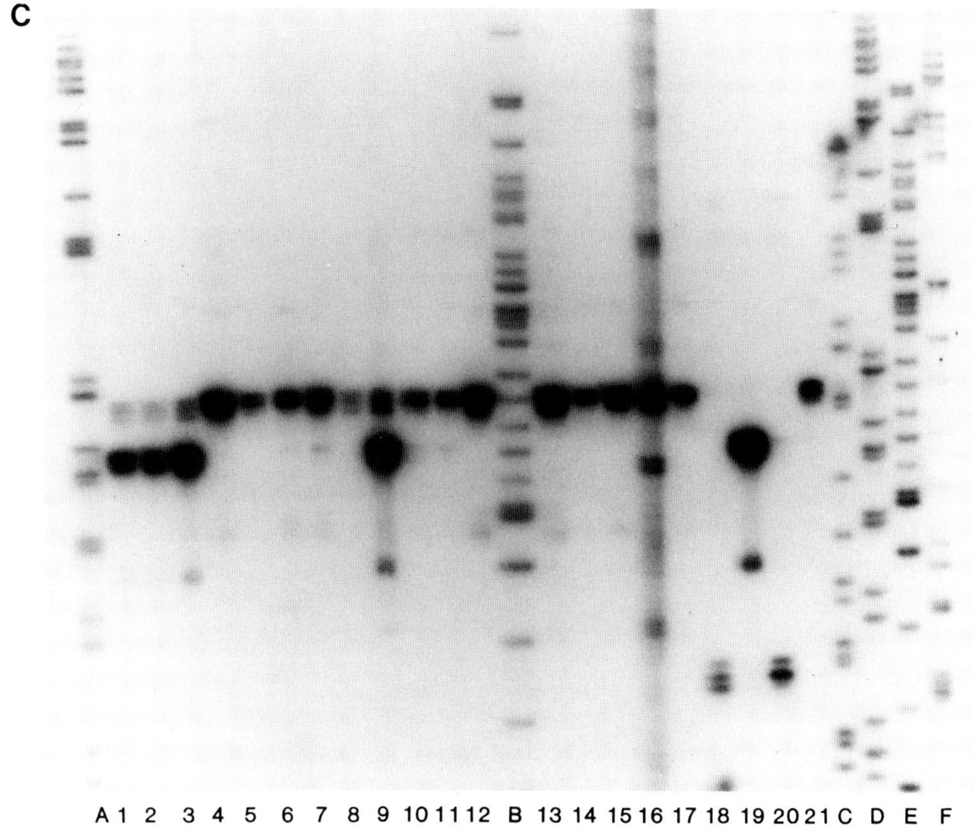

FIG. 1 (*continued*)

the last 3′ nucleotide of the exonic sequence and the first 5′ nucleotide of the next exon, is present in approximately 60% of the intron–exon junctions. By definition, the coding sequence is free of stop codons in at least one reading frame. Therefore, if one makes the assumption that the GG dinucleotide represents the former intron–exon junction at the genomic level, it is safe to assume that regions containing a GG dinucleotide surrounded by stop codons in all three reading frames for one strand, or a CC dinucleotide in all three reading frames for the opposite strand, do not contain an intron at the genomic level. In addition the chance that such a 3′ untranslated region may contain an intron is further diminished by the observation that 80% of the introns are found in coding sequences. Therefore, oligonucleotide primers should be designed for a 3′ untranslated region in which a GG dinucleotide is surrounded by stop codons in all six frames. The oligonucleotide primers should

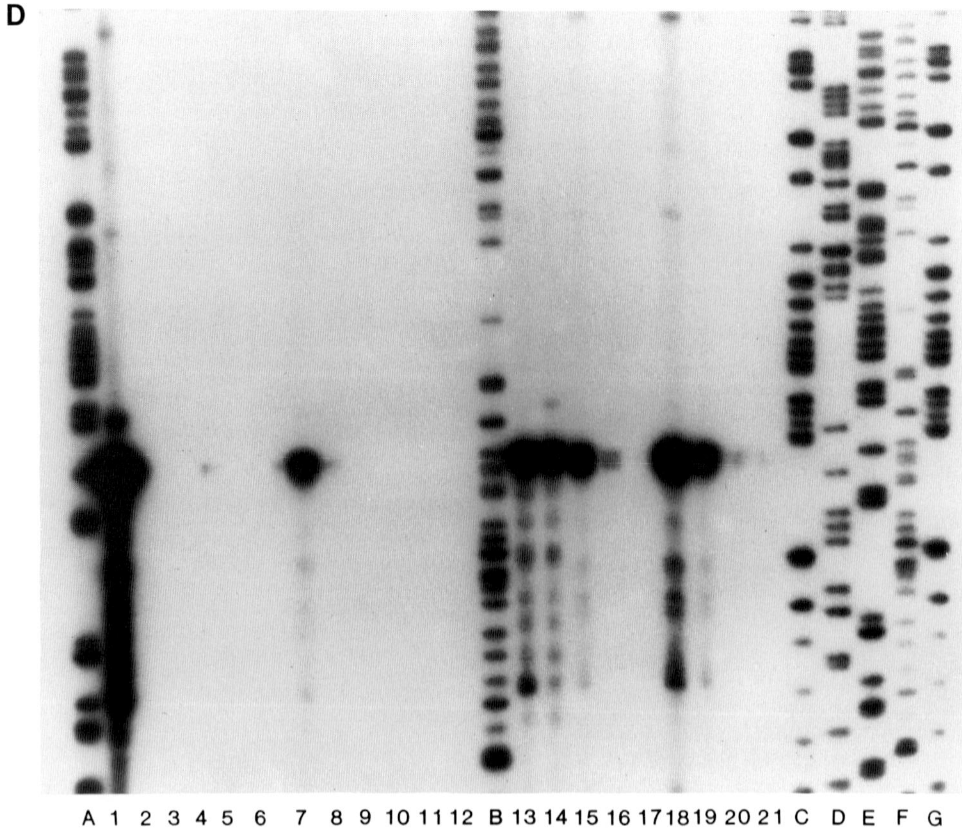

FIG. 1 (*continued*)

be tested for amplification with human genomic templates as well as rodent templates and a mixture of human and rodent templates.

Analysis of Amplified Products

A single-base pair resolution is absolutely necessary for the success of this mapping strategy. Often this single-base pair resolution is required for identification of the expected human product from other human background or more frequently the rodent background amplified products (Fig. 1). Such resolution can be achieved with analysis on standard 6% (w/v) polyacrylamide gel electrophoresis (PAGE) sequencing gels. Sequencing markers should be used, such as M13mp18 sequencing ladders, for an accurate estimation of the length of the amplified products. The labeling protocol discussed above results in an incorporation of the [α-^{32}P]dCTP throughout both strands of the amplified product. Because of the nature of the electrophoretic conditions, the sequencing gels will often resolve two bands. These bands represent the two different strands of the amplified product. Separation of these strands can vary by a distance equivalent to 0 to 3 or 4 bp for products of 80–150 bp in length, and this separation will also vary depending on the sequence composition of each strand. Therefore, differences of migration of up to 5 bp for a given amplified sequence should be expected. Autoradiography of the sequencing gels can be performed at $-70°C$ without prior drying.

The procedures described above will result in successful chromosomal assignments of cDNA clones in approximately 90% of the sequences studied. These procedures based on the PCR are highly transportable between laboratories and will facilitate the physical mapping efforts of the human genome project and the identification of disease genes.

References

1. A. S. Wilcox, A. S. Khan, J. A. Hopkins, and J. M. Sikela, *Nucleic Acids Res.* **19,** 1837 (1991).
2. M. D. Adams, J. M. Kelley, J. D. Gocayne, M. Dubnick, M. H. Polymeropoulos, H. Xiao, C. R. Merril, A. Wu, B. Olde, R. Moreno, A. R. Kerlavage, W. R. McCombie, and J. C. Venter, *Science* **252,** 1651 (1991).
3. M. H. Polymeropoulos, H. Xiao, A. Glodek, M. Gorski, M. D. Adams, R. F. Moreno, M. G. Fitzgerald, J. C. Venter, and C. R. Merril, *Genomics* **12,** 492 (1992).
4. B. Trask and D. Pinkel, *Methods Cell Biol.* **33,** 383 (1989).
5. C. M. Abbot and S. Povey, *Genomics* **9,** 73 (1991).
6. S. Theune, J. Fung, S. Todd, A. Y. Sakaguchi, and S. L. Naylor, *Genomics* **9,** 511 (1991).
7. M. B. Shapiro and P. Senapathy, *Nucleic Acids Res.* **15,** 7155 (1987).
8. J. D. Hawkins, *Nucleic Acids Res.* **16,** 9893 (1988).
9. M. W. Smith, *J. Mol. Evol.* **27,** 45–55 (1988).

Section III

Chromosome Mapping

[10] Cosmid Clones and Generation of Detailed DNA Maps of Large Chromosomal Regions

Alasdair C. Ivens, Timothy F. King, and Peter F. R. Little

Introduction

The generation of maps for large regions of DNA is a central component of any genome mapping project. Particularly attractive within these projects is the construction of maps at the level of overlapping cloned DNAs (so-called cloned DNA maps), because this approach potentially generates DNA clones, contained in yeast artificial chromosome (YAC) or cosmid vectors, that can be incorporated into "reference" clone strategies (1). These, along with expressed sequence tag (EST) and sequence tag site (STS) sequences, can be used as convenient resources for the isolation of regions around loci for which probes already exist.

Genome mapping projects, until relatively recently, have focused almost exclusively on specific genetic and regional targets; a number of analytical techniques, including cell-mediated gene transfer (CMGT), radiation hybrids, YAC cloning, pulsed-field gel electrophoresis (PFGE), and cosmid cloning, have been used to dissect regions into their constituent parts. Many of these techniques use or generate fragments of DNA from one species (transgenomes) on the background of the complete genome of another species.

The construction of cosmid libraries is an important component of the analysis of transgenomes. The ultimate goal is to identify a minimal number of overlapping cosmids that represent the entire linear DNA fragment of the transgenome: the cosmids can then be used for detailed analyses of transcription, rearrangement, and sequencing.

This article deals with the entire process of transgenome analysis by cosmid contiguous sequence (contig) assembly. We have been using the techniques of cosmid fingerprinting and contig mapping (2, 3) to construct a large cloned DNA map of the short arm of chromosome 11. In this technique, cosmid overlaps are identified by comparison of restriction fragment digest patterns: overlap of cosmids is determined by partial or complete similarity of restriction fragment sizes. To date, this has involved preparing over 8500 cosmid DNA samples from small bacterial cultures (micropreparations) and setting up stereotyped restriction and labeling reactions in microtiter trays. Our approach has been to use the strategy originally developed by Coulson *et al.* (2), making extensive use of microtiter trays to handle samples.

Methods in Molecular Genetics, Volume 1

Scope of Article

The construction of cosmid libraries, the selection of clones of interest, DNA extraction, fingerprinting, and data analysis are all detailed. In addition, we specifically detail the application of these procedures to the construction of cosmid libraries from YAC clones. Although a variety of cosmid vector–host systems are currently available (4), the procedures described below are based entirely on the lorist series of vectors (5, 6), although the procedures can be modified simply for use with other vector systems.

The analysis of large numbers of cosmid clones creates the problems of sample and data storage and recording. To overcome some of these, we have developed programs for the data gathering and indexing of microtiter tray contents: DOT and TRAY, respectively. These run on IBM PC compatibles. DOT is designed to record data from autoradiographs of colony array hybridizations with complex probes, whereas TRAY is used to store details of clones and their location. Both programs have greatly facilitated the speed and accuracy of both the acquisition and analysis of clone information.

Preparation of large numbers of plasmid or cosmid DNAs is both time consuming and labor intensive. Two options are available for this kind of analysis: manual or automated. We have written routines for a Beckman (Fullerton, CA) Biomek robotic workstation to prepare DNA in a semiautomated way (7). The Biomek is essentially a programmable pipetting station with user control over the type of pipettor, volumes dispensed, source and destination, how volume is dispensed, and from what height. It is particularly suited to the manipulation of microtiter plates, with each phase controlled by user-defined subroutines. The routines that we have written for the Biomek have proved to be valuable in a project that requires a high throughput of DNAs.

Cosmid Libraries

This section details the necessary procedures for the construction of cosmid libraries, identification, and isolation of positive clones, and general preparations required for clone analysis.

Construction of Cosmid Libraries

Construction of cosmid libraries from cell line DNA has been described (6), and is reproduced in a condensed form below. Similarly, the use of YAC clones as a source of DNA for cosmid libraries is becoming more popular as the average size of cloned

YAC insert DNA increases (8). We are routinely using the procedures detailed below for a number of projects, not all of which involve transgenome analysis.

1. Resuspend or homogenize 10^8–10^9 phosphate-buffered saline (PBS)-washed cells in 5 ml 1× SSC (0.15 *M* NaCl, 0.015 *M* sodium citrate). Add 5 ml 100 m*M* Tris-HCl (pH 7.5), 100 m*M* NaCl, 10 m*M* ethylenediaminetetraacetic acid (EDTA; pH 8.0), 1% (v/v) Sarkosyl. Mix carefully for 2–3 min to lyse the cells.

2. Add proteinase K to 100 μg/ml and incubate at 55°C for 2 hr.

3. Cool to room temperature and carefully extract twice with phenol [TE (10 m*M* Tris-HCl, 1 m*M* EDTA) (pH 8.0) buffered], taking the interface the first time. It is important that the DNA is not shaken vigorously, as this results in shearing. Reextract the aqueous phase with phenol–chloroform (1:1, v/v), and finally extract with chloroform.

4. Dialyze the aqueous phase against at least 5 liters of TNE [TE (pH 8.0), 100 m*M* NaCl] for 16 hr at 4°C. Repeat dialysis against TE (pH 8.0) for 24 hr at 4°C. Do not be tempted to ethanol precipitate the DNA, as this is a primary cause of difficulties when making cosmid libraries: the DNA is broken into approximately 100-kb fragments, which are unsuitable for cloning.

5. Estimate the concentration and size of purified DNA by field-inversion gel electrophoresis against known concentrations of bacteriophage λ DNA and λ DNA concatamers. The size of the isolated DNA *must* be greater than 200 kb.

6. Proceed with partial digestions as detailed below.

Preparation of Yeast DNA

1. Grow YAC cultures at 30°C with vigorous shaking for 48–72 hr in 500 ml yeast medium [1.3% (w/v) yeast nitrogen base, 4% (w/v) glucose, 1× amino acid supplement (20 μg/ml adenine, 20 μg/ml arginine, 20 μg/ml isoleucine, 20 μg/ml histidine, 60 μg/ml leucine, 20 μg/ml lysine, 20 μg/ml methionine, 50 μg/ml phenylalanine, 0.15 mg/ml valine, 30 μg/ml tyrosine)].

2. Pellet cells by centrifugation (5000 *g* for 10 min at 4°C), and resuspend in 30 ml SCE [1 *M* sorbitol, 0.1 *M* sodium citrate (pH 5.8), 10 m*M* EDTA (pH 7.5), 14 m*M* 2-mercaptoethanol].

3. Add 200 μl 10-mg/ml lyticase and incubate with shaking at 37°C for 1 hr.

4. Pellet spheroplasts by centrifugation at 1000 *g* for 10 min at 4°C. Resuspend spheroplasts gently in 25 ml TNE. Add proteinase K to a final concentration of 250 μg/ml and incubate at 55°C for 2 hr.

5. Purify DNA by extraction with an equal volume of phenol–chloroform (twice), followed by a single chloroform extraction, ensuring that all mixing is as gentle as possible to prevent shearing of the DNA.

6. Although it is preferable to dialyze the extracted DNA against TE (pH 8.0) for 24–48 hr at 4°C, it is possible to precipitate DNA by the addition of 0.1 vol 3 *M* sodium acetate (pH 5.5) and 2 vol of ethanol. Recover the DNA by gentle spooling. Wash the pellet in 70% ethanol (v/v), air dry, and resuspend overnight in 500 μl TE.

Partial Digestion

The aim of partial digestions is to generate insert fragments of 45 kb that have the restriction site of interest at each end, which can be subsequently ligated into the cosmid vector. It is important to purify the 30 to 50-kb fraction of partially digested DNA by centrifugation on sucrose gradients, but this luxury is not usually available for DNA isolated from YAC clones, as approximately 300 μg DNA is typically needed per gradient. Gradients *should* be used when constructing libraries from cell line DNA samples. (For a full description of the method, refer to Ref. 6.) In both instances, it is critical to treat the partial fractions of interest with alkaline phosphatase. This prevents coligation events, increases the overall efficiency of the ligation and packaging reactions, and should be done prior to gradient fractionation.

1. Determine the concentration and integrity of the DNA by electrophoresis against a known concentration of bacteriophage λ DNA (e.g., 100 ng). The molecular weight of the DNA *must* be greater than 200 kb.

2. Set up a 10-point partial series on ice. Dispense the appropriate volume of DNA into 10 tubes, such that each tube contains approximately 200 ng DNA in 20 μl of 1× *Sau*3AI reaction buffer. The first tube of the series should contain a double volume of reaction mix. Dilute 2 μl stock *Sau*3AI enzyme [5 units (U)/μl] 10-fold in 1× restriction buffer. Add 2 μl diluted enzyme to the first point of the partial series. Immediately remove 20 μl from the first tube, transfer to the second, and mix thoroughly. Repeat for all points in the dilution series. Incubate at 37°C for exactly 30 min.

3. Place the samples on ice, add EDTA (pH 8.0) to a final concentration of 10 m*M*, and heat the reaction to 68°C for 10 min.

4. Analyze the digestion products by extended electrophoresis on a 0.3% (w/v) agarose gel, using intact bacteriophage λ DNA and uncut YAC DNA as molecular weight markers. Identify the conditions for which the modal molecular weight of the products is 45 kb and calculate the amount of enzyme required to obtain the NEXT LESS-digested sample (9).

5. Scale up the appropriate dilution such that approximately 2 μg of DNA is being restricted; also scale up the dilution points on either side of the "ideal" dilution. Incubate at 37°C for 30 min, add EDTA (pH 8.0) to a final concentration of 10 m*M*, and heat the reaction to 68°C for 10 min.

6. Precipitate the reactions by the addition of 0.1 vol of sodium acetate (pH 5.5) and 2 vol of ethanol. Recover the precipitate by centrifugation. Wash the pellet in 70% ethanol, and dissolve in 50 μl TE.

7. Phosphatase "insert" DNA by treating 30 μl of the "insert" DNA at 37°C for 30 min with 5 U of calf intestinal phosphatase (Boehringer Mannheim, Indianapolis, IN), then inactivate the enzyme by the addition of NTA (nitrilotriacetic acid) to a final concentration of 10 m*M* and incubation at 68°C for 10 min.

8. There is no need to phenol extract the phosphatased DNA. Ethanol precipitate the DNA, wash twice with 70% ethanol, and resuspend the pellet in 50 μl TE (ph 8.0). Analyze a small aliquot (2 μl) by electrophoresis, comparing it to a known concentration of bacteriophage λ DNA.

9. Check both the integrity of the DNA and the efficiency of the phosphatase step by self-ligation tests. Ligate 100 ng of DNA with 1 U T4 DNA ligase (Boehringer Mannheim) in a final volume of 10 μl for 1 hr at room temperature. Analyze the products of ligation by agarose gel electrophoresis with suitable controls and markers. The unphosphatased DNA should self-ligate (reduced mobility), whereas the treated DNA will travel with the same mobility as the unligated control. If the phosphatased DNA ligates, repeat steps 7–9.

10. Ligate insert to vector (prepared as described in Ref. 6) at an 8–10:1 ratio (insert:vector by weight). We typically use 1 μg of insert per ligation (1 U T4 DNA ligase, final volume 10 μl, 1 hr at room temperature).

11. Package 4 μl of the ligation products [e.g., Stratagene (La Jolla, CA) Gigapack Gold] for 2 hr at room temperature. Add 1 ml SM [0.1 *M* NaCl, 5m*M* $MgSO_4$, 50 m*M* Tris-HCl (pH 7.5), 0.01% (w/v) gelatin] and 50 μl chloroform. Store at 4°C. Efficiencies are variable in these experiments and can range from 1×10^3 to $2 \times 10^6/\mu g$ of insert DNA.

Screening of Cosmid Libraries

The strategy employed at this stage will depend entirely on the transgenome being studied. The overall aim is to identify clones of interest in a background of irrelevant clones. If the transgenome makes up 1% or more of the recombinant clones, a low-density screening would be employed (detailed below). The main advantage of this method is that secondary screening is not required and there is little, if any, loss of clones due to variable growth characteristics. If, however, the transgenome makes up significantly less than 1% of the recombinant clones, then a high-density screening, coupled with secondary, and possibly tertiary, screening of positive clones will be required. In the case of screening cosmid libraries derived from YACs, a low-density screening is recommended, independent of the insert size of the starting YAC clone.

1. Before each "round" of plating, the packaging reaction must first be titered against *Escherichia coli* 1046 (the host strain for the lorist X cosmid). Set up a 5-ml overnight culture at 37°C in L broth plus 10 m*M* $MgSO_4$, 0.2% (w/v) maltose (LMM). The next day, inoculate 100 ml LMM with the overnight culture, and incubate with shaking at 37°C for 4 hr (approximately midlog). The cells are now ready to use, but can be stored at 4°C for up to a week.

2. Titer the packaging reaction: we routinely adsorb 100 μl of cells with 2 and 10 μl of the packaging reaction. Incubate the adsorption reactions at 37°C for 30 min, add 200 μl L broth, and incubate an additional 45 min at 37°C prior to plating out on L-agar plates containing 30 μg/ml kanamycin. Once the titer of the packaging reaction has been determined, a simple scale-up is performed. The plating cells prepared for the titration *must* be used for plating out the library.

3. We routinely screen cosmid libraries on large (22 × 22 cm) plates (Nunc, Naperville, IL). L-Agar plates containing kanamycin (30 μg/ml) are poured approximately 1 cm thick (500 ml), flamed to remove bubbles, and allowed to dry for one to 2 hr in a laminar flow hood. They are sufficiently dry when tiny "lines" appear on their surface.

4. The optimum number of colonies for a low-density screen on a large plate is 7000 colonies. Under 5000 colonies is wasteful in terms of both time and resources. If 10,000 colonies are obtained, colony identification becomes harder, but the staining method described below will partially ameliorate matters. Scale up the results obtained for the titration *directly*. Do not be tempted to reduce volumes for the sake of simplicity. After incubation with LB, pellet the cells gently by centrifugation (2000 *g* for 5 min at 4°C) and resuspend the pellet in 3 ml LB. Spreading the absorption reaction on large plates requires patience and should be done in a laminar flow hood, which will also aid drying. Spread the plates until they are completely dry, and continue spreading for approximately 5 min. Invert and incubate the plates for 12 hr at 37°C. (*Note:* This time is critical, as it is important that the colonies do not grow larger than 0.5–0.7 mm in diameter. Larger colonies do not transfer as well as small colonies, and do not give a "better" positive signal.)

Colony Screening

1. At present we are using Pall (Glen Cove, NY) BioSupport membranes (type A). These are labeled in pencil along the edge of the membrane. The dry membrane is gently placed number side up on the colony-covered plate. We have found it easiest to lower the middle of the membrane first and then to lower the sides gently. This must be done in one smooth motion to avoid colony smearing or membrane creasing. The membrane is left on the plate for approximately 1 min, during which time the membrane/plate reference marks are made by stabbing the membrane through to the plate with a needle. Make 10 to 12 widely spaced, asymmetrical

stabs per membrane. Take care when pulling out the needle after each stab, placing tweezers on the membrane around the needle. Briefly flame the needle between plates.

2. We have found that the number of bacteria transferred onto a membrane depends on the speed at which the membrane is removed from the plate. Firmly grip one corner of the membrane with forceps and peel the membrane back approximately 2 cm. "Whip" off the membrane with a quick motion. If the whip-off method has worked well, then the colonies on the membrane should be clearly visible, whereas the plate will be "picked clean."

3. The plate should immediately be placed at 37°C for 3–5 hr. By this stage, the colonies should again be visible as tiny opaque smears in the indentations. Do not let the plates overgrow.

4. Processing of the replica filters follows standard procedures: three sheets of Whatman (Clifton, NJ) 3MM filter paper are cut to fit each tray. Place the membrane colony side up on the wad of denaturing solution (0.5 *M* NaOH, 1.5 *M* NaCl) for 5 min, drain by briefly blotting on dry Whatman 3MM paper, and then place on the wad of neutralizing solution (3*M* sodium acetate, pH 5.5) for 5 min. (*Note:* Each sheet must be laid down separately, covered in solution, and air bubbles gently removed before the next sheet is placed. During denaturation, it is important to have the Whatman 3MM filter paper well soaked but not "swimming," as the DNA from the colonies on the membrane will smear, making subsequent alignment of autoradiographic signals with colonies difficult.)

5. After the membrane has been neutralized, rinse in 2× SSC. If many membranes are being processed, it is more efficient to stagger the procedures accordingly. Change the solutions after every two membranes. Let the membranes air dry for 30 min, wrap in Whatman 3MM, and bake at 80°C for 1 hr.

Hybridization of Membranes

The purpose of the hybridization step is to identify, for example, human clones on a mouse or yeast background. Because complex DNAs are being used as probes, hybridization times must be relatively long.

1. After the membranes have been baked, rinse them in 2× SSC, and place up to four membranes in a polythene bag with 14 ml per membrane of hybridization solution [5× Denhardt's, 5× SSC, 50 m*M* NaP_i (pH 6.5), 0.1% (w/v) sodium dodecyl sulfate (SDS), 250 μg/ml sonicated salmon sperm DNA, 10 μg/ml poly(A), 10 μg/ml poly (C), 50% (v/v) formamide]. Incubate with gentle rocking at 42°C for a minimum of 4 hr.

2. We routinely label 1 μg of probe DNA, using the random primer method (10) to achieve a specific activity of 1×10^9 disintegrations per minute (dpm)/μg. Purify

the radiolabeled DNA probe by chromatography on Sephadex G-50. The purified probe is boiled for 10 min prior to addition to hybridization solution (8 ml/membrane). Hybridization is for 72 hr at 42°C.

3. Wash the membranes at room temperature for 30 min in 1 × SSC, 0.1% (w/v) SDS with several changes, followed by two washes at 60°C in 0.1 × SSC, 0.1% (w/v) SDS. After washing, lay the membranes on Whatman 3MM filter paper to drain. Using a marker pen, mark on the opposite side of the original pencil mark the number of the filter, and the plate/membrane reference marks. Ensure that all excess moisture is removed before sealing the membranes in polythene bags.

4. Expose the membranes to Kodak (Rochester, NY) XAR5 film and intensifying screens at −70°C. Develop the autoradiographs after 3 days, and mark the autoradiographs with membrane number and reference marks.

Colony Picking

Colony picking can be difficult when attempting to line up a small number of positive signals obtained on the autoradiograph with many colonies. Parallax makes this difficult, errors are common, and the process is slow. To overcome this, we developed simple methods for identifying *all* colonies on the filter by staining and then marking the positive colonies on this master image. These can then be easily picked by virtue of their location relative to neighboring colonies.

1. Protein components left by colony debris on the membranes can be stained with Coomassie dye mix, thereby giving, on the membrane, the location of every colony lifted (11). The colonies stain a dark blue, although care is required to avoid a high background staining. One membrane at a time is placed colony side up in a large petri dish. Coomassie dye mix (1 g Coomassie Brilliant blue, 500 ml 2-propanol, 500 ml glacial acetic acid, 1.3 liters water) is added until it covers the membrane. Leave for approximately 5 min. Place the membrane in a plastic dish with water to rinse off excess stain and to reduce background. Dry the membranes completely; the background fades on drying.

2. Membranes are photocopied onto gridded transparencies. (Photocopy the appropriate grid on transparencies; we use a 4-cm grid.) The position of reference marks can easily be identified and highlighted with a green pen. This gives the location of all colonies with respect not only to each other, but also to the reference marks. Mark positives on the autoradiograph with a removable orange marker pen, lay the appropriate transparency over the marked autoradiography, and align the plate reference marks. Adjust the transparency so that the orange dots (i.e., autoradiograph positives) align to colonies. Then, using a red marker, mark on the transparency the colonies that align to the positives. Great care should be taken to avoid sampling bias because the positive clones obtained on the autoradiographs

will range from strong signals to weak signals. As detailed above, clones are thus identified not only by their relation to the key marks, but also to each other: both absolute and relative position are provided in this way.

3. Prior to picking, an Eppendorf repeat dispenser is used to aliquot 500 μl of LB + kanamycin (30 μg/ml) into the wells of a deep-well microtiter tray (Beckman). Picking of positive clones, using sterile toothpicks, is aided by a light box with the gridded transparency under the plate. Deep-well trays are capped with Micronics (Greiner Labortechnik, UK) cap bands, taped at an angle to a rotating shaker, and shaken overnight at 37°C.

Secondary Screening of Colony Arrays

Individual secondary screening of the colonies is to be avoided if the work load is to remain realistic. Consequently, rescreening of whole microtiter trays is carried out by making colony arrays. In essence, all picked clones are gridded out as an array from which replica filters are made. These are rehybridized to confirm the status of the clones. They form the basis of a reference filter set and are made using a replicator. Duplicate arrays are screened with, for example, total human DNA to identify YAC-derived cosmids and total yeast DNA to identify mispicked colonies: colonies hybridizing with both are frequently mixed or, more rarely, ligation artifacts.

It is at this stage that the power of the Biomek automated workstation is appreciated (12). All procedures can be performed manually using a multichannel pipettor, although the sheer repetitiveness of the task greatly increases the risks of both error and clone contamination; the procedures detailed below are for manual operation. Advice on the use of the Biomek workstation can be obtained directly from the authors.

1. Examine the deep-well plate and note which wells did not grow or produced poor growth.

2. Initially, the overnight cultures are used to make two −70°C stocks. The stock trays must be properly labeled. Use a marker pen to write the tray number, date, and stock number (1 or 2) along the top and on the side. The lids placed on the trays should also be labeled, as this enables rapid identification when retrieving trays from the −70°C freezer. Stock tray 1 is used for making colony arrays and for repicking clones from the −70°C stocks for further analyses (*Note:* Stock tray 2 should be used only in emergencies.) Fifty microliters of a sterile 50% glycerol, L-broth solution is dispensed into each well of two microtiter trays for each tray of clones grown. Culture (250 μl) is dispensed into a third microtiter tray for microprepping.

3. All colony array manipulations should be done in a laminar flow hood. Cut

the membranes to 10 × 12 cm and mark, in pencil, the date on the top right, and the colony array number and replica number on the bottom left. Also make cuts in the membrane: the corners are usually easiest. This is for ease of alignment during the array-making process. Many different methods for making replicas exist: we use a simple 96-pin device that is held in a frame that can be offset to 1 of 4 positions, allowing the spotting of 4 microtiter trays into the area normally occupied by 1 (so-called 2 × 2 arrays). Any similar apparatus is suitable.

4. The glycerol stock tray about to be replicated is placed on the replicator and the 96-prong device is dipped a couple of times into the stock tray, and then touched onto the membrane. The membrane should be wetted by this procedure. It is convenient to make four copies of each array: the transgenome "target" and "background" hybridizations can be done at the same time, leaving the other two copies for future use. The process is repeated to give four copies of a particular tray. Three new trays are then spotted, with the appropriate offsets, to generate finally, in our case, 4 sets of 4 × 96 colonies per filter. Between trays, the 96-prong device is sterilized twice by dipping in 95% (v/v) ethanol and flaming. After the colony arrays have been made, remove the membranes from the frames and place them on kanamycin (30 μg/ml) plates and incubate at 37°C overnight.

5. We use transparencies as templates. These are reference transparencies that have the location of all clones for a complete colony array of four trays. They are placed over the dried membrane such that all the colonies on the membrane align with their correct position on the transparency: needle holes are pierced through the membrane at appropriate locations on the transparency, enabling future colony alignments. These alignment marks are particularly important, as probes may detect only a single clone on the membrane.

6. The preparation of the colony array membranes for hybridization is identical to that for library membranes. After baking, cross-link the membranes using UV light for 30 sec. Rinse the membranes in 2× SSC prior to prehybridization. Hybridize one set of membranes with ^{32}P-labeled transgenome "target" DNA and one set with ^{32}P-labeled transgenome "background" DNA for 72 hr at 42°C.

7. The membranes are washed under the same wash conditions as the initial library screen, with the exception of the final wash, which is at 65°C. Autoradiography is against Kodak XAR5 film with intensifying screens at −70°C for 72 hr.

8. After probing, the membranes can be reused; bound probe is removed by soaking the membranes for 10 min in 0.2 *M* NaOH, followed by neutralization in 0.1 *M* phosphate buffer (pH 6.5), and a brief rinse in 2× SSC. The stripped membranes are left to dry before being stored.

Data Collection

The results from the "target" and "background" hybridizations for each tray are compared. Summary data can be compiled for each tray:

how many clones were target-only positive
how many clones were background-only positive
how many clones were target and background positive
and how many clones were neither target nor background positive; these did not hybridize for whatever reason, but grew on the colony array

The above results will indicate the efficiency of the picking of primary positives. Seventy percent target-only positives per tray is satisfactory, whereas anything lower than 60% for a full tray is inefficient both in terms of time and resources. Once the array results have been collected, the clones are assigned consecutive numbers that act as identifiers for the subsequent fingerprinting and contig assembly. We currently include both target-only and target- and background-positive clones. Clones that did not give a definitive array result (even though they did grow) are given a number "*n*" and are included. Clones that are either background only or did not grow on the array are ignored. (The computerized version of data collection, using a flatbed digitizer and DOT, has increased both the accuracy and the speed of this part of the work.) The 2 × 2 arrays of clones on filters can subsequently be used as reference sets for the identification of specific DNA sequences within the otherwise anonymous cosmid clones.

Cosmid Clone Analysis

The subsequent stages of the overall procedure involve the preparation of DNA, fingerprinting of DNA samples, and contig assembly. As for the methodologies described above, we routinely use the Biomek workstation and microtiter trays, although these procedures can be carried out manually either in Eppendorf tubes (4) or in microtiter trays. The manual method for microtiter trays is detailed below. The procedure uses the 250-μl sample set aside at the time of secondary screening (described above).

Micropreparation of DNAs

1. Cover the tray with Saran Wrap and spin at 2500 rpm (930 *g*) in a benchtop centrifuge for 2 min. [We use a Du Pont (Wilmington, DE) T6000B centrifuge, H1000(B) rotor with microtiter plate carriers.] Discard the supernatant and drain the microtiter plate on absorbent tissues.

2. First add 25 μl of solution I [50 m*M* glucose, 10 m*M* EDTA (pH 8.0), 25 m*M* Tris (pH 8.0)]. Tap the side of the tray gently to resuspend the pellets. Add 25 μl freshly made solution II [0.2 *M* NaOH, 1% (w/v) SDS]. Mix by tapping the tray; leave for 5 min, by which time the solution will have cleared. Finally, add

25 μl of cold solution III (3 *M* sodium acetate, pH 5.5). Cap, and invert several times until a precipitate has formed. Spin at 3500 rpm for 10 min at 4°C.

3. While the plates are spinning, dispense 100 μl of 2-propanol into another microtiter tray. Transfer 70 μl of the resulting supernatant into the tray containing 2-propanol, cap, and mix thoroughly.

4. Precipitation of nucleic acid is achieved by placing the tray at −20°C for 1 hr, at which stage the tray is spun down again at 3500 rpm for 10 min at 4°C. The supernatant is discarded and the plate drained on tissues. Sterile water (25 μl) is added, the pellet resuspended, and 25 μl of cold 4.4 *M* LiCl added.

5. Cap the wells, mix thoroughly, and leave at 4°C for at least 1 hr. Spin the plate at 3500 rpm for 10 min at 4°C to pellet debris; while the tray is spinning, dispense 100 μl of 2-propanol into another tray. Transfer the entire supernatant (50 μl) to the 2-propanol tray. This tray is then placed at −20°C for a minimum of 1 hr.

6. When the tray is ready to be fingerprinted, spin down the pellet at 3500 rpm, discard the 2-propanol, gently add 150 μl of 70% ethanol, and drain well. Resuspend the pellets in 10 μl of TE (pH 7.4).

Fingerprinting

Once the target or background status of the clones has been determined, fingerprinting of target clones can commence (13, 14). In summary, cosmid DNAs are restricted with *Hin*dIII, termini labeled with [^{32}P] dATP using reverse transcriptase, restricted with *Sau*3AI, the products denatured and run on 4% (w/v) acrylamide gels, and fragments detected by autoradiography. Fingerprinting reactions are also carried out in microtiter trays although, once again, this can be done manually in Eppendorf tubes if numbers permit.

1. Although a proportion of clones will be categorized as background clones, it is commonly more practical to fingerprint the entire tray. The discrimination between target and background clones can be made later when analyzing the products by electrophoresis. Fingerprint reaction mix sufficient for 1 full microtiter tray (96 samples) is made up in the following manner:

M buffer (10×)	40.0 μl
ddGTP (1 m*M*)	5.0 μl
Sterile double-distilled water	160.0 μl
Avian myoloblastosis virus	
(AMV) reverse transcriptase (10 U/μl)	4.0 μl
*Hin*dIII (10 U/μl)	4.0 μl
[^{32}P]dATP	8.0 μl

M buffer is 50 mM NaCl, 10 mM Tris-HCl (pH 7.5), 10 mM $MgCl_2$, 1 mM dithiothreitol (DTT).

2. Transfer 2 μl of the resuspended DNA to the fingerprint tray. After the 2 μl of DNA has been dispensed, keep the tray covered and on ice.

3. Transfer 2 μl of the reaction mix to each well in the tray, using a Hamilton syringe. The reaction mixture is added to the side of the unmarked wells, just away from the 2 μl of DNA. This should be done on ice.

4. Cover the tray with Saran Wrap (not Parafilm) and place a lid over the tray. Briefly centrifuge the tray, tap the tray a few times, and respin. Incubate the tray at 37°C for 45 min, followed by incubation at 68°C for 25 min to inactivate the reverse transcriptase and *Hin*dIII. Place the tray on ice to cool for 5 min, and spin down any condensation that has formed.

5. Add 2 μl of the *Sau*3AI mixture:

M buffer (10×)	22.5 μl
Sterile double-distilled water	197.0 μl
*Sau*3AI (50 U/μl)	5.7 μl

to the bottom side of the well, using a Hamilton dispenser. Spin briefly, and incubate at 37°C for 2 hr.

6. After incubation, add 4 μl of formamide stop dye [98% formamide, 0.03% (w/v) bromphenol blue, 0.03% (w/v) xylene cyanol, 10 mM EDTA (pH 8.0)]. Spin again and store at −20°C.

Preparation of λ Marker DNAs

Bacteriophage λ DNA restricted with *Sau*3AI and subsequently end labeled with [^{35}S]dATP is used as a molecular weight marker for gel electrophoresis. As the entire procedure of data digitization and analysis is dependent on the markers, their quality is critical.

1. Digest 10 μg of λ DNA with *Sau*3AI as follows:

λ DNA (10 μg)	25.0 μl
M buffer (10×)	50.0 μl
Sterile double-distilled water	415.0 μl
*Sau*3AI (50 U/μl)	10.0 μl

The reaction is incubated for 1 hr at 37°C and then radioactively labeled.

2. Label the DNA in the following reaction:

*Sau*3AI-cut λ DNA	43.5 μl
M buffer (10×)	1.1 μl
dGTP (10 m*M*)	2.0 μl
ddTTP (10 m*M*)	2.5 μl
[^{35}S]dATP	4.0 μl
AMV reverse transcriptase (10 U/μl)	1.0 μl

The labeling reaction is performed at 37°C for 30 min, at which point 55 μl of formamide stop dye mix is added.

Fingerprint Electrophoresis

Electrophoresis of fingerprints is a critical step in the analysis: the fingerprints will be analyzed by computer and this requires rigid adherence to certain procedures. In particular, the bacteriophage λ DNA standards are used by the computer programs to normalize for gel-to-gel electrophoretic variation: without high-quality markers (no partials, clean bands, no gel artifacts), the data on any individual gel are unusable. Accordingly, gels are bonded to one of the glass plates, resulting in the high dimensional stability that is required for absolute migration distances to be accurate (in marked contrast to sequencing gels). We cannot overstress the importance of both marker and gel quality.

1. We use standard Bethesda Research Laboratories (Gaithersburg, MD) sequencing gel tanks (model S2) for the electrophoretic analysis of fingerprinting reactions. Prepare the gel plates as follows: siliconize the larger plate by washing it first with detergent, drying, and then wiping over the ethanol. Apply the siliconizing agent (dimethyldichlorosilane) to the plate, spread evenly, and leave for 10 min. Rinse the plate with water and wipe dry.
2. Bonding of gel to plate: Care is required when applying the bonding solution to the smaller gel plate, as "overbonding" will result in the gel sticking to both plates and splitting when the gel is processed. The smaller plate is washed well with detergent, dried, wiped with ethanol, and then wiped with the bonding solution. [Bonding solution consists of 3 ml ethanol, 50 μl 10% acetic acid, 2.5 μl of WACKER silicone solution (Wacker-Chemie GmbH, Munich, Germany)]. Leave the plate for 1 min after applying the bonding solution, then wash the plate with ethanol. Tape the plates as normal.
3. The acrylamide used for the gel is a 19:1 acrylamide–bisacrylamide 40% stock (w/v) that has been filtered (0.45 μm). For each gel, dissolve 42 g of urea in 45 ml if distilled water plus 10 ml of 10× TBE [108 g Tris base, 55 g boric acid, 40 ml 0.5 *M* EDTA (pH 8.0) per liter]. When the urea has dissolved, add 10 ml of

the 40% acrylamide stock, 800 μl of 10% ammonium persulfate, and 80 μl of *N*, *N*, *N'*,*N'*-tetramethylethylenediamine (TEMED). Mix well for 10 sec and pour the gel. After pouring, insert the comb and clamp the plates around the comb. Leave the gel to set for 1 hr in a horizontal position. Gels can be prepared the day before they are used and wrapped in Saran Wrap to prevent drying. The comb must have lanes 3 mm wide with an interlane separation of 2 mm. The comb thickness is 0.4 mm. There must be uniform spacing across the loaded tracks (two smaller combs cannot be used instead of a single full-width one). Shark's-tooth combs also cannot be used (the software cannot identify lanes if there is no gap between them).

4. Prior to electrophoresis, the combs are removed and sealing tape stripping off the bottom of the plates to enable the migration of the bromphenol dye to be observed. TBE (1 ×) is used for electrophoresis, and the wells are carefully flushed prior to sample loading. Prerunning the gel does not appear to be necessary. Denature the fingerprint reactions at 80°C for 10 min. Markers are denatured by boiling for 5 min. We flush the urea that has leached out of each set of six wells before loading the samples (3.0 μl). The markers (3.0 μl) are then loaded, every seventh lane, eight to a gel. Normally, 42 fingerprints are run per gel. The order of samples to be run on the gel is critical to final data analysis. The first track *must* be a marker followed by six samples and then a marker followed by six samples, and so on across the gel, ending with a marker track.

5. Gels are run at constant power (74 W) until the bromphenol blue dye front is 1 cm from the bottom of the gel. Gel plates are carefully separated by removing the tape and prising the plates apart at one corner with a scalpel. The gel sticks to the bonded plate, which is first placed in a large tray containing 1 liter of 10% acetic acid for 15 min and subsequently in water for 20 min prior to drying on the surface of a hot gel drier for 45 min.

6. Wrap the dried gel (while it is still hot) in Saran Wrap, squeezing out the air bubbles with a tissue. Clean another glass plate. Place this over the gel plate and hinge it on one side with tape. Place a sheet of Kodak XAR5 film between the plates with the film overhanging the wells. Expose for 1–3 days at room temperature in an envelope, taking care not to disturb the plates. When developing the autoradiograph, be careful to mark the film on one corner to ensure that the orientation of the sample loading is known.

Computerized Data Entry

Introduction

The computerized analysis of fingerprints and their subsequent assembly into contigs relies on a critical criterion: that all data can be directly compared. This is achieved by a process of normalization. In outline, the fingerprint autoradiographs are scanned and the position of fragments detected by alterations in the optical density of the film. The scanner generates an image of the autoradiograph as a whole. The software attempts to find lanes on the basis of the variations in densities (bands) within this image. Assuming that the markers are of sufficiently high quality, the first and last lanes are identified, both known to be markers, and every seventh track similarly identified as a marker track. Software then attempts to align ("lock") an internal image of a standardized marker set on the experimental markers, that is, those on the autoradiograph. The expansion or contraction of the experimental markers compared to the standard set is calculated, enabling a direct measure of gel mobility compared to the ideal standardized marker set. An image of the gel with locked markers and sample bands indicated on it is sent to the terminal screen. At this stage, manual editing of the clone fingerprint image is undertaken, whereby any false or partial bands detected by the scanner software or bands derived wholly from the vector can be removed. The editing facility not only enables the user to monitor the entire normalization process manually, but also allows the entry of suboptimal autoradiographic data, as the software is able to compensate for "smiling" gels. When editing is complete, the only information that is stored is the name of the clone in the appropriate sample well, the number of the gel, the number of fragments it contained, and their normalized mobility. In this fashion, migration distances of clones analyzed on different gels can be readily compared to within 0.7 mm.

Contig Assembly

We do not give full details here of how we operate the image analysis programs or of how they work. A full working copy is available on request, along with detailed operating instructions. Analysis of DNA fingerprints involves a sophisticated and specialized set of software packages that were originally developed by Sulston *et al.* (3, 15). B. Loos (Massachusetts General Hospital, Boston, MA) has rewritten the display and editing programs to run on a Vaxstation GPS2. We have modified these further to run on a Vaxstation 3100 running VWS. We use an Amersham autoreader to scan the autoradiographs and output the digitized film images to an Elonex 386 (a 33-MHz IBM clone) running Interactive UNIX. Images are either stored on the Elonex hard disk for batch transfer (multireads mode), or transferred immediately (single-reads mode) to the Vaxstation. In both cases, ethernet and NFS software are used for data transfer. A background program, WAITER, running on the Vaxstation, monitors the incoming files and automatically initiates a batch job to run the track, band, and marker finding and aligning programs (3, 15). Default threshold values for band finding are specified in WAITER and numbers of samples on the digitized autoradiograph are defined by the file name of the incoming data file. The programs expect the first track on a gel to be a marker, followed by a defined number of samples, in our case six, and then another marker followed by six samples, and so on.

Data processing is reasonably fast on this system: total processing time is 12.5 min. Multiple film images can be transferred to the Vaxstation while earlier images files are being processed, although only a single image is processed by WAITER at a time. Each gel is then edited using the mouse-driven editor to accept or reject bands as described above, prior to entry into the clone database.

Data Entry

1. When the autoradiograph has been developed, check the quality of the markers. Imperfect markers require the entire set of samples to be rerun.
2. Choose the exposure considered best for entering: clear markers and good fingerprint bands. In our experience, larger fragments are often weakly labeled, whereas smaller fragments may be so close together that the signals merge. The merged signals will be digitized by the scanner as single band instead as two. A compromise is required; when editing the scanned image (see below), fragments can only be confirmed or deleted: it is *not* possible to add new fragments.
3. Draw a line on the autoradiograph over the origin of the samples. Feed the gel through the scanner, and the image analysis programs will be run on the Vaxstation. Once the film has been scanned and the image generated by the software, the lock of the markers must be carefully examined.
4. The lock *must* be perfect before proceeding with entry of clones into the database. Attaining this is dependent on the quality of the markers: not only their preparation, but also the quality of the autoradiograph. Although it is possible to improve the chances of markers locking by altering the scanning threshold levels, many problems can be avoided by attention to marker quality. If a gel has failed the first time, use a range of thresholds to obtain the best lock.
5. Once the data are acceptable, the appropriate records are entered into a database. Clones are then compared using the programs detailed by Sulston *et al.* (3, 15).

The overall aim of this kind of project is to join clones into overlapping clusters or contigs and then to join contigs to each other. Thus, any new clone will do one of four things: (1) remain unattached, (2) form new contigs; (3) be buried into established contigs; or (4) extend already established contigs.

It is also important to keep a good record of what is happening in the database. After each new set of data has been analyzed, make a record of the following: number of new clones entered, number of new contigs, number of old contigs extended, number of clones that are buried into old contigs, and number of new clones entered that are unattached.

The above records will provide a continuous record of the project and will give updated information on how the project is progressing. In the early stages one expects a higher proportion of unattached clones. In the middle of the project we would expect a higher proportion of clones to extend contigs and to join unmatched clones into new contigs. At a later date we expect the number of unattached clones to be reduced, and the number of new clones that are buried into contigs to increase dramatically.

Conclusion

In this article we have detailed the construction of cosmid libraries for large-scale analysis of transgenomes. The cosmids isolated by these methods can be used either as resources for further analysis without any attempts to establish their relationships directly (reference clones), or they can be aligned to make a cloned DNA map. This latter area has been only briefly discussed, and further information can be obtained by contacting the authors directly. A large project that relies on fingerprint analysis is not a trivial proposition to set up successfully; fortunately, a considerable body of experience now exists both in our own laboratory and those of fellow genome mappers.

Acknowledgments

Much of the work detailed in this article was supported by the Cancer Research Campaign and by special project grants of the UK Human Genome Mapping Project-directed program of the UK Medical Research Council.

References

1. H. Lehrach, D. Drmanac, J. Hoheisel, Z. Larin, G. Lennon, T. Monaco, D. Nizetic, G. Zehetner, and A.-M. Poustka, *in* "Genome Analysis. Vol. I: Genetic and Physical Mapping:" (K. E. Davies and S. H. Tilghman, eds.), p. 39. Cold Spring Harbor Lab. Press, Cold Spring Harbor, New York,
2. A. Coulson, J. Sulston, S. Brenner, and J. Kahn, *Proc. Natl. Acad. Sci. U.S.A.* **83**, 7821 (1986).
3. J. Sulston, F. Mallet, R. Staden, R. Durbin, T. Horsnell, and A. Coulson, *CABIOS* **4**, 125 (1988).
4. J. Sambrook, E. F. Fritsch, and T. Maniatis, "Molecular Cloning: A Laboratory Manual," 2nd Ed. Cold Spring Harbor Lab. Press, Cold Spring Harbor, New York, 1989.
5. S. H. Cross and P. F. R. Little, *Gene* **49**, 9 (1986).
6. P. F. R. Little, *in* "Gene Cloning" (D. M. Glover, ed.), "Vol. III, p. 19. IRL Press, Oxford, 1987.
7. I. J. J. P. Heding, A. C. Ivens, J. Wilson, M. Strivens, S. Gregory, J. M. N. Hoovers,

M. Mannens, B. Redeker, D. J. Porteous, V. van Heyningen, and P. F. R. Little, *Genomics* **13**, 89 (1991).

8. C. L. Smith, S. R. Klco, and C. R. Cantor, *in* "Genome Analysis: A Practical Approach" (K. E. Davies, ed.), p. 41. IRL Press, Oxford, 1988.
9. B. Seed, R. C. Parker, and N. Davidson, *Gene* **10**, 249 (1982).
10. A. P. Feinberg and B. Vogelstein, *Anal. Biochem.* **132**, 6 (1983).
11. I. J. J. P. Heding and P. F. R. Little, *Technique* **2**, 218 (1990).
12. I. J. J. P. Heding, M. Strivens, S. Gregory, A. C. Ivens, and P. F. R. Little, *Technique* **3**, 129 (1991).
13. A. Coulson and J. Sulston, *in* "Genome Mapping: A Practical Approach" (K. E. Davies, ed.), p. 19. IRL Press, Oxford, 1988.
14. K. J. Harrison-Lavoie, R. M. John, D. J. Porteous, and P. F. R. Little, *Genomics* **5**, 501 (1989).
15. J. Sulston, F. Mallet, R. Staden, R. Durbin, and T. Horsnell, *CABIOS* **5**, 101 (1989).

[11] Physical Mapping of Human Chromosomes: Generation of Chromosome-Specific Sequence-Tagged Sites

Eric D. Green

Introduction

The process of constructing a complete physical map of the human genome is an enormous challenge that is currently being faced as part of the Human Genome Project. One common strategy used to establish physical maps of DNA involves the isolation of the genomic region of interest in cloned form and the localization of landmarks within the recombinant clones. For historic and practical reasons, the human genome has (in most instances) been divided into smaller portions, typically individual chromosomes or parts thereof, with each portion being examined in detail by one or more laboratories. Thus, reliable methods for developing chromosome-specific DNA landmarks and/or clones have become essential for many physical mapping projects, especially those aiming to develop long-range maps of whole chromosomes.

An important unifying principle that has emerged during the early phase of the Human Genome Project is the utilization of sequence-tagged sites (or STSs) as the landmarks on which to base the chromosome maps (1). An STS is defined as a short tract of DNA that can be specifically detected by a corresponding polymerase chain reaction (PCR) assay. Thus, the physical landmark is the DNA sequence itself, with PCR representing the method used to detect the landmark. Sequence-tagged sites offer at least two important advantages. First, the use of PCR as the front-line tool for detecting physical landmarks in cloned DNA is highly desirable because of its high sensitivity, specificity, and potential for automation. Second, the defining features of an STS can be reduced to information (primer sequences, STS size, and PCR conditions) that can be stored in a computer database and made readily available to any investigator. By synthesizing the primers and performing PCR under the described conditions, any STS can thus be "transported" from one laboratory to another without any exchange of materials, as demonstrated by Green *et al.* (2). The ability to access readily the DNA landmarks on which the physical maps are based, without the expense, inconvenience, and error-prone nature of storing and distributing recombinant DNA clones, represents a major rationale for using STSs as the "common language" for the large-scale physical mapping of human DNA (1).

Our approach to the establishment of STS-based physical maps of human DNA

Methods in Molecular Genetics, Volume 1

has emphasized the use of yeast artificial chromosomes (YACs) as the source of the cloned DNA. The YAC cloning system represents a powerful tool for obtaining large (100- to >1000-kb) segments of DNA in cloned form (3, 4). In our paradigm for mapping human chromosomes, STS-specific PCR assays are used to isolate (5) and characterize YACs (6), with the resulting data providing the relative "content" of STSs within YAC clones. Overlap relationships among the YACs can then be deduced based on the sharing of common STSs among the different clones. This information can in turn be used to construct "STS-content maps" that reflect the relative locations of STSs within intervals of cloned DNA (6, 7). A detailed review of the theory and practice of constructing YAC-based STS-content maps of human chromosomes has been published (7).

Intrinsic to an STS-content mapping project on the scale of a whole human chromosome is the need to develop significant numbers of PCR assays, each amplifying a specific STS. For example, we are developing a YAC-based STS-content map of human chromosome 7, which contains roughly 160 Mb of DNA. Based on computer simulations (7), 1000 or more STSs will need to be generated in order to assemble YAC contigs (sets of overlapping clones) averaging 2 Mb in size that cover most of the chromosome. Given the paucity of preexisting, sequenced DNA markers and their nonrandom distribution, virtually all of the STSs utilized in this endeavor will be most efficiently derived from random (anonymous) segments of DNA. Here are described several of the approaches we have implemented for generating random STSs from a specific (targeted) region of the human genome, such as that of a whole chromosome. These strategies have thus far been used to generate over 500 chromosome 7-specific STSs (2; unpublished observations, 1993).

Of note, most of the discussion that follows is geared toward the mapping of a whole human chromosome; however, the same approaches can be readily adapted for mapping more limited subchromosomal regions. For example, rodent cell lines containing a single human chromosome can be irradiated and used to create a set of new cell lines (8). Each of the resulting "radiation hybrid" cell lines typically contains one or more smaller segments of the initial human chromosome and can serve as a more limited source of human DNA for STS development (8, 9). Similarly, microdissection of human chromosomes (10, 11) can be used to limit the target DNA region used for STS generation to a defined subchromosomal region (discussed in the next section).

Overview of Sequence-Tagged Site Generation

The process of generating chromosome-specific STSs can be broadly divided into the steps shown in Fig. 1. The first step, generating chromosome-specific DNA templates, is the key aspect of the overall scheme. The strategies utilized for this step are actually quite similar to those employed for the development of chromo-

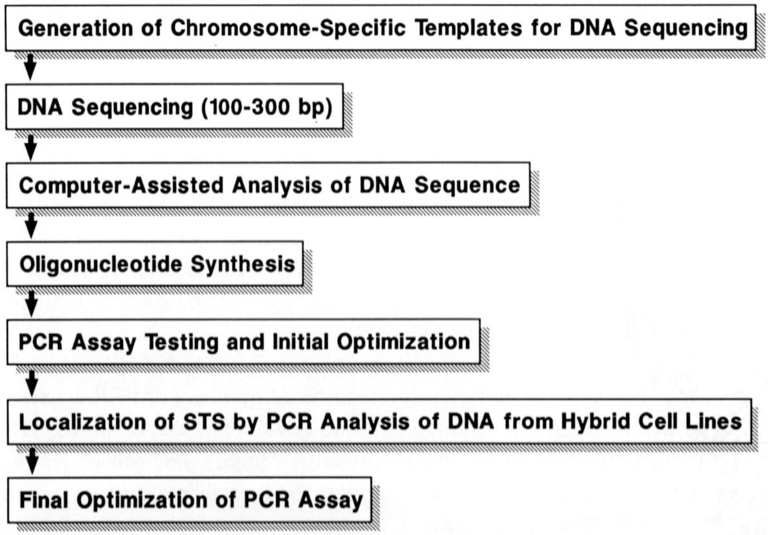

FIG. 1 Overview of the sequential steps in the generation of PCR assays corresponding to chromosome-specific STSs. Details for each step are provided in the text.

some-specific hybridization probes. However, for STS generation the DNA segments are isolated in a form suitable for sequencing rather than for testing as a hybridization probe.

Several approaches for generating chromosome-specific preparations of DNA have been described. A major group of these utilizes somatic hybrid cell lines, such as those made with human and rodent cells. Such cell lines can provide a valuable source of human DNA from a targeted region of the genome (e.g., a single chromosome or portion thereof). The advantage of hybrid cell lines is that the human DNA of interest is essentially isolated from the remainder of the genome; however, this DNA remains within a background of total rodent DNA (which has roughly the same complexity as total human DNA). The relative amount of human DNA in a hybrid cell line depends on the size of the chromosome of interest, the percentage of cells in the population that actually contain human DNA, and the copy number of the endogenous rodent chromosomes. Of note, some human–rodent hybrid cell lines are tetraploid with respect to the rodent chromosomes, which results in a proportional decrease in the relative amount of human DNA in the cells. For example, ~3% of the DNA within a human chromosome 7-containing hybrid cell line would be of human origin (assuming that all of the cells contain one copy of human chromosome 7 in an otherwise diploid genome). The challenge is to employ methods that, in effect, "extract" the human DNA from the rodent DNA background. The two approaches described below both rely on the ubiquitous presence of repetitive sequences (e.g., *Alu*) in human DNA. These repetitive sequences can be exploited by various means to isolate segments of human DNA from the remainder of the cell line DNA.

Another important source of chromosome-specific DNA segments consists of the insert ends of genomic clones, particularly YACs derived from a chromosome of interest. Such YACs can either be isolated from a total genomic library using a chromosome-specific STS (or probe) or from a library constructed with DNA from an appropriate hybrid cell line. A reliable protocol for isolating the insert ends of YACs is provided below, along with a discussion about the potential pitfalls of such an approach for generating chromosome-specific STSs.

Other strategies for developing chromosome-specific DNA preparations have also been described. One such approach employs flow-sorting technology to "enrich" for DNA from a particular chromosome (10, 12). Unfortunately, the extent of enrichment varies depending on the chromosome of interest and the cells being utilized as the source of DNA (e.g., human vs hybrid cell lines), with unacceptably high amounts of irrelevant DNA segments obtained in many instances. Furthermore, this approach requires specialized equipment that is not widely available to many investigators. Another approach for obtaining DNA from specific genomic regions utilizes the technique of microdissection (10, 11). In this method, cytogenetically distinct subchromosomal regions are physically excised and isolated from the remaining genomic DNA. This material is then amplified by PCR to generate sufficient amounts of DNA for subcloning and subsequent analysis. The desirable features of this method include the use of human cell lines (which tend to be more karyotypically stable than hybrid cell lines) and the ability to generate highly targeted DNA preparations from subchromosomal regions. However, like flow sorting, this approach relies on equipment that is not widely available. Finally, a variant source of targeted DNA templates is represented by sequences contained in expressed genes. In this strategy, random clones from a tissue-specific cDNA library are sequenced, and particular portions of the resulting sequence (e.g., the 3′ untranslated regions) are used to develop gene-specific STSs (13), which in turn are mapped to individual chromosomes. The application of this approach to whole chromosome mapping projects is hampered by the difficulty in selecting cDNA clones that originate from a particular region of the genome. Although flow sorting, microdissection, and cDNA cloning methods represent viable strategies for generating STSs, they are not discussed further here.

Isolation of Chromosome-Specific DNA Sequences by Subcloning

One approach for isolating the human DNA contained in human–rodent hybrid cell lines involves subcloning the DNA of a hybrid cell line into an appropriate vector and identifying the subclones that contain human-specific repetitive sequences (14),

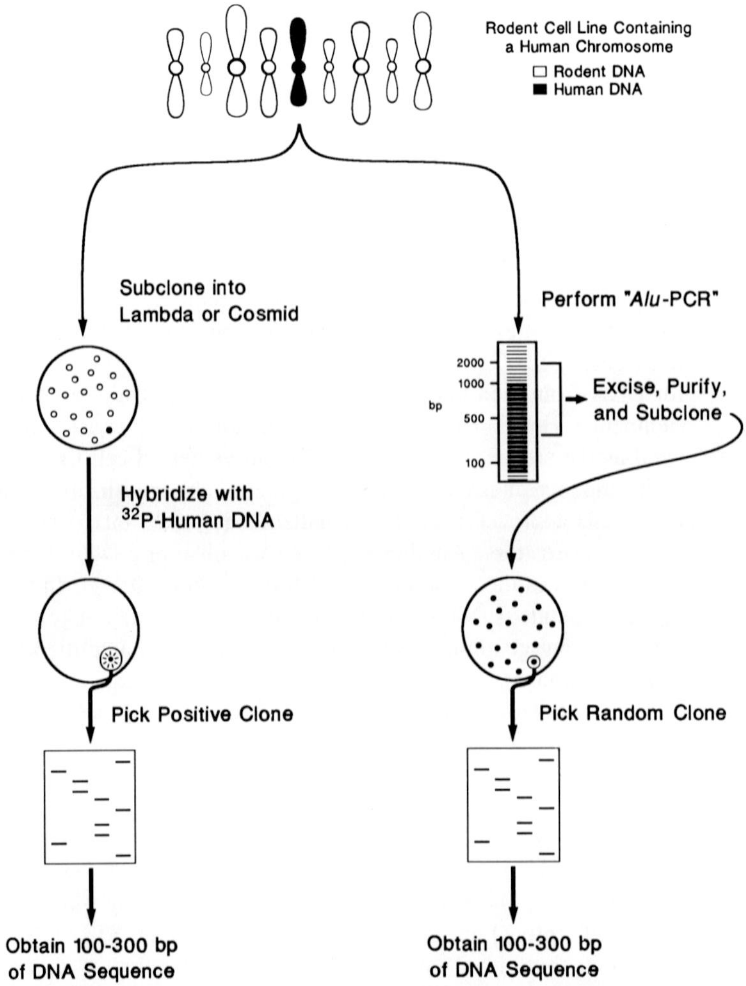

FIG. 2 Strategies for generating chromosome-specific DNA sequences from human–rodent hybrid cell lines. The two approaches described in the text for using human–rodent hybrid cell lines to develop random, human chromosome-specific DNA sequences are depicted. In the strategy on the left, DNA from a human–rodent hybrid cell line containing the chromosome of interest is subcloned into a bacteriophage λ or cosmid vector. The resulting plaques or colonies are hybridized with radiolabeled total human DNA, hybridizing (positive) clones are selected, and 100–300 bp of DNA sequence of the cloned insert is determined. In the strategy on the right, *Alu*-PCR is used to amplify a heterogeneous population of human DNA fragments, which in turn are gel purified and subcloned. Random subclones are selected, and 100–300 bp of DNA sequence of the cloned insert is determined. For both strategies, the resulting DNA sequence is used in the scheme shown in Fig. 1 to develop a corresponding STS-specific PCR assay.

as illustrated in Fig. 2. To ensure recognition of a high fraction of the subclones containing human DNA, large-insert bacteriophage λ (15–25 kb) or cosmid (45 kb) cloning systems should be employed, because DNA inserts of this size typically contain sufficient repetitive DNA (e.g., *Alu* sequences) to allow their detection by hybridization with radioactive probes generated from total human DNA (15). Once isolated, such human DNA-containing subclones provide an excellent source of randomly located DNA sequence.

Generation of Human DNA-Containing Subclones

1. Purify total genomic DNA from a human–rodent hybrid cell line of interest, using standard methods (16). Because all or some of the human DNA can be deleted and/or rearranged within hybrid cell lines during prolonged growth, it is advisable to assess the human DNA content (e.g., by karyotype analysis) of cell populations used for preparing DNA.

2. Partially digest the purified DNA with an appropriate restriction enzyme, size fractionate the DNA (typically by pulsed-field gel electrophoresis), and subclone the resulting DNA into either a large-insert bacteriophage λ (15–25 kb) or cosmid vector. The particular restriction enzyme and size range used for preparing the insert DNA is dictated by the particular cloning system employed. The choice between λ or cosmid cloning vectors can largely be based on personal preference and familiarity. We have had good success in cloning *Sau*3AI-digested DNA into the *Bam*HI site of the λ vector EMBL3 (Stratagene, La Jolla, CA).

3. Grow recombinant subclones on solid medium, and immobilize DNA from λ plaques or cosmid colonies on nitrocellulose or nylon membranes (16). The phage or bacteria should be plated at relatively low densities (200–400 plaques or colonies per 150-mm diameter petri plate) to avoid the need for any additional purification of positive subclones.

4. Hybridize the membranes with radiolabeled total human DNA. Although other hybridization probes (e.g., Cot-1 DNA, cloned *Alu* segments) can be employed to identify human repetitive sequences, we have found total human DNA to provide excellent discrimination between human DNA- and rodent DNA-containing subclones. Similarly, a number of hybridization solutions and washing conditions can be utilized for this purpose. We routinely hybridize membranes overnight at 65°C in 10% (w/v) dextran sulfate, 1 *M* NaCl, 1% (w/v) sodium dodecyl sulfate and wash the membranes twice in 2× SSC [20× SSC is 3 *M* NaCl, 0.3 *M* trisodium citrate dihydrate (pH 7.0) (16)] for 10 min at room temperature followed by twice in 0.5× SSC, 1% sodium dodecyl sulfate, for 30 min at 65°C.

5. Following detection by autoradiography, isolate all positive (hybridizing) subclones, prepare duplicate membranes containing lysed plaques or colonies, and hybridize each membrane with either radiolabeled total human DNA or radiolabeled total rodent DNA (hamster or mouse, depending on the background genome of the

hybrid cell line), as above. The subclones that hybridize to human but not rodent DNA are used for additional analysis. The fraction of positive subclones will depend on both the characteristics of the target DNA of interest and the particular hybrid cell line utilized. For example, ~3% of λ subclones derived from a hybrid cell line containing a single copy of human chromosome 7 (in >90% of cells examined) were found to contain human DNA (2).

DNA Sequencing of Subclones

Once human DNA-containing λ or cosmid subclones are available, a number of different approaches can be employed to derive a small amount (100–300 bp) of DNA sequence from each insert. The choice of strategy is largely dependent on the type of sequencing template that is preferred and the general sequencing method being utilized (e.g., radioactive- vs flourescent-based sequencing). One approach is to sequence directly purified λ (17) or cosmid (16) DNA using oligonucleotide primers specific to vector sequences. Alternatively, one (or both) vector–insert junction of a subclone can be amplified using a ligation-mediated ("one-sided") PCR technique and the resulting PCR product sequenced either directly or after subcloning into a plasmid or M13 vector (2). A third option is to digest purified λ or cosmid DNA with an appropriate restriction enzyme, subclone a random, gel-purified restriction fragment into a plasmid or M13 vector, and obtain small amounts of DNA sequence from the resulting subclone using a vector-specific oligonucleotide primer(s).

Isolation of Chromosome-Specific DNA Sequences by *Alu*-Polymerase Chain Reaction

Another approach for recovering human DNA segments from hybrid cell lines utilizes the technique of interspersed repetitive sequence PCR (IRS-PCR), specifically *Alu*-PCR (18, 19), as illustrated in Fig. 2. In this strategy, total DNA from the hybrid cell line is used as template in PCR reactions containing oligonucleotide primers that are specific to consensus *Alu* sequences and that point outward from the repeat units. The resulting PCR products are thus enriched for DNA segments that reside immediately adjacent to *Alu* motifs. Typically, one or two *Alu*-specific oligonucleotide primers are employed, and PCR conditions are selected to ensure virtually no amplification of rodent DNA. Following PCR, the amplified DNA fragments are cloned into a suitable vector, and individual subclones are randomly selected and sequenced. This approach is relatively simple and rapid to perform. However, only a subset of the human DNA sequences is recovered from the hybrid cell line, and the DNA segments themselves are relatively short (i.e., intervals that can be amplified by PCR).

Alu-Polymerase Chain Reaction Amplification and Subcloning of Polymerase Chain Reaction Products

1. Purify total genomic DNA from a human–rodent hybrid cell line of interest using standard methods (16). As discussed above, the content of human DNA in the final cell preparation should be assessed to ensure the absence of any gross karyotypic changes.

2. Perform PCR using one or two *Alu*-specific oligonucleotide primers and purified hybrid cell line DNA as template. A number of different oligonucleotide primers for performing *Alu*-PCR have been reported (reviewed in Ref. 19). The choice of primer(s) and PCR conditions employed should be carefully determined for each experiment. Test the DNA from the hybrid cell line and the parental rodent cell line in parallel and analyze the PCR products on a 1–2% (w/v) agarose gel. Ideally, a heterogeneous smear of DNA (ranging from 100 to >2000 bp) is obtained following amplification of hybrid cell line DNA, with virtually no detectable DNA amplified from the rodent DNA. We have been particularly pleased with the results obtained using the oligonucleotide primers (ALE34 and ALE1) and PCR conditions described by Cole *et al.* (9). However, other reported *Alu*-PCR assays (19) could undoubtedly be used for this purpose as well.

3. Following preparative PCR amplification of the hybrid cell line DNA, using established conditions, separate the PCR products on a 1–2% (w/v) agarose gel and excise DNA segments in the 300 to ~2000-bp size range. Isolation of PCR products smaller than ~300 bp should be avoided, because only a small amount of useful (nonrepetitive) DNA sequence can be derived from such fragments due to the presence of residual *Alu*-specific sequences.

4. Purify the above agarose-embedded DNA [e.g., using Geneclean (Bio 101, San Diego, CA)], subclone the purified PCR products into a plasmid (or M13) vector containing a blue/white selection system [e.g., pBluescript (Stratagene)], and pick individual recombinant (i.e., white) subclones for additional analysis. Many strategies for subcloning PCR products are available. For example, we have successfully utilized the method of Padgett *et al.* (20) to clone *Alu*-PCR products into pBluescript.

DNA Sequencing of *Alu*-Polymerase Chain Reaction Products

Provided with a cloned *Alu*-PCR product, two approaches can be readily employed to derive a small amount (100–300 bp) of DNA sequence from the insert. One strategy is to purify plasmid (or M13) DNA from individual subclones and sequence the purified DNA using vector-specific oligonucleotide primers. A simpler alternative is to amplify the cloned insert by PCR using bacterial cells (or phage particles) and vector-specific oligonucleotide primers that flank the cloning site (21). We have achieved excellent results using the following protocol.

Incubate a mixture of 2 μl of bacterial culture (containing a pBluescript transformant) and 3 μl of water at 95°C for 5 min. Add the appropriate buffer [10 m*M* Tris-HCl (pH 8.3), 50 m*M* KCl, 2.5 m*M* $MgCl_2$ final concentrations], dNTPs (200 μM final concentration of dGTP, dCTP, dTTP, and dATP), thermal-stable DNA polymerase (0.05 units/μl), and vector-specific oligonucleotide primers (5′ GTTTTCCCAGTCACGAC 3′ and 5′ GGATAACAATTTCACACAGG 3′, each at 1 μM final concentration) to the mixture to bring the total volume to 50 μl. Subject the sample to 30 cycles of 92°C for 10 sec, 55°C for 1 min, and 72°C for 1 min in a Perkin-Elmer/Cetus (Norwalk, CT) 9600 thermal cycler. Purify the resulting PCR product using a QIAwell-8 column (Qiagen, Chatsworth, CA) and sequence as described by Halloran *et al.* (22) using fluorescently labeled M13 "forward" and "reverse" oligonucleotide primers, cycle-sequencing reactions with *Taq* polymerase, and an Applied Biosystems (Foster City, CA) 373A DNA sequencer.

Of note, most of the resulting *Alu*-PCR products contain a stretch of A's at one or both ends, which are derived from the characteristic poly(A) region of *Alu* motifs (23). Obtaining DNA sequence beyond such poly(A) tracts is notoriously problematic because of the difficulty of the polymerase to extend faithfully nascent strands beyond this motif. However, when DNA sequence cannot be read distal to this stretch of A's, repeating the sequencing reactions using an oligonucleotide primer from the opposite side of the cloning site often yields the necessary sequence data. This latter capability represents an important advantage of using a plasmid vector (which yields double-stranded sequencing templates) rather than M13 (which yields single-stranded sequencing templates) for subcloning *Alu*-PCR products.

Isolation of Yeast Artificial Chromosome Insert Ends

Yeast artificial chromosome clones containing DNA from a particular chromosome can be obtained either by screening a total genomic YAC library with known chromosome-specific STSs (or probes) or by constructing a chromosome-specific YAC library. The latter can be derived either from a hybrid cell line containing a single human chromosome [with the identification of human DNA-containing clones by hybridization (as described above for λ and cosmid subclones)] or from flow-sorted DNA preparations. Any isolated, chromosome-specific YAC represents a potential source of other chromosome-specific DNA sequences that can be used for STS generation. In principle, any segment of DNA from the YAC insert can be used to develop an STS. However, in practice, the most desirable YAC segments for STS generation are the insert ends, because these are separated by the maximal physical distance within the YAC and thus are optimal starting points for subsequent chromosome walking.

The major problem associated with using YAC insert ends for generating STSs is the common occurrence of YAC clones that contain DNA segments from more than one region of the source genome (24). The frequency of such "chimeric" YACs varies considerably among libraries made from human DNA, with reported rates ranging from 10 to 60% (24). The source of the "unrelated" DNA segment in a chimeric YAC is usually either other human DNA (in the case of clones derived from total human DNA) or rodent DNA (in the case of clones derived from hybrid cell lines). Thus, with current YAC libraries, there is no guarantee that an insert end will correspond to the same chromosomal region as the remainder of the YAC insert. Nonetheless, the ability to generate STSs from YAC insert ends represents a critical component of both large-scale STS-content mapping projects (7) and more conventional YAC-based chromosome walking efforts.

Several methods are available for isolating the insert ends of YACs, including those involving subcloning, ligation-mediated PCR, and the recovery of bacterial plasmids using sequences present in the YAC vector (reviewed in Refs. 4 and 7). Described below is a modest modification of the elegant technique developed by Riley *et al.* (25) that we use for the routine isolation of YAC insert ends. This method, referred to as "vectorette"- (25) or "bubble"-PCR, involves the ligation of a double-stranded linker containing an internal region of mismatch (the "vectorette" or "bubble") to yeast DNA that has been digested with a compatible restriction enzyme. Using this ligated DNA as template, PCR is performed using one oligonucleotide primer that anneals to the YAC vector and one oligonucleotide primer that anneals to the newly synthesized DNA complementary to the region of mismatch in the bubble. In the first thermal cycle, only the YAC vector-specific primer can prime DNA synthesis, producing a complementary strand from which the bubble-specific primer can prime. Amplification of the other sequences ligated to the bubble is avoided, because specificity is provided by the prior synthesis of DNA originating from YAC vector sequences. The resulting PCR products consist of a small piece of the YAC vector arm, a variably sized stretch of the DNA insert adjacent to the cloning site, and a small piece of the bubble. A number of restriction enzymes and compatible bubbles can be used to ensure that a PCR product of adequate size is obtained. Of note, the protocol described below is designed for the routine recovery of the left and right insert ends of YAC clones constructed with the vector pYAC4 (3). For other YAC vectors, different vector-specific oligonucleotide primers, restriction enzymes, and reaction conditions (e.g., buffer, thermal cycling parameters) would likely be required.

Preparation of Bubble-Ligated Yeast DNA

1. Purify total yeast DNA from a YAC clone using standard methods (5). Alternatively, high molecular weight yeast DNA can be prepared in agarose plugs (26), which can then be melted and diluted three- to fourfold with water to prevent resolidification.

2. Digest ~1-μg aliquots of the purified yeast DNA with *Rsa*I or *Hin*fI in 50-μl reactions at 37°C for 2–4 hr, and then incubate the samples at 65°C for 15 min to inactivate the restriction enyzme.

3. In a total volume of 50 μl, combine 200 ng of *Rsa*I- or *Hin*fI-digested yeast DNA, ligase buffer [preferably containing 5% (final concentration) polyethylene glycol 8000], 2 units of ligase, and 2 pmol of annealed *Rsa*I- or *Hin*fI-compatible bubble. [The annealed bubble is formed by combining equal amounts of bottom and appropriate top oligonucleotide strands (Table I), heating to 68°C for 15 min, and slow cooling to room temperature.]

4. Incubate the ligation reactions at 37°C for 2 hr (or room temperature overnight), and then add 150 μl of water (to bring the DNA concentration to 1 ng/μl).

In a modification of the above protocol developed by Arnold and Hodgson (27), the bubble utilized bears terminal sequences that are compatible for ligation to the enzyme-digested DNA but fails to reform the restriction site on ligation. This bubble is added along with ligase and ATP directly to the restriction enzyme reac-

TABLE I Sequences of Bubble Components, Polymerase Chain Reaction Primers, and Sequencing Primers for Bubble-Polymerase Chain Reaction

Bubble components	
Universal bottom strand of bubble[a]	5′ CTCTCCCTTCTCGAATCGTAACCGTTCGTACGAGAATCGCTGTCCTCTCCTTC 3′
*Rsa*I-compatible top strand of bubble (creates blunt-ended bubble)[a]	5′ GAAGGAGAGGACGCTGTCTGTCGAAGGTAAGGAACGGACGAGAGAAGGGAGAG 3′
*Hin*fI-compatible top strand of bubble[a]	5′ A(ACGT)TGAAGGAGAGGACGCTGTCTGTCGAAGGTAAGGAACGGACGAGAGAAGGGAGAG 3′
Bubble-specific primers	
224 PCR primer[b]	5′ CGAATCGTAACCGTTCGTACGAGAATCGCT 3′
Bubble-specific sequencing primer[b]	5′ CGCTGTCCTCTCCTTC 3′
YAC vector-specific primers	
HYAC-C	5′ GCTACTTGGAGCCACTATCGACTACGCGAT 3′
LS-2	5′ TCTCGGTAGCCAAGTTGGTTTAAGG 3′
HYAC-D	5′ GGTGATGTCGGCGATATAGGCGCCAGCAAC 3′
RA-2	5′ TCGAACGCCCGATCTCAAGATTAC 3′

[a] Adapted from Ref. 25 by A. MacMurray and E. Lander (personal communication, 1992).

[b] From Ref. 25.

tion (without prior inactivation of the restriction enzyme). The sample is then incubated at 20°C for 60 min and then at 37°C for 30 min, and this incubation cycle is repeated a total of three times. The incubation at 37°C results in the cleavage of sites formed by the ligation of the target DNA segment with other pieces of yeast DNA, but not that formed between the target DNA segment and the bubble. This leads to the increased formation of the latter and improved specificity of subsequent PCR amplification.

Bubble-Polymerase Chain Reaction Amplification of Yeast Artificial Chromosome Insert Ends

The different sequences (and restriction sites) on the left (centromere-containing) and right (non-centromere-containing) vector arms of pYAC4 (3) create a slight problem in formulating a single strategy for recovering both insert ends by bubble-PCR. As a result, two different approaches are described below. In general, the first tends to be more robust, but is limited in its recovery of right insert ends and is more labor intensive (requiring two rounds of PCR amplification). Although the second approach is easier (requiring a single round of PCR amplification) and more likely to yield both insert ends of a YAC, it tends to be less robust. As a result, it is recommended that the first approach be utilized as the front-line method, with the second approach employed when difficulty is encountered in recovering a particular (more often right) insert end of a YAC.

Heminesting Bubble-Polymerase Chain Reaction

This strategy involves two sequential rounds of PCR. In the first round, the above bubble-ligated yeast DNA is used as template to perform PCR using a bubble-specific primer and a YAC vector-specific primer. In the second round, a small amount of the PCR product from the first round is used as template to perform PCR using the same bubble-specific primer and a new YAC vector-specific primer that primes internal (i.e., nested) to the primer used in the first round (thus, the name "heminesting"). This approach can be used to recover the left insert ends with either *Rsa*I- or *Hin*fI-digested DNA. However, for the recovery of the right insert ends, only *Rsa*I-digested DNA can be used (because *Hin*fI cleaves between the two YAC vector-specific primers on the right pYAC4 vector arm).

1. Prepare 5–50 μl PCR reactions containing 10 mM Tris-HCl (pH 8.3), 50 mM KCl, 1.5 mM $MgCl_2$, 200 μM concentrations of each dNTP (dGTP, dCTP, dTTP, and dATP), 0.05 units/μl of thermal-stable DNA polymerase, 0.2 μM concentra-

tions of each oligonucleotide primer [primers 224 and HYAC-C for the left side and primers 224 and HYAC-D for the right side (Table I)], and 1 μl of a diluted ligation reaction from above (at 1 ng/μl) per 5 μl of total reaction volume.

2. Perform 20 cycles of 92°C for 1 min, 62°C for 2 min, and 72°C for 2 min (Perkin-Elmer/Cetus thermal cycler).

3. Dilute each sample 1:100 with sterile water.

4. Prepare 5- to 50-μl PCR reactions containing 10 m*M* Tris-HCl (pH 8.3), 50 m*M* KCl, 1.5 m*M* $MgCl_2$, 200 μM concentrations of each dNTP (dGTP, dCTP, dTTP, and dATP), 0.05 units/μl of thermal-stable DNA polymerase, 0.2 μM concentrations of each oligonucleotide primer [primers 224 and LS-2 for the left side, and primers 224 and RA-2 for the right side (Table I)], and 1 μl of a 1:100 diluted sample from the first round of PCR per 5 μl of total reaction volume.

5. Perform 30 cycles of 92°C for 1 min, 65°C for 2 min, and 72°C for 2 min (Perkin-Elmer/Cetus thermal cycler).

6. Analyze products by gel electrophoresis.

Nonnesting Bubble-Polymerase Chain Reaction

This strategy can be used to recover both the left and right insert ends of a YAC, utilizing a single round of PCR and either *Rsa*I- or *Hin*fI-digested DNA. Of note, for recovery of the right insert end, a PCR primer specific to the region of pYAC4 derived from the yeast *SUP4* gene is used. This can lead to amplification of spurious bands derived from endogenous yeast sequences. The latter can often be overcome by addition of a commercially available "specificity enhancer" to the PCR reactions.

1. Prepare 5- to 50-μl PCR reactions containing 10 m*M* Tris-HCl (pH 8.3), 50 m*M* KCl, 1.5 m*M* $MgCl_2$, 200 μM concentrations of each dNTP (dGTP, dCTP, dTTP, and dATP), 0.05 units/μl of thermal-stable DNA polymerase, 0.2 μM concentrations of each oligonucleotide primer [primers 224 and HYAC-C for the left side and primers 224 and RA-2 for the right side (Table I)], a 1:15 dilution of Perfect Match DNA polymerase enhancer (Stratagene), and 1 μl of a diluted ligation reaction from above (at 1 ng/μl) per 5 μl of total reaction volume.

2. Perform 35 cycles of 92°C for 1 min, 65°C for 2 min, and 72°C for 2 min (Perkin-Elmer/Cetus thermal cycler).

3. Analyze products by gel electrophoresis.

Characterization of Bubble-Polymerase Chain Reaction Products

For the great majority of YACs, the above methods result in the successful amplification of both insert ends using yeast DNA digested with one or the other of the restriction enzymes (*Rsa*I or *Hin*fI). The amount of YAC vector sequences contained in the amplified products differs depending on the primer utilized in the final round of PCR (25): 225 bp for HYAC-C, 50 bp for LS-2, and 25 bp for RA-2. Bubble sequences also comprise about 42 bp of each PCR product. In most instances, it is preferable to have at least 100–300 bp of actual YAC insert DNA in the PCR product for additional analysis. If the PCR products obtained with *Rsa*I- and *Hin*fI-digested DNA are too small, then an additional panel of blunt-cutting restriction enzymes can be used, with the subsequent ligation of the blunt-ended (*Rsa*I) bubble (Table I). For example, individual yeast DNA samples can be digested with *Eco*RV, *Hin*cII, *Ssp*I, *Stu*I, *Alu*I, *Pvu*II, or *Sca*I and ligated to the blunt-ended bubble. The PCR is then performed as above, preferably using the nonnesting approach. Most of the potential products will be too large to amplify by PCR. However, typically one of the above restriction enzymes will yield a small enough product (<2 kb) to allow amplification by PCR.

All authentic bubble PCR products derived from YACs constructed with pYAC4 should contain an *Eco*RI site (representing the cloning site). This can be verified by digesting the resulting PCR product with *Eco*RI and analyzing the products by gel electrophoresis. Two major DNA fragments are typically seen: one of a constant size (corresponding to the amplified segment of the YAC vector) and one of a variable size (corresponding to the amplified segment of the insert end). Occasionally, two variably sized fragments are seen, reflecting the presence of an *Eco*RI site within the amplified insert-end segment. Such an analysis is recommended to confirm that the bubble-PCR product indeed represents the YAC insert end.

DNA Sequencing of Bubble-Polymerase Chain Reaction Products

The approach taken for sequencing bubble-PCR products is, once again, largely dependent on the preferred sequencing template (PCR product vs cloned DNA) and the method being utilized (e.g., radioactive- vs fluorescent-based sequencing). For direct sequencing of bubble-PCR products, sequencing primers specific to the YAC vector (LS-2, RA-2) and the bubble (see Table I) have been successfully employed. Once subcloned into a plasmid vector, this choice expands to include sequencing primers specific to the cloning vector itself.

Development and Testing of Sequence-Tagged Site-Specific Polymerase Chain Reaction Assays

Following the generation and sequencing of chromosome-specific DNA templates by any of the above strategies, each DNA sequence is processed through a standard series of steps tailored for the development of a large number of robust PCR assays. These steps are outlined in Fig. 1, with important details provided below.

Computer-Assisted Analysis of DNA Sequences

Each tract of newly generated DNA sequence is subjected to two types of computer-assisted analysis. First, each sequence is scrutinized for the presence of known repetitive DNA sequences (*Alu*, L1, and alphoid repeats), using a standard homology identification program [e.g., Fasta (28)]. Regions of DNA sequence with significant similarity (typically >75%) to any of these repeats are discarded. Remaining sequences are then analyzed to select appropriate PCR primers, using the program OSP (29). With this program, we have obtained functional PCR assays in >90% of instances in which primers are recommended (unpublished observations, 1993). The recommended primers are then synthesized using standard chemistries.

Polymerase Chain Reaction Assay Testing and Initial Optimization

Most of the myriad protocols and modifications for performing PCR cannot be considered when applied to the development of large numbers of independent PCR assays. As a result, we have attempted to develop a systematic and generic approach for batch testing and optimizing new PCR assays. This scheme minimizes the number of different parameters that are tested, making the process relatively routine and convenient. In our experience, the greatest gains in optimizing PCR assays are achieved by varying the temperature during the annealing phase of the thermal cycle and by altering the magnesium concentration in the reaction buffer. Our standard approach (2) involves the initial testing of PCR assays using two annealing temperatures that differ by 10°C (either 50 vs 60°C, or 55 vs 65°C) and, at first, 2.5 m*M* Mg^{2+} in the reaction buffer. Both positive (total human) and negative (either total yeast or rodent) control DNA templates are tested. Each of these two types of negative control templates has its relative merits. The analysis of yeast DNA early in the testing scheme helps to detect PCR assays that amplify products from yeast DNA of the same size as the STS (such assays cannot be easily used for characterizing YACs), whereas the early analysis of rodent DNA can be used to eliminate STSs corresponding to rodent sequences that were inadvertently isolated from hybrid cell line DNA.

The initial results obtained by the above testing strategy reveal a significant amount of information about the nature of the PCR assay (weak vs robust, specific vs nonspecific). Most often, PCR assays that will ultimately be judged as "acceptable" require one of these initial conditions or some minor modification thereof (such as raising the annealing temperature by 5°C to increase stringency, lowering the annealing temperature by 5°C to decrease stringency, or using 1.5 m*M* Mg^{2+} in the reaction buffer to improve specificity). It is important to stress that, in general, we are not concerned about the amplification of spurious DNA fragments that differ in size from the STS itself. In fact, we make no attempt to optimize a PCR assay for the sole purpose of eliminating the amplification of other products, unless they hinder the ability to detect the STS. Of note, we find that the use of high-resolution acrylamide gels (5) provides good size discrimination between an STS and most irrelevant PCR products.

Occasionally, a new PCR assay will be functional (i.e., amplify the appropriately sized product from human DNA) but not robust. In these instances, we turn to a small number of alternate strategies for improving assay performance (2). These include the use of "specificity enhancers" (such as that described above for performing bubble-PCR), different thermal cyclers, and/or alternate protocols for performing PCR (such as the addition of more DNA polymerase after half of the thermal cycles) (2). It is not obvious where the balance lies between the desirability of using a minimal set of PCR conditions and that of maximizing the probability that a particular pair of oligonucleotide primers will yield a robust PCR assay. Undoubtedly, some problem assays could be improved by simply synthesizing a new pair of PCR primers from the same DNA sequence. The decision to perform the latter must be compared with the effort needed to optimize further the initial assay as well as the impact of completely abandoning any attempt to develop an STS from that particular DNA sequence.

Using the above scheme, over 90% of the >500 STS-specific PCR assays that we have developed require 1 of 5 annealing temperatures (45, 50, 55, 60, 65°C) and one of two Mg^{2+} concentrations (1.5 or 2.5 m*M*) in the reaction buffer (2; unpublished observations, 1993). This relatively small set of conditions likely reflects the use of the computer program OSP (29) to select oligonucleotide primers for these PCR assays. We likely could generate an even more uniform collection of PCR assays by narrowing the range of acceptable values for the various parameters considered by OSP. However, this would result in a larger fraction of DNA sequences for which no acceptable primers could be found, and thus require a proportional increase in the overall sequencing effort to maintain a given rate of STS generation. The balance between DNA sequencing and uniformity in the behavior of the resulting PCR assays must be considered when setting the parameters for any program that selects PCR primers.

Genomic Localization of Sequence-Tagged Sites

Once a PCR assay is operational, it must be used to confirm that the corresponding STS was derived from the genomic region of interest. This can be efficiently accomplished by using the PCR assay to analyze a small panel of hybrid cell lines, some of which contain the chromosome (or chromosomal region) of interest and some of which do not. It is advisable to include in this panel a hybrid cell line that contains a number of other human chromosomes but not the one of interest. Polymerase chain reaction assays that amplify the STS exclusively from DNA of the appropriate cell lines are utilized further. The two most common negative results at this stage are (1) amplification of the appropriate product from all human DNA-containing cell lines but not from the control rodent (parental) cell line, suggesting that the STS represents a human repetitive sequence (perhaps less common than *Alu,* L1, or alphoid sequences), and (2) amplification of the appropriate product from all hybrid cell lines as well as the control rodent (parental) cell line, suggesting that the STS is conserved between humans and rodent. The latter result is particularly problematic, because the location of the STS in the human genome cannot be assessed without analysis by some other technique (e.g., *in situ* hybridization using a YAC, cosmid, or λ clone containing the STS). Occasionally, the latter two types of negative results can be overcome by additional optimization of the PCR assay, as described above.

Final Optimization of Polymerase Chain Reaction Assays

The proposed utility of any developed STS-specific PCR assay dictates the strategy used for any additional optimization steps. Our approach is geared toward the development of PCR assays capable of screening a YAC library, and thus we strive to ensure that each PCR assay is sufficiently robust to allow detection of STS-containing YACs within pools of thousands of clones (5). In our experience, many PCR assays that appear robust when tested with total human DNA perform in a notably weaker manner when amplifying target sequences from yeast-derived DNA preparations. We have found that a good approach for mimicking a YAC library screen is to use DNA purified from pools of 384 YAC clones (at 33 ng/μl) that have been spiked with 5 ng/μl of total human DNA (2). Amplification of the appropriate product with such a DNA preparation correlates extremely well with the subsequent success of identifying YAC clones containing that STS during a PCR-based library screen (5). Unsatisfactory performance at this stage of testing (which is rare) can often be overcome by slight alterations in the reaction conditions along the lines outlined above. For other utilities of the developed PCR assays, it is advisable to design analogous testing schemes that will help to ensure their satisfactory performance.

Concluding Remarks

The large-scale physical mapping of the human genome is underway, and there appears to be an increasing commitment to the use of PCR-based landmarks (STSs) as the common language on which the physical maps will ultimately be based. As a result, there is an enormous amount of activity being directed at the development of STS-specific PCR assays, most often in a targeted (typically chromosome-specific) manner. In this article, several distinct strategies have been presented for efficiently isolating and sequencing random pieces of chromosome-specific DNA. Also discussed was our scheme for progressing from raw DNA sequence to a functional, authenticated PCR assay. The outlined approaches have been extensively applied to the development of human chromosome 7-specific STSs. Overall, our success rate has been good, with a PCR assay having the requisite features [functional, amplifying a chromosome 7-specific STS, and suitable for YAC library screening (2)] obtained in >75% of the instances in which PCR primers were synthesized.

One aspect of the described strategies that remains untested is the chromosomal representation provided by the developed STSs (i.e., whether the STSs reflect a random sampling from the targeted genomic region). The two methods described for generating STSs from hybrid cell lines both exploit the presence of human repetitive sequences, either as targets for hybridization probes or as sites for annealing of PCR primers. It is likely that such repeats are not randomly distributed across a human chromosome. The potential impact of any such nonrandom distribution of STSs on a YAC-based STS-content mapping strategy remains to be determined, but has prompted us to use several additional approaches for deriving chromosome 7-specific DNA templates (including flow sorting, microdissection, and recovery of YAC insert ends). The extent to which such issues affect long-range physical mapping efforts will become apparent over the next several years as the quality and representation of the resulting STS-content maps are revealed.

Acknowledgments

I wish to thank David Bentley for advice about *Alu*-PCR, Armand MacMurray and Eric Lander for help with bubble-PCR, Maynard Olson for support and guidance in the development of the described strategies, and Rose Tidwell, Jackie Idol, Valerie Braden, Regina Lim, Mindi Ueltzen, and Jim Landgraf for technical assistance. I also thank Phil Green, Rick Wilson, and Gabriela Adelt Green for critical review of this article. This work was supported by NIH Grant P50-HG00201. E.D.G. is a Lucille P. Markey Scholar and this work was supported in part by a grant from the Lucille P. Markey Charitable Trust.

References

1. M. Olson, L. Hood, C. Cantor, and D. Botstein, *Science* **245,** 1434 (1989).
2. E. D. Green, R. M. Mohr, J. R. Idol, M. Jones, J. M. Buckingham, L. L. Deaven, R. K. Moyzis, and M. V. Olson, *Genomics* **11,** 548 (1991).
3. D. T. Burke, G. F. Carle, and M. V. Olson, *Science* **236,** 806 (1987).
4. P. Hieter, C. Connelly, J. Shero, M. K. McCormick, S. Antonarakis, W. Pavan, and R. Reeves, *in* "Genome Analysis" (K. E. Davies and S. M. Tilghman, eds.), p. 83. Cold Spring Harbor Lab. Press, Cold Spring Harbor, New York, 1990.

5. E. D. Green and M. V. Olson, *Proc. Natl. Acad. Sci. U.S.A.* **87,** 1213 (1990).
6. E. D. Green and M. V. Olson, *Science* **250,** 94 (1990).
7. E. D. Green and P. Green, *PCR Methods Appl.* **1,** 77 (1991).
8. D. R. Cox, M. Burmeister, E. R. Price, S. Kim, and R. M. Myers, *Science* **250,** 245 (1990).
9. C. G. Cole, P. N. Goodfellow, M. Bobrow, and D. R. Bentley, *Genomics* **10,** 816 (1991).
10. J. C. Hozier and L. M. Davis, *Anal. Biochem.* **200,** 205 (1992).
11. R. D. C. Saunders, *BioEssays* **12,** 245 (1990).
12. L. L. Deaven, M. A. Van Dilla, M. F. Bartholdi, A. V. Carrano, L. S. Cram, J. C. Fuscoe, J. W. Gray, C. E. Hildebrand, R. K. Moyzis, and J. Perlman, *Cold Spring Harbor Symp. Quant. Biol.* **51,** 159 (1986).
13. M. D. Adams, J. M. Kelley, J. D. Gocayne, M. Dubnick, M. H. Polymeropoulos, H. Xiao, C. R. Merrill, A. Wu, B. Olde, R. F. Moreno, A. R. Kerlavage, W. R. McCombie, and J. C. Venter, *Science* **252,** 1651 (1991).
14. J. F. Gusella, C. Keys, A. Varsanyi-Breiner, F.-T. Kao, C. Jones, T. T. Puck, and D. Housman, *Proc. Natl. Acad. Sci. U.S.A.* **77,** 2829 (1980).
15. R. K. Moyzis, D. C. Torney, J. Meyne, J. M. Buckingham, J.-R. Wu, C. Burks, K. M. Sirotkin, and W. B. Goad, *Genomics* **4,** 273 (1989).
16. J. Sambrook, E. F. Fritsch, and T. Maniatis, "Molecular Cloning: A Laboratory Manual," 2nd Ed. Cold Spring Harbor Lab. Press, Cold Spring Harbor, New York, 1989.
17. C. Helms, J. E. Dutchik, and M. V. Olson, *in* "Recombinant DNA," Part D (R. Wu and L. Grossman, eds.), Methods in Enzymology, Vol. 153, p. 69. Academic Press, Orlando, Florida, 1987.
18. D. L. Nelson, S. A. Ledbetter, L. Corbo, M. F. Victoria, R. Ramirez-Solis, T. D. Webster, D. H. Ledbetter, and C. T. Caskey, *Proc. Natl. Acad. Sci. U.S.A.* **86,** 6686 (1990).
19. D. L. Nelson, *Methods* **2,** 60 (1991).
20. K. A. Padgett, J. E. Mullersman, and B. K. Saha, *BioTechniques* **11,** 56 (1991).
21. D. Gussow and T. Clackson, *Nucleic Acids Res.* **17,** 4000 (1989).
22. N. Halloran, Z. Du, and R. K. Wilson, *in* "DNA Sequencing: Laboratory Protocols" (H. G. Griffin and A. M. Griffin, eds.). Humana Press, Clifton, New Jersey, in press, 1993.
23. C. W. Schmid and W. R. Jelinek, *Science* **216,** 1065 (1982).
24. E. D. Green, H. C. Riethman, J. E. Dutchik, and M. V. Olson, *Genomics* **11,** 658 (1991).
25. J. Riley, R. Butler, D. Ogilvie, R. Finniear, D. Jenner, S. Powell, R. Anand, J. C. Smith, and A. F. Markham, *Nucleic Acids Res.* **18,** 2887 (1990).
26. E. M. Southern, R. Anand, W. R. A. Brown, and D. S. Fletcher, *Nucleic Acids Res.* **15,** 5925 (1987).
27. C. Arnold and I. J. Hodgson, *PCR Methods Appl.* **1,** 39 (1991).
28. W. R. Pearson and D. J. Lipman, *Proc. Natl. Acad. Sci. U.S.A.* **86,** 2444 (1988).
29. L. Hillier and P. Green, *PCR Methods Appl.* **1,** 124 (1991).

[12] Nonradioactive Multiplex Procedure for Genotyping of Microsatellite Markers

Alain Vignal, Gabor Gyapay, Jamilé Hazan, Simon Nguyen, Chantal Dupraz, Nathalie Cheron, Noelle Becuwe, Martine Tranchant, and Jean Weissenbach

Introduction

About 3000 human polymorphic markers have been isolated to date, but 90% of these do not have a high frequency of variation in the population and are unevenly distributed throughout the genome. The lack of better markers impedes considerably exclusion mapping of genetic diseases, especially the less frequent ones. Mapping of multifactorial diseases will require the establishment of a high-resolution map to detect involvement of loci with minor effects. There is thus a need to develop both numerous and highly polymorphic systems. In the past such markers have been difficult to obtain. This situation has dramatically changed with the use of simple sequence repeats (microsatellites) developed by Litt and Luty (1) and Weber and May (2), which can be readily isolated from genomic libraries, sequenced, and tested by the polymerase chain reaction (PCR) for polymorphism. The polymorphism of microsatellites results from variations in the number of repeat units within a block. These variations are analyzed by PCR amplification of the sequence containing the polymorphic block, followed by electrophoretic separation (1, 2). However, because the size differences between alleles can be as small as one or two nucleotides, conventional agarose separations cannot be used as classical restriction fragment length polymorphism (RFLP) markers. Instead, PCR products must be separated on denaturing polyacrylamide gels as in sequencing. To reduce the number of sequencing gels and gel loadings, a procedure derived from the multiplex sequencing technique of Church and Kieffer-Higgins (3) was developed by Hazan *et al.* (4). In this multiplexing technique the PCR amplifications of different markers from the same genomic DNA sample are pooled together and loaded into a single well of a sequencing gel. After migration, the products are transferred onto a nylon membrane and successively hybridized to one of the primers of each pair used for the PCR amplification (Fig. 1). The initial procedure (4) has been modified and is now adapted to the large-scale mapping project performed at Généthon (Evry, France). Carried out on a large scale, this multiplexing procedure requires significant quantities of radioisotopes. We therefore employed a nonradioactive probing technique that is safer, facilitates working conditions, and improves the resolution of the amplification patterns on film. The different steps of this genotyping procedure are described below.

Methods in Molecular Genetics, Volume 1

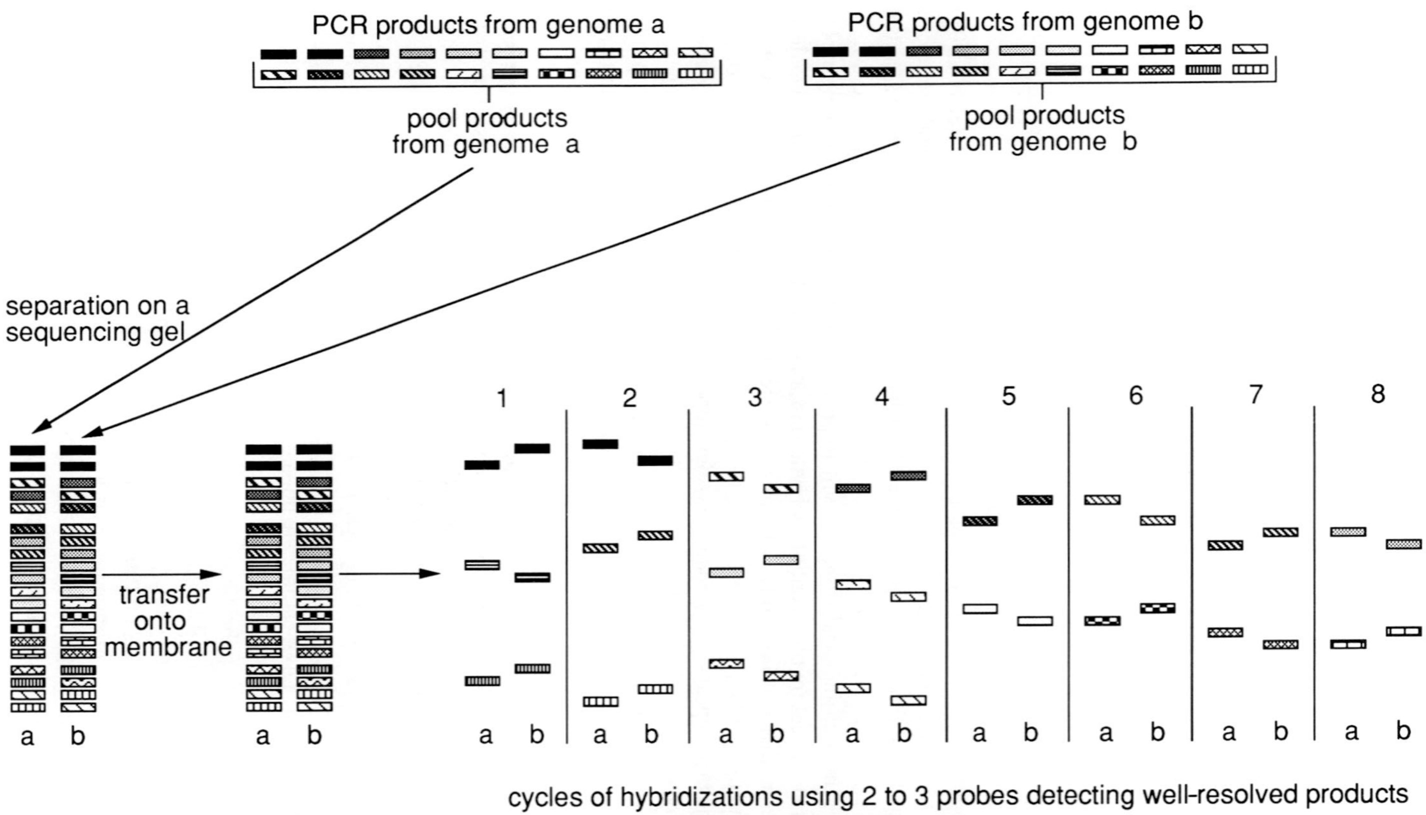

FIG. 1 The multiplex procedure. Products from PCR performed separately on the same DNA sample, with primer pairs corresponding to different markers, are pooled together and loaded into a single well of a denaturing polyacrylamide gel. After electrophoresis, PCR products are transferred onto a nylon membrane. Simultaneous hybridization, using one of the PCR primers from two to three well-resolved amplification products, allows detection of all the markers in several hybridization cycles. One band represents a marker with its two alleles.

Polymerase Chain Reaction Amplifications

Polymerase chain reactions are performed on microtiter plates, using a single set of reaction conditions: this allows processing of several markers on the same plates. Addition of DNA samples and various reagents (reaction mix, primers, etc.) is carried out using robotic equipment to reduce the number of possible errors, especially with the DNA samples.

Choice of Primers

Polymerase chain reaction primers were chosen using the computer program OLIGO of Rychlik and Rhoads (5) and additional criteria such as base composition and avoidance of homologies with repeated sequences (*Alu, Kpn,* etc.). Their T_m (melting temperature) ranges from 55 to 70°C. Special care has been taken to avoid self-priming, particularly at the 3′ end.

Polymerase Chain Reaction Conditions

The standard conditions described here permit successful amplification of 85% of the markers we have studied. These conditions represent a compromise probably remote from the optimum for most primer pairs.

Polymerase chain reactions are performed in a total volume of 50 μl, containing 40 ng of genomic DNA, 50 pmol of each primer, 1.25 m*M* concentrations of dNTPs, and 1 unit of *Taq* polymerase. Amplification buffer (1 ×) contains 10 m*M* Tris base (pH 9), 50 m*M* KCl, 1.5 m*M* $MgCl_2$, 0.1% (v/v) Triton X-100, and 0.01% (w/v) gelatin. The reactions are performed using a "hot-start" procedure: *Taq* polymerase is added only after an initial denaturation step of 5 min at 96°C. Amplification is carried out during 35 cycles of denaturation (94°C for 40 sec) and annealing (55°C for 30 sec). An elongation step (72°C for 2 min) ends the process after the last annealing.

This hot-start procedure provides an efficient primary denaturation step and minimizes enzyme degradation. Moreover, addition of *Taq* polymerase at a high temperature considerably reduces mispriming. Because the amplification products to be obtained are short (90 to 350 bp long) and the time interval to raise the temperature from 55 to 94°C (obtained with a ramping rate of 1°C/sec) is rather long, completion of DNA elongation can be achieved without an elongation step at 72°C. This speeds up the overall amplification and allows a greater number of reactions to be performed on the same machine in the same day.

For many markers, the number of cycles can be reduced to 30 or even 25 cycles. For such markers, this may improve the quality of the amplification pattern and subsequent film reading, with less ladder effect and spurious bands.

Preparing Polymerase Chain Reactions on a Large Scale Using Robotic Equipment

Samples of 40 ng of DNA in 10 μl [dissolved in 10 m*M* Tris base (pH 7.6), 1 m*M* EDTA (pH 8.0)] are distributed in the microplates with a Hamilton AT plus robotic equipment (Hamilton, Bonadaz, Switzerland), followed by addition of 30 μl of mix containing the primers, dNTPs, and 1 × buffer. The samples are then overlaid with heavy mineral oil (Sigma, St. Louis, MO). At this stage, the microplates can be stored up to 24 hr at 4°C.

Amplifications are performed using the hot-start procedure. After an initial denaturing step of 5 min at 96°C, 1 unit of *Taq* polymerase, dissolved in 10 ml of 1 × amplification buffer, is added manually on the thermal cycler, through the oil layer, while the samples are kept at 94°C. The cycling then proceeds.

Attempts to add *Taq* polymerase using robotic equipment resulted in a dramatic decrease in successful amplification. In this instance, reaction mixes were prepared at room temperature, probably favoring mispriming and consecutive nonspecific elongation, resulting in depletion of the original primer(s). The hot-start procedure resolves this problem and allows preparation ahead of time of a great number of microplates that can be stored at 4°C before temperature cycling.

Coprecipitation of Polymerase Chain Reaction Products

Eight, 12, or 16 amplification products of different markers from aliquots of the same DNA sample are coprecipitated using robotic equipment, before loading into a single gel well. Twenty microliters of PCR products from different microplates is taken out directly through the oil, and pooled in another microplate. To remove residual mineral oil carried over with the samples when using robotic equipment, one-fourth of the pooled samples is transferred manually to another microplate. Precipitation is then performed by adding 0.2 vol of 3 *M* sodium acetate and 2 vol of cold 100% ethanol. The samples are left for 45 min at −20°C and centrifuged in the microplate at 2500 *g* for 1 hr at 4°C. The pellets are rinsed twice with cold 70% (v/v) ethanol, with intermediate spinning at 2500 g for 1 min at 4°C, dried, and resuspended in 5 μl of 10 m*M* Tris base (pH 7.6), 1 m*M* EDTA (pH 8.0).

Polyacrylamide Gel Electrophoresis

As the differences in allele sizes of microsatellite markers may be as small as one or, more usually, two nucleotides, allele separation is performed on 0.4-mm thick 6% (w/v) polyacrylamide gels in 8.3 *M* urea, identical to sequencing gels (6), with some slight modifications. For running the gels, the TBE used is 9 m*M* Tris base, 9 m*M* boric acid, 1 m*M* ethylenediaminetetraacetic acid (EDTA).

A

B

C

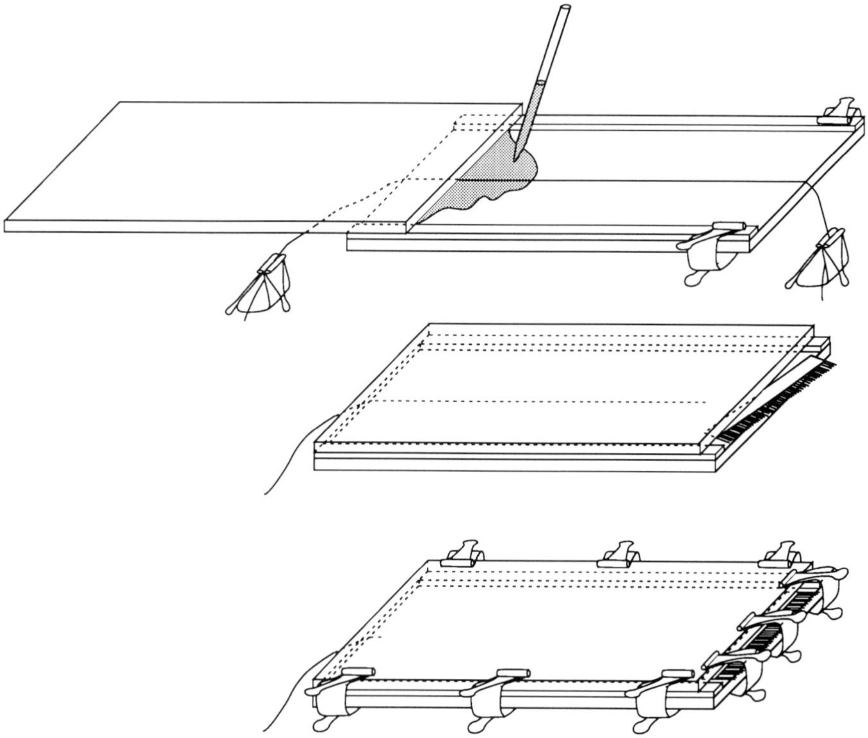

FIG. 2 Casting of polyacrylamide gels. (A) The bottom plate is placed flat on a support table and the two spacers are arranged in place along the sides, held in place with binding clips. A few drops of water help to keep the spacers in position for the next steps. If large plates are used, a fishing thread is arranged across the middle of the plate. Then the top plate is laid in position. The acrylamide solution is poured on the bottom plate, while sliding the top plate along. (B) Once the acrylamide solution is poured, and the top plate is in position, the nylon thread is pulled carefully, until enough space is left to insert the comb. (C) The nylon thread is pulled out, until only a few millimeters are left at the bottom of the gel. This acts as an internal spacer and prevents the gel from being too thin, due to the bending of the top plate under its own weight.

Gel Pouring

Instead of sealing the assembled gel molds by binding the assembled plates and spacers with tape, we use a procedure derived from a method described for the Macrophor sequencing system of Pharmacia (Uppsala, Sweden). In this method, the gel is cast flat by pushing the top plate in place, while the acrylamide is poured over the bottom plate. The acrylamide is held in place simply by capillary action (see Fig. 2).

If large amounts of gel are poured, the top glass plate may bend slightly under its own weight, causing difficulties in inserting the comb and/or reducing gel thickness in the middle. In such instances, migration of the samples is altered, with a strong "antismile" pattern across the gel and a poor transfer of the DNA from the gel to the nylon membrane during blotting. Such "antismile" patterns can interfere with the genotyping of large pedigrees. Therefore, for 0.4-mm thick gels, a nylon thread 0.35 mm in diameter, such as a fishing thread, is used as an additional spacer. This helps, when inserting the comb, by lifting the top plate slightly, and keeps the thickness of the gel even (see Fig. 2).

Gradient Gels

To load as many different-sized samples at a time on one gel, wedge gradient or buffer gradient gels can be used that slow down the migration of the smaller products (6). However, loading of the former is difficult and casting of the latter takes too long. A simpler method consists of casting the gel and performing the prerun normally in 1 × TBE, then adding a $\frac{1}{3}$ vol of 3 *M* sodium acetate in the lower buffer chamber of the electrophoresis apparatus, as the samples are loaded. A salt gradient then forms during electrophoresis (7).

Loading and Running Gels

One microliter of the coprecipitated PCR products is mixed with 2 vol of deionized formamide and 0.5 vol of loading mix containing 0.2% (w/v) xylene cyanol, 0.1% (w/v) bromphenol blue, and 50% (w/v) sucrose, and loaded on the gels.

According to the size of the markers to be separated, gels are run for approximately 3 to 4 hr at a constant power of 50 W, to keep the temperature around 50°C during the run. With 6% acrylamide gels, xylene cyanol migrates as do 110-base DNA strands.

Size Standards

Whenever the $(AC)_n$-containing M13 DNA template cognate with a marker is available, it is used as a template for PCR, and the amplification product is loaded on the gel to help define the allele sizes in the genotyped individuals. In addition, a ladder showing jumps of one nucleotide, composed of sonicated commercial poly(dA), is loaded on the gels. This ladder will be revealed by a poly(dT) probe (Fig. 3).

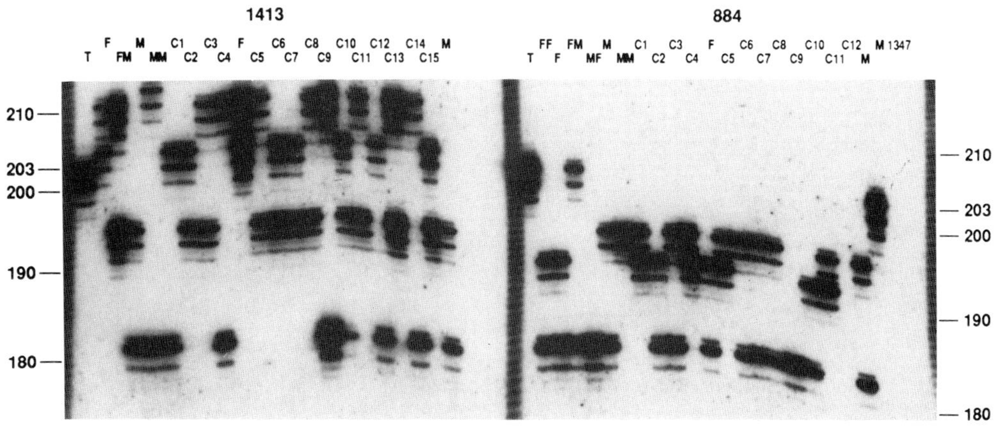

FIG. 3 Autoradiograph illustrating a microsatellite marker amplified on samples from families 1413 and 884 from the collection of the Centre d'Etude du Polymorphism et Humain (CEPH, Paris) families 1413 and 884, along with the sonicated poly(dA) ladder used as an internal scale in determining allele sizes. FF, Father's father; F, Father; FM, father's mother; MF, mother's father; M, mother; MM, mother's mother; C1 to C15, children; T, PCR product of the M13 template cognate with the revealed microsatellite marker; M1347, PCR product of the DNA sample from which the M13 human genomic library used to develop the microsatellite markers was derived.

Transfer of Polymerase Chain Reaction Products onto Nylon Membranes

Electrophoresed amplification products are transferred onto nylon membranes. Several transfer procedures have been compared. Electrotransfer (4) is the most efficient procedure in terms of quantity of DNA, but results in irregular transfers with large areas of poorly transferred material. Vacuum blotting gives more reproducible results. However, both procedures require special equipment that takes much laboratory space. The most satisfactory technique is simple capillary blotting, in which the membrane is left in contact with the gel so that the DNA can be transferred. The amount of DNA transferred is not as quantitative as with the two other techniques but, if performed carefully, capillary blotting yields an even transfer of a sufficient amount of material onto the membrane for detection by hybridization.

Once the migration is completed, the gel apparatus is cooled down with tap water. This prevents the gel from drying too quickly after unmolding. The top plate is taken off and the gel is carefully lifted onto Whatman (Clifton, NJ) 3 MM filter paper, as for sequencing gels (6). The filter paper, with the gel facing upward, is then laid over another Whatman paper soaked in 1× TBE (Fig. 4). This prevents the gel from drying too quickly, which otherwise results in a poor transfer. A piece of Hybond N^+ membrane, about 5 mm smaller than the gel in both dimensions, is soaked in 1× TBE and placed on top of the gel. Any trapped air bubbles are

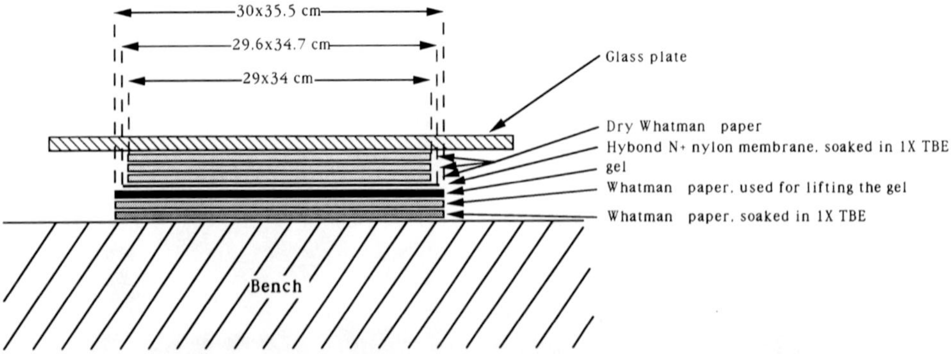

FIG. 4 Capillary transfer of DNA from sequencing acrylamide gels onto nylon membranes.

removed by rolling over the paper with a pipette and any excess buffer must be carefully wiped away. Finally, three sheets of dry Whatman 3MM paper cut about 5 mm smaller than the nylon membrane are superimposed and a glass plate is placed on top to hold the assembly (Fig. 4). Cutting the membrane and the three top sheets of Whatman 3MM paper smaller than the gel prevents irregular transfer toward the edges. An alternative to this contact transfer method is to leave the gel on the bottom plate of the migration apparatus and to lay the wet membrane, the three sheets of Whatman 3MM paper, and the glass plate directly on top. After 90 min of transfer, the membrane is peeled off, rinsed in 2× SSC (sodium saline citrate), and the DNA is cross-linked by exposure to UV light.

Oligonucleotide Labeling

Each PCR product is detected by hybridization to one of the labeled primers. These primers can be labeled by tailing with radioactive dNTP precursors, using calf thymus terminal deoxynucleotidyltransferase (Tdt) or by a nonradioactive labeling procedure. We have used the ECL system (Amersham, UK) based on a horseradish peroxidase-mediated chemiluminescent reaction (8, 9). Covalent linking of peroxidase to oligonucleotides can be achieved by reacting the enzyme molecule with a thiol group located on the 5′ end nucleotide of the nucleic acid chain. However, preparation of such modified oligonucleotides is difficult and expensive. We therefore used the more classical cross-linking procedure using glutaraldehyde (8). A protocol involving a random dNTP tail addition using Tdt allows the addition of tails approximately 300 nucleotides long, as observed on polyacrylamide gels. Such tailed oligonucleotides form a substrate long enough for successful binding of peroxidase molecules, which normally attach approximately every 30 nucleotides.

Tailing of Oligonucleotides

Oligonucleotide (160 pmol) is incubated for 3 hr at 37°C with 75 units of terminal deoxynucleotidyltransferase, in a total volume of 100 μl, containing 200 μM concentration of each dNTP, 200 mM potassium cacodylate (pH 7.6), 100 μM dithiolhreitol (DTT), and 1 mM $CoCl_2$. The $CoCl_2$ must be added last, just before the enzyme. The reaction is stopped with 4 μl of 250 mM EDTA. Unincorporated nucleotides, which might interfere with the peroxidase in the cross-linking step, are separated from the tailed oligonucleotides by gel filtration using a Sephadex G-25 spin column (6). Absorbance at 260 nm is then measured to estimate the average length of the nucleotide tail. An approximate absorbance increase of 15- to 20-fold is typically obtained for a 20-mer primer.

Covalent Binding of Horseradish Peroxidase

Binding of the horseradish peroxidase to tailed oligonucleotides is performed just before using the probe for hybridization. Typically, for a volume of 20 ml of hybridization buffer, 300 ng of tailed oligonucleotide in 30 μl of water is mixed with 30 μl of horseradish peroxidase and 30 μl of gluteraldehyde, according to the protocol of Amersham. The mixture is vortexed gently and incubated for 10 min at 37°C. If two or more probes are hybridized together to the same membrane, they are labeled in a single reaction: 30 μl of glutaraldehyde and 30 μl of horseradish peroxidase are added to 30 μl of the mixed oligonucleotides for each 300 ng of tailed oligonucleotide.

The sonicated poly(dA) ladder, used to help determine allele sizes (Fig. 3), is detected using a poly(dT) probe. Poly(dT) DNA (300 ng) is labeled as above and hybridized together with the specific elongated oligonucleotide probes.

Hybridization

Hybridizations are carried out at 42°C overnight in Gold buffer (Amersham), with 5% (w/v) blocking agent (Amersham), 0.5 M NaCl, and 0.16 pmol/ml of probe in glass cylinders fixed on a hybridization rotor in a thermostatically controlled oven. Blots are washed once in 3× SSC, 0.1% sodium dodecylsulfate (SDS) for 15 min, and twice in 0.2× SSC for 20 min at 36°C. Hybridized probes are revealed using the ECL detection reagents (Amersham), and blots are exposed to autoradiographic X-ray film for 10 min to 3 hr (Figs. 3 and 5). A first round of hybridization using peroxidase-labeled oligo(CA) allows visualization of the overall transfer (Fig. 5A) and saturates nonspecific binding sites of the bound DNA, prior to the primer-specific hybridizations (Fig. 5B).

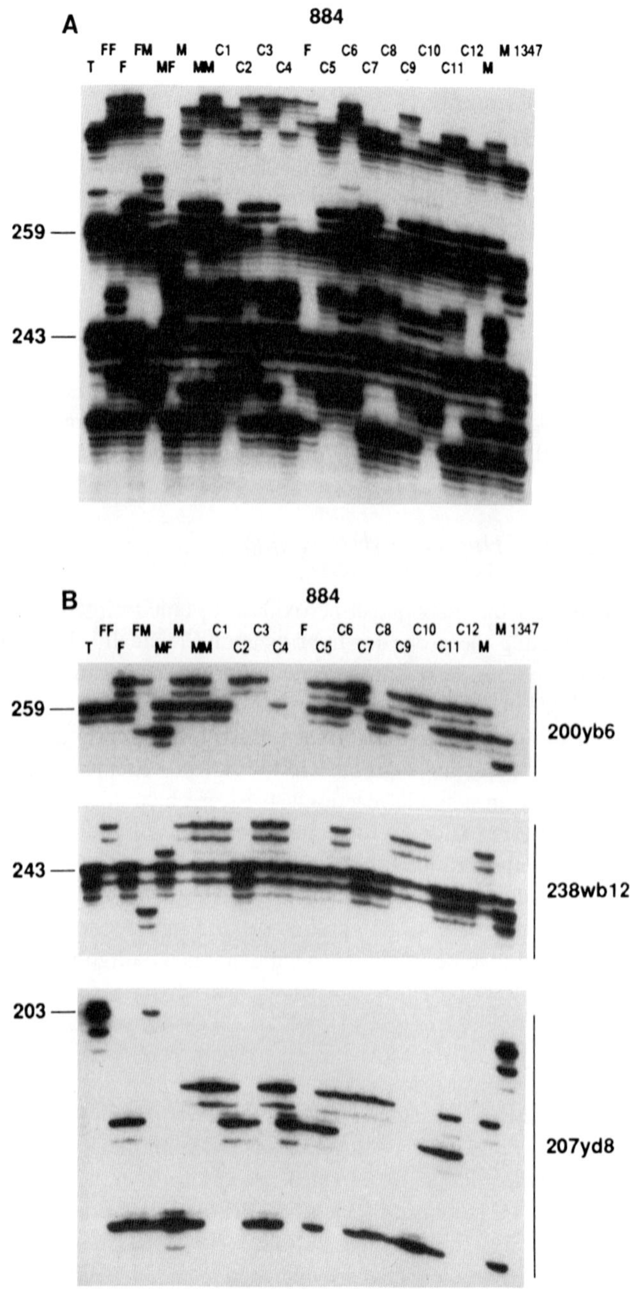

FIG. 5 Autoradiographs illustrating microsatellite markers amplified on DNA samples from a CEPH family, coprecipitated and comigrated on a single gel, transferred by contact blotting on a nylon membrane, and revealed by probes labeled with peroxidase. (A) Hybridization with a poly(dA-dC) probe, revealing all markers bound to the membrane. (B) Successive hybridizations with one of the oligonucleotides used for the PCR. Abbreviations as in Fig. 3.

References

1. M. Litt and J. A. Luty, *Am. J. Hum. Genet.* **44**, 397 (1989).
2. J. L. Weber and P. E. May, *Am. J. Hum. Genet.* **44**, 388 (1989).
3. G. M. Church and S. Kieffer-Higgins, *Science* **240**, 185 (1988).
4. J. Hazan, C. Dubay, M.-P. Pankowiak, N. Becuwe, and J. Weissenbach, *Genomics* **12**, 183 (1992).
5. W. Rychlik and R. E. Rhoads, *Nucleic Acids Res.* **17**, 8543 (1989).
6. J. Sambrook, E. F. Fritsch, and T. Maniatis, "Molecular Cloning: A Laboratory Manual," 2nd Ed. Cold Spring Harbor Lab. Press, Cold Spring Harbor, New York, 1989.
7. F. M. Ausubel, R. Brent, R. E. Kingston, D. D. Moore, J. G. Seidman, J. A. Smith, and K. Struhl, "Current Protocols in Molecular Biology," Vol. 1, p. 7.4.13 (Suppl. 2). Green Publ. Assoc. and Wiley (Interscience), New York, 1992.
8. M. Renz and C. Kurz, *Nucleic Acids Res.* **12**, 3435 (1984).
9. T. P. Whitehead, G. H. G. Thorpe, T. J. N. Carter, C. Groucutt, and L. J. Kricka, *Nature (London)* **305**, 158 (1983).

[13] Prenatal Cytogenetics

Gail Stetten

Introduction

For over 20 years the hallmark of a cytogeneticist has been preparation of banded metaphase chromosomes, and a trypsin–Giemsa method has become standard practice in most clinical cytogenetics laboratories. This method makes accurate chromosome identification and karyotyping a straightforward and reliable means for detecting a wide range of chromosome abnormalities. A new molecular method, fluorescence *in situ* hybridization (FISH), has emerged, permitting the rapid visualization of DNA probes hybridized to complementary DNA sequences on metaphase or interphase chromosomes. This has dramatically increased the level of resolution possible on metaphase chromosomes and has also made it possible to identify accurately specific chromosomes in the interphase nucleus. These new techniques are already a useful adjunct to classical clinical cytogenetics (1) and gene mapping (2, 3) and are beginning to change how we do prenatal cytogenetic testing.

Banded Chromosomes

Banded metaphase chromosomes are prepared for prenatal diagnosis from cultured amniotic fluid, chorionic villi, or fetal blood. Amniocytes are routinely grown as clones on glass coverslips and are harvested after about 7 days of culture (4). Chorionic villi can be prepared directly for cytogenetic analysis about 5 hr after sampling (5). Unfortunately, the chromosome morphology and mitotic index are often poor in these rapid preparations. For this reason dispersed cells are also cultured for about 6 days prior to harvesting. Fetal blood cells are grown in the presence of a mitogen (phytohemagglutinin) for a minimum of 48 hr before a satisfactory mitotic index can be obtained.

The harvesting procedure is similar for all three cell types. Cell division is first blocked by exposure to a mitotic poison (Colcemid at 0.05–0.1 μg/ml) followed by treatment of cells with a hypotonic solution (0.075 *M* KCl or 0.8% sodium citrate) to swell the nuclei and an acid fixative [3:1 (v/v) methanol–acetic acid] to fix the basic protein of the chromatin.

Metaphase nuclei are physically ruptured by dropping fixed cells onto slides and/or blowing a stream of air on the slide or coverslip at high humidity. Slides are heated (e.g., for 6 hr at 45°C) and then dipped in a trypsin solution (0.4 μg/ml; pH

Methods in Molecular Genetics, Volume 1

7.2) and stained for bright-field microscopy (e.g., Giemsa or Wright's stain). The lateral banding pattern produced, G bands, reflects the stage of condensation of the chromosomes in each particular spread. The longer the chromosomes, the greater the number of bands that can be visualized. It is often easy to identify the chromosomes belonging to a particular spread by their unique stage of condensation. Chromosomes from nuclei in late prophase will be more elongated and contain a larger number of distinct bands than will midmetaphase chromosomes (Fig. 1).

The level of resolution of traditional chromosome banding can be estimated from the number of bands visualized on the most elongated chromosomes as follows:

$$\frac{3 \times 10^9 \text{ bp/haploid genome}}{1000 \text{ bands/haploid genome}} = 3 \times 10^6 \text{ bp/band} \quad \text{or} \quad 3000 \text{ kb/band}$$

This implies that small (<3000 kb) duplications, deletions, or rearrangements of chromosomal material might not be detected. It should also be noted that the banding pattern, although unique for each chromosome, will not permit detection of rearrangements involving similar light–dark sequences of bands.

Interphase chromosome identification has never been as accurate as metaphase analysis. Although methods for identification of the inactive X chromosome (the Barr body) and the fluorescent heterochromatin in the long arm of the Y chromosome do exist, identification of autosomes in the interphase nucleus has not been possible with these traditional approaches. (The one exception to this rule is chromosome 9, which can be identified by Giemsa 11 staining.) It is in part for this reason that the newer molecular techniques appear so promising.

Aneuploidy Detection

The introduction of molecular cytogenetics to the clinical laboratory has made aneuploidy detection possible in interphase nuclei. Aneuploidies of chromosomes X, Y, 13, 18, and 21 account for the majority of chromosome abnormalities diagnosed prenatally. (In our laboratory the aneuploidy rate is about 2%). Repetitive DNA sequences specific for the centromeric regions of chromosomes X, Y, and 18 produce compact fluorescent signals that can be counted to determine copy number in interphase and metaphase (Fig. 2). Similarly, a repetitive sequence common to the centromeres of chromosomes 13 and 21 is useful in detecting these two chromosomes.

Direct preparations of blood, chorionic villi, and amniocytes can all be used for

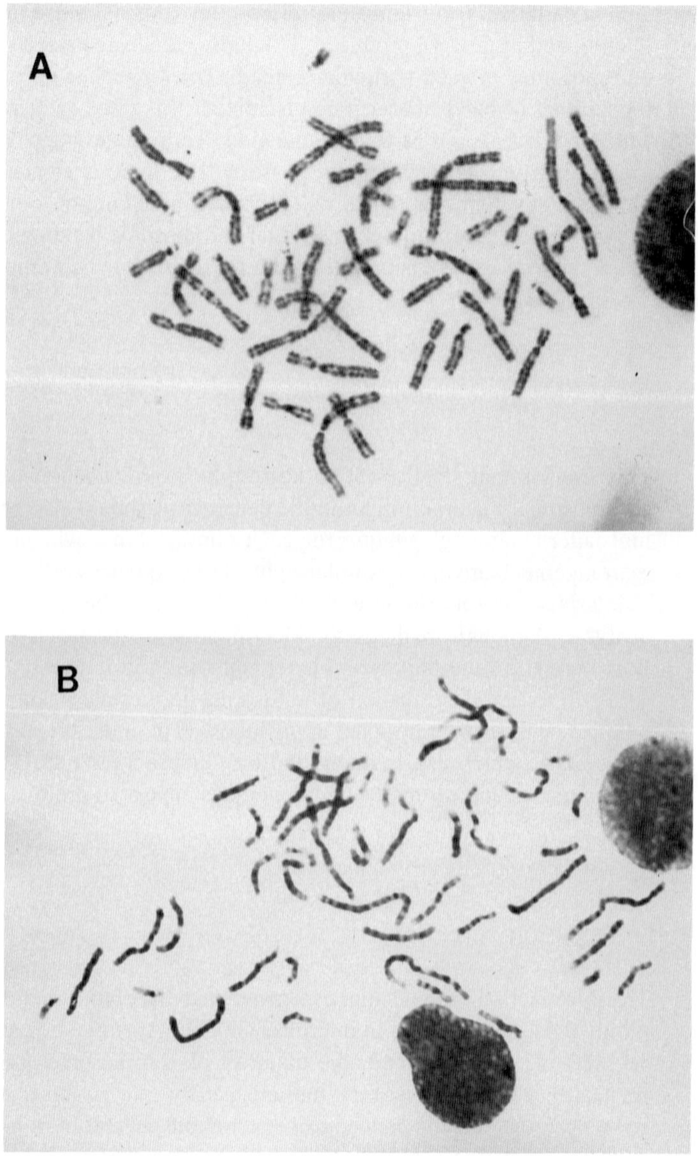

FIG. 1 G-banded chromosomes form lymphocyte nuclei in (A) midmetaphase (~400-band stage) and (B) late prophase (~800-band stage).

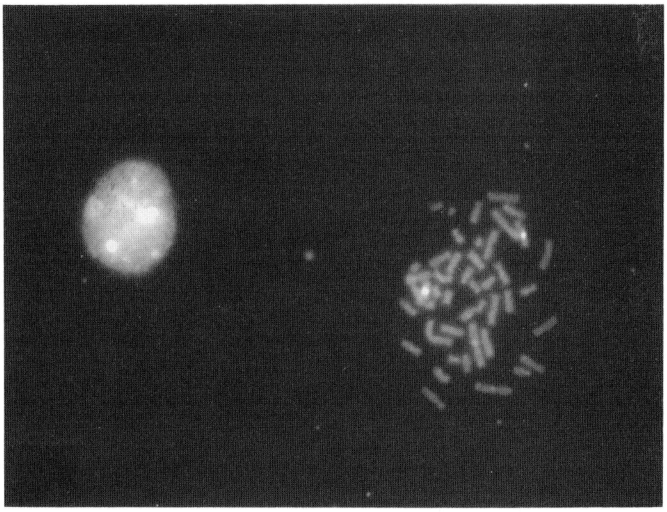

FIG. 2 Fluorescence *in situ* hybridization with the α-satellite probe DXZ1 (Oncor, Inc.), showing the X chromosome centromeres in an interphase nucleus and a metaphase spread from cultured amniotic fluid.

interphase analysis. Cells are treated with a hypotonic solution followed by fixative in a manner identical to that used in preparing metaphase chromosomes. In our laboratory, hybridization is successful in about 80% of lymphocytes, 98% of chorionic villi, but only 5–25% of amniocytes (Fig. 3A–C). In this last instance, the poor hybridization efficiency to amniocyte nuclei may reflect the relatively high proportion of nonviable cells present in uncultured amniotic fluid. Cells that do not show a signal can be characterized as having a massive cytoplasm enclosing a compact nucleus or lacking entirely a well-defined nucleus (6).

The first report of trisomy 18 detected by FISH (7) emphasized the problem of signal heterogeneity and the necessity for a statistical analysis of the data. Often the signals lie in different focal planes or background signals cannot be easily distinguished from a true hybridization signal. Using probes under proper stringent conditions in both the hybridization step and the wash step is critical. It is always necessary to run control samples and code the slides for microscope analysis. It is also useful to have the slides read by two microscopists.

Aneuploidy is sometimes suspected in the third trimester of a pregnancy based on information obtained by ultrasound studies. Direct preparations of fetal cells or extraembryonic tissue can be analyzed immediately by FISH for a specific trisomy. A prenatal detection of trisomy 18 in a twin pregnancy was made in our laboratory on direct preparations of amniotic fluid obtained at 33 weeks gestation. Only one

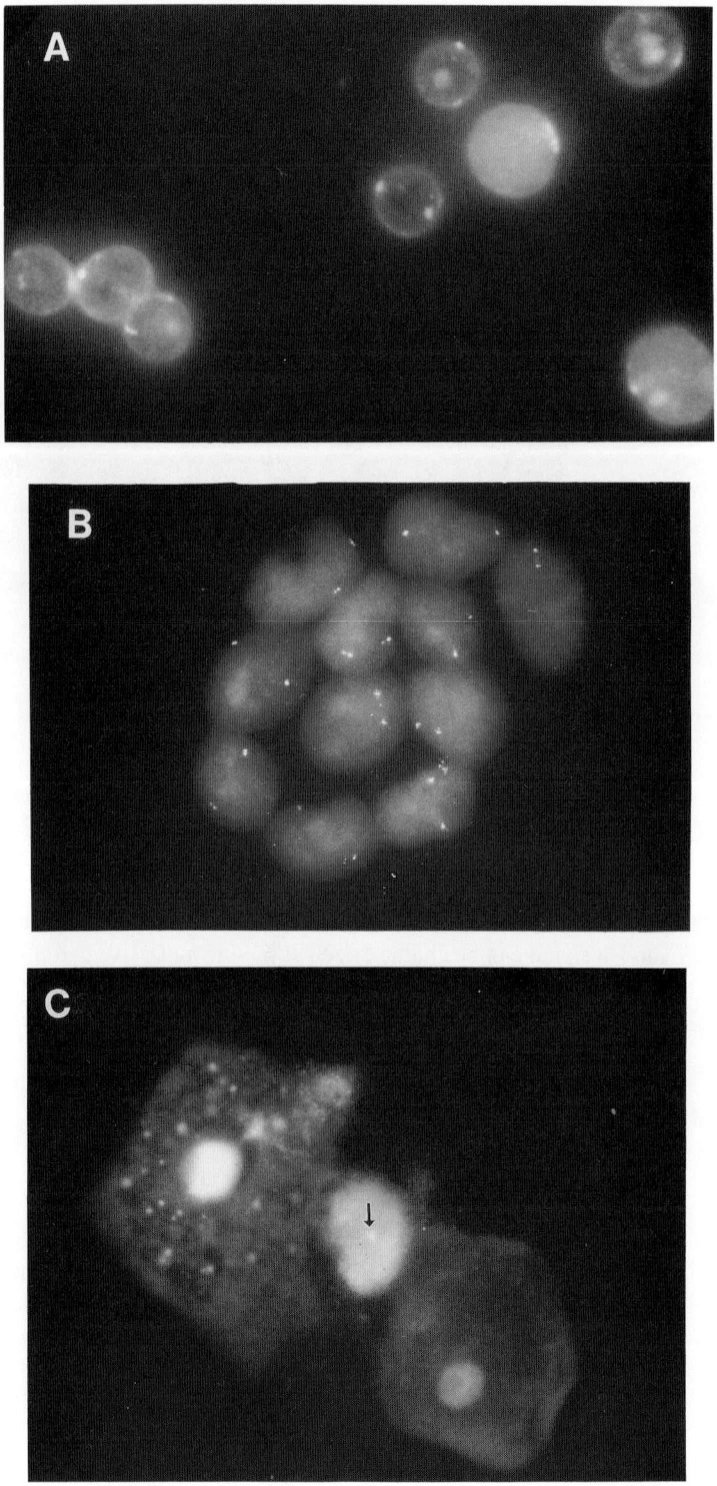
A
B
C

twin was affected (Fig. 4). X and Y centromere-specific probes were used to show sexual discordance of the twins, thus ruling out the possibility of low-level mosaic trisomy 18 in the normal fetus. Our results were complete in 24 hr and were an essential part of the obstetrical management of this case (8).

Using this general approach a few cytogenetics laboratories are developing rapid (24-hr) analysis of amniotic fluid for identification of sex and the major autosomal trisomies (trisomy 13, 18, and 21). Chromosome-specific partial libraries (cosmid contigs) (9) and whole-chromosome probes (10) have been developed for this purpose. These tests are being followed by analysis of metaphase chromosomes to confirm the diagnosis made on interphase cells.

Identification of a chromosome abnormality with a chromosome-specific probe (centromeric or partial library) is especially useful when the abnormality is thought to be mosaic by metaphase chromosome analysis. Interphase nuclei are usually readily available in large numbers and can be rapidly screened for trisomy vs disomy or disomy vs monosomy of a specific chromosome. For example, if a mosaic trisomy is ascertained in metaphase spreads from cultured chorionic villi, the direct preparation (often with only a few metaphase spreads of poor quality) can be rapidly checked for the presence of the trisomy. Several hundred nuclei can be examined in the time it takes to analyze five metaphase spreads. However, interpreting mosaic results from interphase nuclei can be difficult, and simultaneous control samples must always be run.

Analysis of Structural Rearrangements

Structural rearrangements in chromosomes are detected by G banding in about 1% of prenatal specimens in our laboratory. However, many undoubtedly go undetected because they involve rearrangements below the level of microscopic resolution or regions of chromosomes with similar banding patterns. Common chromosome rearrangements seen in the prenatal laboratory include balanced and unbalanced translocations, duplications, deletions, inversions, ring chromosomes, and marker chromosomes. Identification of *de novo* unbalanced structural abnormalities can be especially difficult with banding methods alone, because the pattern of bands is often insufficient to unambiguously identify a region of a particular chromosome. If an abnormality is familial and can be identified as "balanced" (with no effect on

---

FIG. 3 Fluorescence *in situ* hybridization with probe DXZ1 (Oncor, Inc.) showing direct preparations of (A) white blood cells, (B) chorionic villi, and (C) amniocytes. In most instances only one signal per chromosome is seen with a centromere-specific probe. However, occasionally a double signal is seen representing slight separation of the replicated chromatids in a single chromosome. In (C) the arrow indicates the hybridization signal in an amniocyte nucleus.

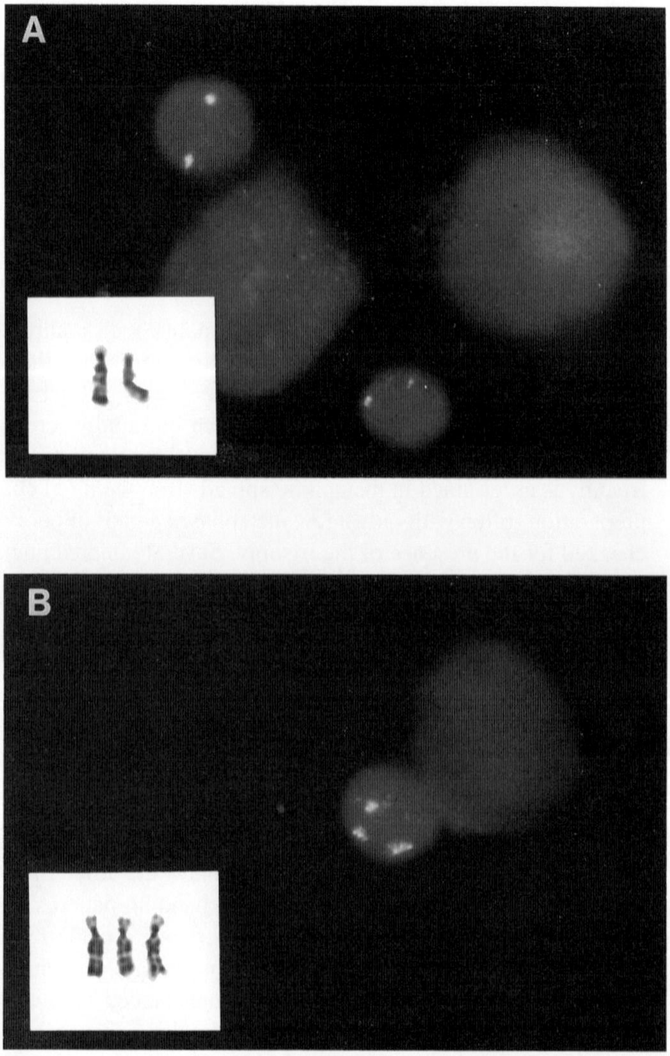

FIG. 4 Fluorescence *in situ* hybridization with probe L1.84 (7) showing direct preparations of amniocytes from (A) the normal twin with two copies of chromosome 18 and (B) the trisomic twin with three copies of chromosome 18. The insets at lower left shows G-banded chromosome 18 from each case.

the phenotype) in one generation, then the unbalanced form can usually be described. It is increasingly clear that many structural rearrangements are simply not detectable with current chromosome banding methods. For example, microdeletions in syndromes such as Prader–Willi, Angelman, DiGeorge, Wolf–Hirschhorn, and Miller–Dieker can only be visualized in chromosome preparations on a fraction of patients with the disease, even though molecular studies using somatic cell hybrid or restriction fragment length polymorphism (RFLP) analysis often can demonstrate a molecular deletion. In addition, apparently balanced translocations are occasionally seen in association with a clinical phenotype in families where it is highly suggestive that an undetectable loss or gain of chromosomal material, probably at one of the translocation breakpoints, is the underlying problem (11). In many of these situations the new molecular techniques can be useful.

The origin of accessory marker chromosomes can also now be successfully determined prenatally with chromosome-specific centromeric probes. We have identified two cases of similar small mosaic marker chromosomes that were both monocentric and bisatellited with nucleolar organizer region DNA at both ends. In one case the marker chromosome contained a 14 centromere, in the other case a 22 centromere (12). In both cases the pregnancies have resulted in phenotypically normal children. Hopefully, specific identification of these previously unidentified chromosomes will help to uncover any associated abnormal clinical phenotypes.

Chromosome-specific libraries for all human chromosomes have been constructed in Bluescribe plasmids from flow-sorted chromosome preparations (13). These libraries have been successfully used for FISH to identify chromosomes or parts of chromosomes involved in chromosomal rearrangements (14, 15). The method has been appropriately called "chromosome painting" (15). Chromosome translocations that are visible only in elongated G-banded chromosomes can often be more easily seen when one chromosome involved in the translocation is painted.

In prenatal diagnosis it is important to be able to distinguish *de novo* balanced translocations from unbalanced ones. If a chromosome appears to carry a deletion, painting with the DNA library of that chromosome may help to identify a translocation chromosome. In our laboratory a balanced chromosome rearrangement t(3;15)(q29;q26.1), seen only in well-banded, long chromosomes, was painted with the chromosome 15 library and could then be readily distinguished in shorter chromosomes (Fig. 5) (16). The fetus in this case had an unbalanced translocation karyotype, having inherited the der(15) chromosome but not the der(3). This was confirmed by both traditional cytogenetics and FISH in the chromosomes obtained from a placental biopsy.

We have worked on a case in which analysis of a complex translocation with chromosome painting revealed interstitial chromosome rearrangements not detectable with traditional banding methods. An amniotic fluid that appeared to have a *de novo* three-way translocation plus an inversion by G banding was found by painting to have a fourth chromosome involved, and not five but at least nine chromosomal breakpoints have been identified (Fig. 6) (17). Thus it is clear that chromosome painting cannot only increase resolution but also, in some cases, uncover rearrangements in metaphase chromosome structure not previously detected.

In situ hybridization has been used to detect submicroscopic deletions and, in some cases, to reveal the submicroscopic (cryptic) translocations from which they

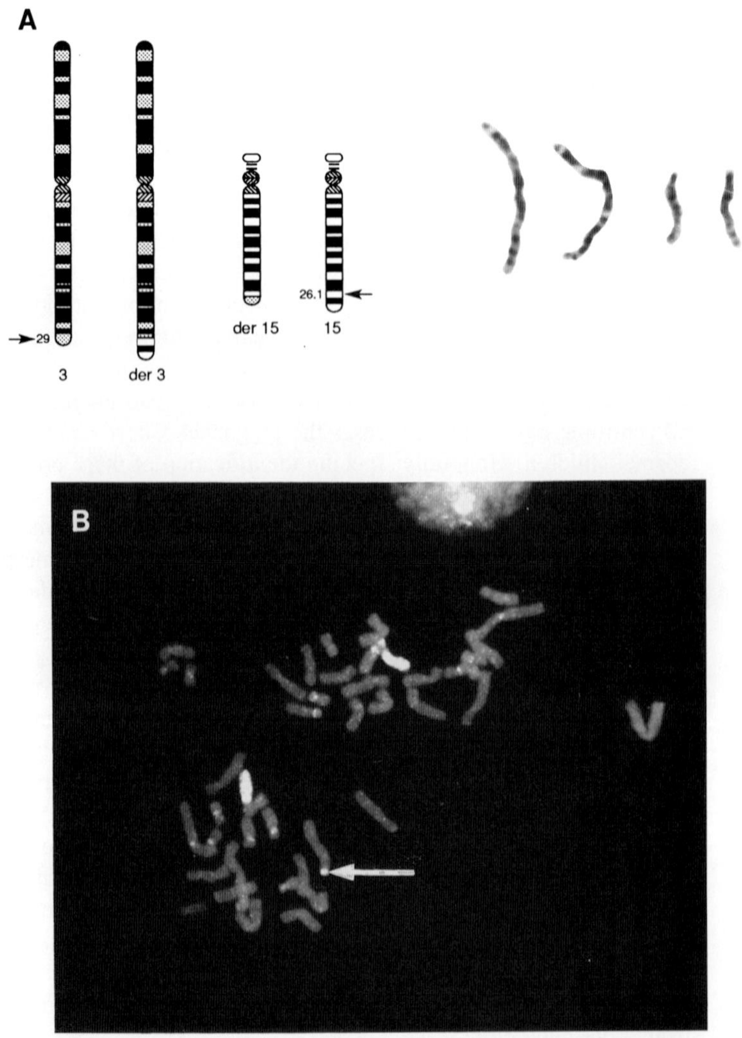

FIG. 5 A balanced reciprocal translocation involving chromosomes 3 and 15 (A) in G-banded chromosomes (right) and an idiogram (left) and (B) in a metaphase spread after painting with the chromosome 15 library. The arrow indicates the der(3) chromosome.

arose (18, 19). Fluorescent *in situ* hybridization with cosmid probes from critical regions of genes is used to localize specific sequences directly on chromosomes. Deletions are detected by the absence of such sequences. It is important to score only replicated chromosomes in which hybridization signals can be discerned on

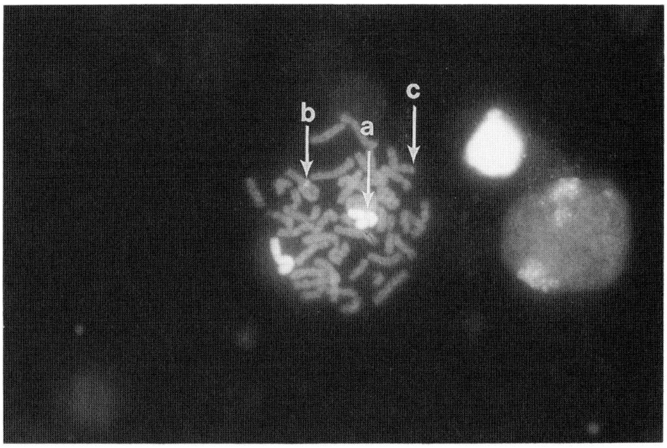

FIG. 6 Fluorescence *in situ* hybridization of a complex chromosomal translocation painted with the chromosome 4 library. The arrows mark a deletion in one chromosome 4 (a) and its interstitial insertion into a derived chromosome 9 (b), and its presence at the telomere region of an inverted derived chromosome 1 (c).

both chromatids, in order to distinguish a true hybridization signal from any background fluorescence. This method should be applicable to amniotic fluid and chorionic villi samples and make it possible to detect specific deletions prenatally. This will be an improvement over high-resolution G banding, which is not always as sensitive. It is also important to detect cryptic translocation carriers because their recurrence risk and risk of spontaneous abortion are different from individuals with normal karyotypes. Over the next few years probes will become available permitting detection by FISH of many of the deletion syndromes. This will be important for diagnosis, and in familial cases for prenatal detection as well.

Prospects

The new molecular approach to cytogenetics has both strengths and limitations. Fluorescent *in situ* hybridization has been shown to improve the resolution of metaphase chromosomes and to permit chromosome-specific identification in interphase nuclei. However, sometimes chromosome-specific libraries fail to show a particular translocation (16) because not all libraries have equal specificity and sensitivity in detecting different chromosomal regions. The uneven representation of DNA from different areas of the chromosome may be an artifact of cloning and subsequent amplification of the library, but can also result from preferential suppression of some areas of the target chromosomes. Interphase chromosome identification, although useful as a rapid means of identifying specific chromosomes, is not a substitute for complete metaphase chromosome analysis. Great care in interpreting data from patients must be taken if screening for specific chromosomal trisomies in interphase is to be offered as a prenatal diagnostic test in the future.

Methods

Protocols using the biotin–avidin system for probe visualization are outlined below. In our laboratory, the chromosome-specific repetitive DNA probes for the α-satellite regions and for the heterochromatic region of the Y chromosome have been obtained from Oncor, Inc. (Gaithersburg, MD), the American Type Culture Collection (Rockville, MD), or from the laboratory of P. L. Pearson (Johns Hopkins University School of Medicine, Baltimore, MD). Chromosome-specific libraries were a gift to P. L. Pearson from J. W. Gray (Lawrence Livermore National Laboratory, Livermore, CA). Methods have been adapted from Pinkel (15, 20).

Probe DNA is obtained already biotin-labeled or it may be biotinylated using biotin-dATP and a nick translation kit (Bethesda Research Laboratories, Bethesda, MD). The fluorescent detection of biotinylated DNA is dependent on the high affinity of the glycoprotein avidin for the vitamin biotin. The avidin is chemically linked to fluorescein isothiocyanate (FITC). Chromosomes are counterstained with propidium iodide to display chromosome morphology (orange-red). The probe signal can be amplified with biotinylated antiavidin and another layer of avidin–FITC. The components of the biotin–avidin system can be purchased from Vector Laboratories, Inc. (Burlingame, CA). Detection requires an epifluorescence microscope equipped with an FITC broad bandpass filter (excitation, 450–490 nm; emission, LP 520 nm). If a second filter set is available (excitation, 365 nm; emission, LP 420 nm) then chromosome bands can be simultaneously produced with 4,6-diamidino-2-phenylindole (DAPI). Slides are photographed with Kodak (Rochester, NY) Ektachrome 400 film at ISO 400.

Slide Preparation

Direct interphase FISH analysis of nuclei from a chorionic villus sample or from amniocentesis is accomplished as follows: cleaned villi (1–2 mg) or amniocytes (5 ml of amniotic fluid) are suspended in hypotonic solution [0.8% (w/v) sodium citrate] for 15–20 min and then fixed in 3:1 (v/v) methanol–acetic acid three times for 10 min each. Chorionic villi are dissociated in 50% aqueous acetic acid as the slides are prepared (21). Amniocytes are dropped onto slides directly from fixative. Cells can be concentrated on one central area of a slide because the hybridization is set up to take place under an 18 × 18 mm square coverslip.

Direct interphase preparations are also possible from white blood cells. Heparinized whole blood is allowed to settle in a syringe and the "buffy coat" fraction is separated from the plasma and red cells. These cells are suspended in hypotonic solution (0.075 *M* KCl) for 15 min, followed by fixation in 3:1 (v/v) methanol–acetic acid three times for 10 min each. Cells are dropped onto slides directly from fixative and slides are air dried. Flaming slides or extensive heating is thought to alter DNA structure, so this is intentionally omitted.

Standard metaphase preparations of chorionic villi, amniocytes, fibroblasts, lymphocytes, or lymphoblast lines all work well for FISH analysis. If slides are not used immediately they are stored at −70°C. However, we have had successful hybridizations with slides that have been stored at room temperature for several weeks. It is also possible to use previously G-banded slides that have been destained in 1:1 (v/v) methanol–acetic acid and dehydrated with an ethanol series.

Sperm can also be used for FISH (22). A sample of semen is washed three times with Hanks' balanced salt solution, fixed with 3:1 (v/v) methanol–acetic acid three times, and then dropped onto slides and air dried. Slides can be stored at this point at −70°C. Excess acetic acid is removed by washing the slides in 2× SSC and dehydrating through a 4°C ethanol series (70, 90, 100%). Slides are air dried and treated with 50 m*M* dithiothreitol in 0.1 *M* Tris-HCl, pH 8.0, at room temperature. We have found that a 12-min treatment is optimal for most samples. Slides are then washed twice in 2× SSC and finally dehydrated with a 4°C ethanol series and air dried.

Fluorescence in Situ Hybridization with Centromere-Specific Repetitive (α-Satellite) DNA Probes

Day 1

1. Dehydrate slides in a 4°C ethanol series (75, 85, 100%). This is especially important if slides have been freshly prepared.
2. Prepare the hybridization mixture:[*,†] 10–20 ng biotinylated probe, 60% formamide, 2× SSC (pH 7.0). The Oncor probes are at a concentration of 0.5 μg/50 μl: use 1 μl of probe/slide. Formamide should be spectral grade: use 6 μl/slide. The 20× SSC (3 *M* NaCl, 0.3 *M* sodium citrate) is adjusted to pH 7.0: use 1 μl / slide. Add deionized water to bring the final volume to 10 μl: use 2 μl/slide.
3. Put 10 μl of probe mixture on each slide, cover with an 18 × 18 mm coverslip and seal with rubber cement.

[*] Small volumes should be mixed in microfuge tubes and centrifuged briefly to collect the total volume.

[†] Store individual reagents at −20°C unless otherwise indicated by the manufacturer. Some laboratories advise making fresh stock solutions every month.

4. Denature the target DNA and the probe simultaneously by placing the slides in an 80°C oven for 5 min.

5. Incubate the slides in a moist chamber overnight at 42°C. Moisten the chamber with a 60% (v/v) formamide solution. We reuse formamide for this step from that used first in the slide washing (step 7).

Day 2

6. Remove the slides from the hybridization chamber and carefully remove the dried rubber cement. The coverslip will fall off in the wash.

7. Wash the slides three times for 5 min each in a Coplin jar in a freshly prepared solution of 50% formamide, 2× SSC (pH 7.0) at room temperature with gentle agitation on a rotating platform. All subsequent washes are at room temperature with agitation on a rotating platform.

8. Wash the slides three times for 5 min each in 2× SSC, pH 7.0.

9. Wash the slides once for 3 min in 4× SSC, 0.05% (v/v) Tween 20 (pH 7.0).

10. Put 100 μl of 4× SSC, 5% (v/v) nonfat dry milk (NFDM pH 7.0 on a 24 × 50 mm coverslip and touch the slide to the fluid.* Place the slides in a moist chamber) at room temperature for a 10-min incubation.

11. Shake off the coverslips.

12. Wash the slides three times for 5 min each in 4× SSC, 0.05% (v/v) Tween 20 (pH 7.0).

13. Put 100 μl of avidin–FITC [5 μg/ml 4× SSC, 5% (w/v) NFDM (pH 7.0)] on each slide as described in step 10. Incubate for 20 min. To prepare avidin–FITC we use a 5-mg/ml solution from Vector Laboratories and dilute 1 μl into 1 ml of 4× SSC, 5% (w/v) NFDM (pH 7.0).

14. Wash three times for 5 min each with 4× SSC, 0.05% (v/v) Tween 20 (pH 7.0).

15. Put 100 μl of biotinylated goat anti-avidin [2.5 μg/ml 4× SSC, 5% (w/v) NFDM] on each slide as described in step 10. Incubate for 20 min. To prepare biotinylated goat anti-avidin, we use a 0.5-mg/ml solution from Vector Laboratories and dilute 5 μl into 1 ml of 4× SSC, 5% (w/v) NFDM (pH 7.0).

16. Wash three times for 5 min each with 4× SSC, 0.05% (v/v) Tween 20 (pH 7.0).

17. Repeat step 13.

18. Wash three times for 5 min each in 4× SSC, 0.05% (v/v) Tween 20 (pH 7.0).

19. Wash once for 5 min in phosphate-buffered saline, pH 7.2.

*For steps 10, 13, 15, and 17 it is necessary to centrifuge the tubes of mixed reagents in a microfuge at 14,000g for 10 min at room temperature to remove any precipitate before applying 100 μl of supernatant to each slide. Incubation in steps 13, 15, and 17 should be carried out in the dark in the same moist chamber used in step 10.

20. Mount the slides with 20 μl antifade solution (23) with 0.4 μg/ml propidium iodide.

Antifade solution is prepared from 100 mg *p*-phenylenediamine dihydrochloride dissolved in 10 ml of phosphate-buffered saline (PBS). The pH is adjusted to 8.0 with 0.5 *M* carbonate–bicarbonate buffer. The solution is filtered through a 0.22-μm syringe filter and added to 90 ml of glycerol. Store in the dark at $-20°C$.

Propidium iodide stock solution is 1 mg/ml H_2O and is also stored at $-20°C$.

A mixture of 2 μl of propidium iodide stock in 5 ml of antifade solution is prepared and stored in the dark at $-20°C$.

The DAPI stock solution is 1 mg/ml H_2O. Two microliters can be added to the propidium iodide–antifade solution to permit simultaneous chromosome banding. The final concentration of DAPI is 0.4 μg/ml.

21. Keep the slides in the dark until ready to view.

22. Slides will last a week or two if a coverslip is added (24 $\times$ 30 mm), sealed with nail polish, and the slides stored in the refrigerator.

Fluorescence in Situ Hybridization with Whole-Chromosome Libraries

Day 1

1. Precipitate the labeled probe with sonicated total human placental DNA to inhibit hybridization of nonspecific repeated sequences. Mix 200 ng biotinylated probe, 26 μg sonicated total human placental DNA, 0.1 vol 3 *M* sodium acetate (pH 5.6), 2.5 vol 100% ice-cold ethanol.

Probes are stored at a concentration of 20 μg/ml; use 10 μl of probe per slide. Human placental DNA is at a concentration of 10 mg/ml: use 2.6 μl of placental DNA per slide. Use 1.3 μl 3 *M* sodium acetate (pH 5.6) per slide. Use 35 μl 100% ice-cold ethanol per slide.

2. Vortex the tubes for 30 sec.
3. Place the tubes in an ice bath for 30 min.
4. Centrifuge the tubes for 30 min in a microfuge at 14,000*g*, 4°C.
5. Remove most of the supernatant, invert the tubes, and air dry the pellets.
6. Add 10 μl of the master mix to the precipitated probe. The master mix [10% (w/v) dextran sulfate, 50% (v/v) formamide, 2$\times$ SSC (pH 7.0)] is prepared as:

Formamide	5 ml
Dextran sulfate (sodium slat)	1 g
SSC (20$\times$)	1 ml

Heat to 70°C for several hours to dissolve the dextran, cool, adjust the pH to 7.0 and the volume to 7 ml. Freeze the stock.

7. Dissolve the pellets by placing the tubes in a 37°C water bath for 30 min.
8. Denature the DNA at 75°C for 5 min.
9. Chill on ice for 1 min.

10. Allow the DNA to preanneal for 90 min at 37°C.

11. Denature the target DNA (metaphase chromosomes) by adding 100 μl of 70% (v/v) formamide, 2× SSC (pH 7.0), to each slide, cover with 24 × 50 mm coverslips, and incubate in a 75°C oven for 5 min.

12. Dehydrate the slides in a 4°C ethanol series (50, 70, 100%).

13. Add 10 μl of hybridization mix with probe to each slide, cover with 18 × 18 mm coverslips, and seal with rubber cement.

14. Incubate the slides in a moist chamber at 37°C for 3 days. Moisten the chamber with a 50% (v/v) formamide solution.

Day 4

15. Remove the slides from the hybridization chamber and carefully remove the dried rubber cement.

16. Wash the slides three times for 10 min each in 50% (v/v) formamide, 2× SSC (pH 7.0), at 42°C.

17. Wash the slides three times for 10 min each in 2× SSC, pH 7.0, at 42°C.

18. Wash the slides once for 5 min in 4× SSC, 0.05% (v/v) Tween 20 (pH 7.0), at room temperature.

19. The detection system is identical to that described for the centromere-specific DNA probes, with the exception of the time of incubation in avidin-FITC and anti-avidin, which is extended from 20 to 30 min in this protocol.

Acknowledgments

I would like to thank D. A. S. Batista, G. Prabhakar, and V. Jaswaney for help with these protocols, and F. Karas for secretarial assistance.

References

1. K. Dyer and J. Meyne, *in* "The ACT Cytogenetics Laboratory Manual" (M. J. Barch, ed.), 2nd Ed., p. 525. Raven, New York, 1991.
2. B. J. Trask, *Trends Genet.* **7**, 149 (1991).
3. J. C. Hozier and L. M. Davis, *Anal. Biochem.* **200**, 205 (1992).
4. D. C. Peakman, M. F. Morton, and A. Robinson, *J. Med. Genet.* **14**, 37 (1977).
5. G. Simoni, B. Brambati, C. Danesino, F. Rossella, G. L. Terzoli, and M. Fraccaro, *Hum. Genet.* **63**, 349 (1983).
6. G. Prabhakar, G. Stetten, V. L. Jaswaney, and J. J. Blakemore, unpublished.
7. T. Cremer, J. Landegent, A. Bruckner, H. P. Scholl, M. Schardin, H. D. Hager, P. Devilee, P. Pearson, and M. van der Ploeg, *Hum. Genet.* **74**, 346 (1986).
8. K. J. Blakemore, C. Rosenberg, V. L. Jaswaney, E. K. Pressman, W. G. Kearns, P. L. Pearson, and G. Stetten, submitted (1993).

9. K. W. Klinger, G. Landes, D. Shook, R. Harvey, L. Lopez, P. Locke, T. Lerner, R. Osathanondh, B. Leverone, T. Houseal, K. Pavelka, and W. Dackowski, *Am. J. Hum. Genet.* **51**, 55 (1992).
10. W.-L. Kuo, H. Tenjin, R. Segraves, D. Pinkel, M. S. Golbus, and J. Gray, *Am. J. Hum. Genet.* **49**, 112 (1991).
11. S. C. Raimondi, F. W. Luthardt, R. L. Summitt, and P. R. Martens, *Hum. Genet.* **63**, 310 (1983).
12. G. Stetten, K. J. Blakemore, A. M. Courter, C. A. Coss, and E. W. Jabs, *Prenatal Diagn.* **12**, 83 (1992).
13. C. Collins, W. L. Kuo, R. Segraves, J. Fuscoe, D. Pinkel, and J. W. Gray, *Genomics* **11**, 997 (1991).
14. P. Lichter, T. Cremer, J. Borden, L. Manuelidis, and D. C. Ward, *Hum. Genet.* **80**, 224 (1988).
15. D. Pinkel, L. Landegent, C. Collins, J. Fuscoe, R. Segraves, J. Lucas, and J. Gray, *Proc. Natl. Acad. Sci. U.S.A.* **85**, 9138 (1988).
16. C. Rosenberg, K. J. Blakemore, W. G. Kearns, R. A. Giraldez, C. S. Escallon, P. L. Pearson, and G. Stetten, *Am. J. Hum. Genet.* **50**, 700 (1992).
17. D. A. S. Batista, C. M. Tuck-Muller, J. E. Martinez, W. G. Kearns, P. L. Pearson, and G. Stetten, submitted (1993).
18. A Kuwano, S. A. Ledbetter, W. B. Dobyns, B. S. Emanuel, and D. H. Ledbetter, *Am. J. Hum. Genet.* **49**, 707 (1991).
19. M. R. Altherr, U. Bengtsson, F. F. B. Elder, D. H. Ledbetter, J. J. Wasmuth, M. E. McDonald, J. F. Gusella, and F. Greenberg, *Am. J. Hum. Genet.* **49**, 1235 (1991).
20. D. Pinkel, T. Straume, and J. W. Gray, *Proc. Natl. Acad. Sci. U.S.A.* **83**, 2934 (1986).
21. S. W. Cheung, J. P. Crane, H. A. Beaver, and A. C. Bergsers, *Prenatal Diagn.* **7**, 535 (1987).
22. M. H. E. C. Pieters, J. P. M. Geraedts, H. Meyer, J. C. M. Dumoulin, J. L. H. Evers, R. J. E. Jongbloed, P. M. Nederlof, and S. van der Flier, *Cytogenet. Cell. Genet.* **53**, 15 (1990).
23. G. D. Johnson, and G. M. de C. Nogueira Araujo, *J. Immunol. Methods* **43** 349 (1981).

Section IV

DNA and DNA–Protein Interactions

[14] Local Structures in Supercoiled DNA

Richard Bowater, Fareed Aboul-ela, James McClellan, Alastair I. H. Murchie, and David M. J. Lilley

DNA Structural Polymorphism

The basic structure of DNA in solution is the right-handed B-form double helix, but this may be altered in a variety of ways as a function of local base sequence, conditions such as ionic strength, temperature, and pH, and topology. Deformation from the standard B conformation may be caused by relatively minor changes at the level of the single base pair (1), or by more major alterations involving large regions of DNA. At the extreme, almost no feature of the B structure is incapable of being changed. This has been the subject of two detailed reviews (2, 3).

Z-DNA formed by alternating purine–pyrimidine sequences (4) is left handed, and contains *syn* purine bases. Runs of oligoadenine phased with the period of the helix generate a curved axial trajectory (5–10); the bases in the A tract exhibit high propellor twist, leading to an array of bifurcated hydrogen bonds in the major groove (11, 12). A number of altered DNA structures involve extensive changes in base pairing, the simplest of which is the formation of a melted region of helix in an (A + T)-rich region (13–15). Cruciform structures are formed from inverted repeat sequences by formation of paired intrastrand hairpins (16–18), whereas H-triplex formation by mirror-repeat sequences generates a three-stranded structure based on conventional A-T or G-C base pairs to which are fused third bases (T or protonated C, respectively) on the major groove side by Hoogsteen bonding (19–23). The number of strands may be further increased to four, based on the guanine tetrad (24–27). These structures may be intramolecular and parallel (24), or intermolecular and antiparallel (25, 26). Conformational transitions have been described between B-DNA and other structures, and their thermodynamic and kinetic properties analyzed.

Not all these structures have been ascribed clear biological roles. Protein binding may involve either flexural (28, 29) or torsional (30) distortion of DNA, and binding sites may be predisposed to these deformations. Regions to be bent around proteins, including histones, frequently have a phased arrangement of A · T and G · C base pairs (8) to permit compression of the minor groove on the inside of the bent region. A-tract bending, first discovered in kinetoplast network DNA, appears to be important in a number of situations (8, 31–33), and sequences with a propensity to adopt G-tetraplex structure are found at the telomeres of eukaryotic chromosomes (25, 26). Some structures may have an importance as models for species that enjoy a brief existence only as transient intermediates. For example, although no biologi-

Methods in Molecular Genetics, Volume 1

cal role has been found for a cruciform, its structure is closely related to the four-way junction believed to be important in recombination processes (34–44), and the structure has been extensively investigated (45–49).

DNA Supercoiling and Local DNA Structure

The formation of many perturbed DNA structures, such as melted regions, Z-DNA, cruciforms and H-triplexes, involves a net underwinding of the helix, and as a consequence these are stabilized in negatively supercoiled DNA molecules. For this reason we frequently want to study them embedded in supercoiled circular DNA molecules; this also provides the opportunity to apply powerful topological methods for their study.

Most natural DNA species potentially exist as supercoiled molecules (50). The most fundamental property of DNA supercoiling is the linking number (Lk), which measures the number of times one strand is linked with the other in a circular DNA molecule. This is related to two geometric properties of the molecule, the twist (Tw, rotation of the strands about the helical axis) and the writhe (Wr, which measures the path of the helix axis in space). These three properties are related (51) by Eq. (1):

$$\mathrm{Lk} = \mathrm{Tw} + \mathrm{Wr} \tag{1}$$

A relaxed closed circular DNA molecule has a linking number (Lk°) given by Eq. (2),

$$\mathrm{Lk}^\circ = N/h \tag{2}$$

where N is the number of base pairs, and h is the helical repeat under the experimental conditions. A negatively supercoiled molecule has a linking deficit ($\Delta\mathrm{Lk}$) relative to the relaxed species, that is,

$$\Delta\mathrm{Lk} = \mathrm{Lk} - \mathrm{Lk}^\circ < 0 \tag{3}$$

whereas for a positively supercoiled molecule $\Delta\mathrm{Lk}$ is positive. The degree of supercoiling is frequently expressed in terms of the specific linking difference, or superhelix density (σ) given by

$$\sigma = \Delta\mathrm{Lk}/\mathrm{Lk}^\circ \tag{4}$$

For most circular DNA molecules isolated from cells this takes a value of about -0.06, that is, the DNA is negatively supercoiled equivalent to an unwinding of about 1 turn in 17. In the absence of strand breakage the Lk value is constant, and therefore the sum of twist and writhe changes is constant in any structural change that maintains strand integrity. Thus the linking deficit is partitioned between geometric alterations in the molecule of torsional and flexural character:

$$\Delta\text{Lk} = \Delta\text{Tw} + \Delta\text{Wr} \tag{5}$$

Both deformations are energetically unfavorable, and a supercoiled DNA molecule has a higher free energy compared to its relaxed isomer (52–54),

$$\Delta G^\circ_{sc} = (1050RT/N)\Delta\text{Lk}^2 \tag{6}$$

where ΔG°_{sc} is the free energy of supercoiling, R is the gas constant, and T the absolute temperature. Thus any process that brings about a relaxation of superhelical stress will be favored by negative supercoiling, and this is the basis of the stabilization of a number of structural variants.

If the formation of a perturbed DNA structure contributes a local negative twist change, this brings about a partial relaxation of the superhelical stress. This reduction in the free energy of supercoiling offsets the free energy of formation of the new DNA structure, and helps stabilize the altered conformation. Because the free energy of supercoiling increases quadratically with linking difference [Eq. (6)], there exists a level of negative supercoiling above which the new structure has a stable existence, and this has been experimentally demonstrated for structures such as cruciforms (16–18), Z-DNA (55, 56), and H-triplex DNA (20). These arguments apply in reverse for a structure that brings about an overwinding, which should therefore be stabilized by *positive* supercoiling. The first example of this has been demonstrated experimentally (57).

Preparation of Supercoiled Plasmid DNA

Supercoiled DNA must be prepared and handled with care because cleavage anywhere within the entire circular molecule completely changes the topology and releases the superhelical constraint. Moreover the molecules are more sensitive to damage than normal DNA because of their topology, and conditions that lower helix stability should be avoided. When preparing DNA for structural studies we adopt procedures based on gentle lysis of cells, followed by isopycnic gradient centrifugation of the supercoiled plasmid DNA in cesium chloride/ethidium. The separation depends on the topology of the DNA, and specifically separates constrained from nicked DNA circles; it therefore gives a higher yield of supercoiled DNA than other methods. The DNA is taken from the gradient as positively supercoiled DNA in the presence of a high concentration of the intercalator, and is free of cruciform and other structures that require negative supercoiling. Subsequent careful removal of the ethidium can leave the DNA free of some (such as most cruciform), but not all, perturbed structures.

The following is the procedure used in this laboratory to prepare supercoiled plasmid DNA for physical studies. This is based on pBR332-based plasmids transformed into a common laboratory *Escherichia coli* strain such as HB101.

Growth and Lysis of Bacterial Cells

Exact conditions for growth will depend on the strain and plasmid required; the following applies to a pBR322-type plasmid grown in HB101. Prepare 500 ml M9 medium in a 2-liter culture flask, containing the following:

KH_2PO_4	1.5 g
Na_2HPO_4	3.0 g
NaCl	0.25 g
NH_4Cl	0.5 g

and autoclave. When cool, add the following, using aseptic technique:

Glucose (40%)	2.5 ml
Casamino acids (10%)	25.0 ml
Thiamine (1 mg/ml)	1.0 ml
$MgSO_4$ (1 *M*)	0.5 ml
Thymidine (0.25%)	2.0 ml

Antibiotic for selection is added at this stage. Innoculate with 5 ml of a fresh overnight culture of the plasmid-carrying strain, grown in LB medium plus antibiotic. Shake the flask with good aeration at 37°C until the cell density gives an A_{660} of 0.6 (usually 3–4 hr after innoculation). If chloramphenicol amplification is required, 75 mg chloramphenicol is added at this stage, and shaking continued for a further 16–18 hr. Longer growth can result in lower levels of supercoiling in the extracted plasmid.

The cells are collected by centrifugation at 7500 rpm for 10 min at 4°C. The supernatant is disposed of according to local safety regulations. The cellular pellet is resuspended in 7.5 ml 25% (w/v) sucrose in 50 m*M* Tris, pH 8.0, and held on ice for 10 min. The following additions are made, with constant swirling to ensure good lysis:

1. 1.5 ml of 10 mg/ml lysozyme in 25 m*M* Tris, pH 8.0; hold on ice 10 min
2. 3 ml 250 m*M* ethylenediaminetetraacetic acid (EDTA); hold on ice 10 min

3. 3.1 ml 5 *M* NaCl; hold on ice 10 min
4. 1.75 ml 10% (w/v) sodium dodecyl sulfate (SDS)

At this point cell lysis occurs, and the suspension becomes viscous. The lysate is held on ice for 4–20 hr.

Isopycnic Gradient Centrifugation of Plasmid DNA

The lysate is cleared by centrifugation at 20,000 rpm for 45 min at 4° C, the supernatant recovered, and its volume measured. To this is added 0.97 g of CsCl and 25 μl of a 10-mg/ml solution of ethidium bromide, per milliliter of supernatant. This mixture is centrifuged at 10,000 rpm for 10 min at 4° C, and the clear solution carefully removed from the dark pellicle and placed in ultracentrifuge tubes. Centrifugation is performed at 54,000 rpm at 15° C for 7–16 hr using a vertical rotor [Beckman (Fullerton, CA) VTi 65]. The (lower) plasmid band is recovered by side puncture (it is usually quite visible in daylight; a UV light source can be employed, but exposure should be minimized to reduce the danger of UV-induced damage of supercoiled DNA), and combined in a fresh centrifuge tube, which can be filled if necessary with a solution of 0.97 g/ml CsCl, 250 μg/ml ethidium bromide. This is recentrifuged at 54,000 rpm at 15° C for at least 10 hr. The plasmid DNA is again recovered by side puncture, transferred to a microcentrifuge tube, and placed on ice.

Preparation of Cruciform-Free Supercoiled DNA

At this stage the DNA is still in the presence of a high concentration of ethidium bromide, and as a consequence it is positively supercoiled, and free of any structures that are underwound, including cruciforms. To regenerate negatively supercoiled DNA the ethidium bromide must be removed. This may be done using conditions that do not result in the formation of cruciform structures, at low temperature, and with avoidance of helix-destabilizing conditions. We employ repeated extraction into 1-butanol to remove ethidium bromide, followed by dialysis.

An equal volume of 1-butanol at 0° C is added to the plasmid solution, and vortexed. The layers are separated, and the organic (upper) phase removed. A further volume of 1-butanol is added, and the entire procedure repeated seven times. Care should be taken throughout to maintain a low temperature. Following extraction of ethidium bromide, the negatively supercoiled DNA is extensively dialyzed against 10 m*M* Tris (pH 7.5), 0.1 m*M* EDTA at 4° C, to remove CsCl. On removal from the dialysis sack the DNA is not manipulated further, and can be used directly for kinetic or other measurements.

The yield of DNA depends on the plasmids and strains used, but typical yields are 200 μg to 2 mg DNA from 1 liter of culture. The concentration after the second gradient is normally in the range 200–1000 μg/ml. The DNA should be >95% supercoiled at the end of the preparation. We store the supercoiled DNA in aliquots at -70°C, and gently thaw the frozen DNA at 4°C just before use.

Two-Dimensional Gel Electrophoresis

Gel electrophoresis has proved to be an immensely powerful method for the study of nucleic acid conformation and DNA–protein interactions. Polyacrylamide and agarose gels have great resolving power, and are remarkably sensitive to the shape and sizes of nucleic acids. At one extreme sequencing gels will clearly distinguish on the basis of size between single-stranded DNA fragments differing by a single base, whereas at the other extreme topological differences between plasmids of several kilobases in size may be distinguished by electrophoresis in agarose. Polyacrylamide gels are extremely sensitive to curvature (5–7, 9) or other kinds of kinking in DNA fragments (45–47), and may even distinguish DNA, RNA, and hybrid molecules of identical size and sequence (58).

The frictional properties of supercoiled DNA molecules are strongly influenced by the topology of the molecule. In general as a DNA molecule becomes more supercoiled (i.e., when the linking difference becomes either more negative or more positive), the molecule adopts a more compact shape, and this may be readily revealed by a greater electrophoretic mobility in an agarose gel. If the molecule undergoes a local structural transition that is coupled to the topology, this may therefore be seen as a discontinuous trend in the mobility of topoisomers.

DNA supercoiling can stabilize altered DNA structure that leads to a local change in twist. On formation of the new structure, the change in twist leads to a compensatory change in writhe, because the linking number of the molecule remains constant throughout. The change in writhe and consequent altered shape of the molecule leads to a change in the electrophoretic mobility; migration becomes slower due to the partial relaxation of writhing. Thus the molecule carrying the new structure will migrate more slowly than a molecule of the same linking difference that has not undergone the structural transition. By measuring the amount by which the migration is retarded, an estimate of the twist change for the transition may be made. Two-dimensional gel electrophoresis (59) provides a simple way to reveal topology-dependent structural transitions and measure some parameters for the process. A more detailed description of two-dimensional gel electrophoresis may be found in Bowater *et al.* (60).

In two-dimensional gel electrophoresis a complete set of topoisomers is electrophoresed from a single well of an agarose gel (first dimension, D1) in the absence of intercalating agents. The gel is then soaked for several hours in an intercalating drug, usually chloroquine. The effect of the intercalation is a local unwinding, but because of the invariance of the linking number, the negative twist change must be compensated by a positive writhe, that is, the molecule becomes more positively supercoiled. The extent of positive supercoiling is directly proportional to the amount of chloroquine added. The gel is then electrophoresed for a second time (second dimension, D2), in a direction 90° to the first direction, with the same chloroquine concentration present in the running buffer. Those topoisomers with the lowest negative linking difference become the most positively supercoiled species in the second dimension; in other words, the topoisomers that were most negatively supercoiled (and therefore migrated fastest) in the first dimension, become the least positively supercoiled (hence slowest) in the second dimension. The first few spots (greatest linking number) are topoisomers that were actually slightly positively supercoiled in the first dimension. We then pass through the position of relaxation (and therefore slowest migration in D1) in the first dimension. This is the position where Lk $= N/h$, where h is the helical periodicity in the absence of chloroquine. Moving to increasing linking difference the migration in D1 increases with $-\Delta$Lk, and this continues until a plateau is reached—this is the point at which the migration as a function of ΔLk has saturated in the first dimension.

The second dimension may be run with a lower concentration of chloroquine, giving improved resolution of topoisomers of lower linking difference. Because the concentration of chloroquine is now lower, not all the topoisomers are positively supercoiled in the second dimension. The topoisomers of higher $-\Delta$Lk remain negatively supercoiled throughout the experiment, although their negative supercoiling is proportionately reduced in D2. At the point at which Lk $= N/h'$, where h' is the helical periodicity in the presence of the prevailing chloroquine concentration, topoisomers are relaxed in D2, and the spots appear to "turn around" at this position.

If the plasmid studied has a sequence that can undergo a structural change giving a local negative change in twist, this will generate a relaxation in the level of supercoiling. However, if the transition requires a threshold level of supercoiling to occur (i.e., if the transition would be unfavorable in relaxed DNA), the altered structure is absent in topoisomers of lower $-\Delta$Lk. At low $-\Delta$Lk the topoisomer spots migrate equivalently to those of an equivalent plasmid that cannot undergo the transition. Above the threshold level of $-\Delta$Lk the perturbed structure may form stably in the molecule, and hence there is a relaxation in the first dimension of the gel. However the presence of the chloroquine in the second dimension effectively reduces the level of negative supercoiling, and hence the alternative conformation is destabilized and these topoisomers migrate with their normal mobility in D2. Thus a structural transition is revealed by topoisomer spots having their normal second dimension mobility, but a retarded mobility in the first dimension. This is seen as a discontinuous "jump," whereby the topoisomer spots describe a new line of slower mobility. An example of the use of two-dimensional gel electrophoresis to observed cruciform formation by an $(AT)_{34}$ sequence is shown in Fig. 1.

From these gels three pieces of information may be obtained. First, by counting

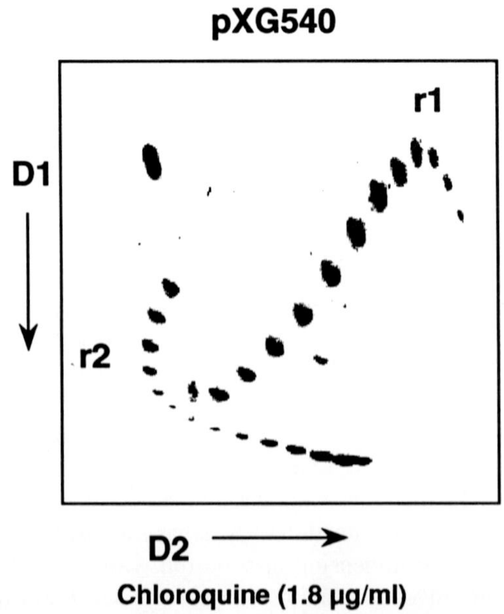

FIG. 1 Two-dimensional gel electrophoresis of a plasmid exhibiting cruciform formation by an $(AT)_{34}$ sequence. A complete set of topoisomers of the plasmid pXG540 [D. R. Greaves, R. K. Patient, and D. M. J. Lilley, *J. Mol. Biol.* **185,** 461 (1985)] of specific linking difference from 0.01 to -0.07 was electrophoresed from a single well at the top left-hand side of the gel as shown. The first dimension (D1) was electrophoresed in TBE at room temperature. The gel was then soaked in TBE containing 1.8 μg/ml chloroquine before electrophoresis was resumed in direction D2 in TBE, 1.8 μg/ml chloroquine. This relatively low concentration of chloroquine induces positive supercoiling in only the topoisomers of lower negative linking difference ($-\Delta Lk < 12$), and topoisomers of greater $-\Delta Lk$ remain negatively supercoiled throughout the experiment. For this reason the mobility of the topoisomers "turn around" in the second dimension at the point r2, corresponding to $Lk = N/h'$, where h' is the helical periodicity in the presence of the intercalator. In the first dimension the topoisomers have their lowest mobility at r1, corresponding to $Lk = N/h$, where h is the helical periodicity in the absence of chloroquine. There is a cooperative transition due to the formation of a cruciform by the $(AT)_{34}$ sequence, corresponding to a relaxation of approximately six helical turns, at a $-\Delta Lk$ of 9. This is clearly seen as a discontinuous "jump" in electrophoretic mobility.

the topoisomer spots from the relaxed (first dimension) position, the lowest value of $-\Delta Lk$ that supports the new structure can be measured. Second, by counting back from a given shifted topoisomer to the equivalent unshifted position with the same mobility in the first dimension, the twist change for the transition may be measured (assuming that the presence of the new local DNA structure does not itself alter the overall gel mobility of the circular molecule; this may be a better approximation for some structures than others). With a reasonable separation of topoisomers, this can be measured with an accuracy of ± 0.2. Finally, by combining these two pieces of information, we may calculate the free energy of formation of the new structure, ΔG_f, according to:

$$\Delta G_f^\circ = (1050RT/N)(L_j^2 - L_{j-t}^2) \quad (7)$$

where L_j is the linking difference of topoisomer j, and L_{j-t} is the linking difference of the unshifted topoisomer of equal mobility. The constant is taken from Refs. 52–54. Ideally a statistical mechanical model of the transition can be used to model the transition as a function of superhelix density, and from the fit between the model and the experimental data the thermodynamic parameters for the transition may be estimated (15, 61–66).

Two-dimensional gels may also be used as a method for spreading out a wide distribution of topoisomers. For example, this has been used in the study of the effects of transcription on the topology of cellular DNA (67, 68). The superhelix density of a narrower distribution of topoisomers may also be assigned by loading the distribution 1–2 hr ahead of the full distribution on a two-dimensional gel.

Variables in Two-Dimensional Gel Electrophoresis

The conditions of gel electrophoresis can be varied to suit the structure it is intended to study. Because the perturbed structure is present only during the first dimension, it is not necessary to run the second dimension under any special conditions, beyond the choice of intercalator concentration. Potentially variable parameters include the following.

First Dimension

DNA structures respond to environmental conditions in different ways, and the following conditions may be readily varied.

pH

Structures that require base protonation are more stable at lower pH. For example, the H-triplex requires cytosine protonation for the formation of $C \cdot G \cdot CH^+$ triple base pairs (19–23).

Buffer Composition

Many perturbed structures require certain cations for their stability, such as the requirement for divalent metal ions for the folded conformation of cruciform structures (47). Z-DNA is strongly stabilized by hexamminecobalt(III) (55), and this may be included in the buffer for the first dimension. Some transitions, such as helix melting in (A + T)-rich regions, are promoted by low ionic strength. With all such experiments the electrophoresis buffer should be recirculated at >1 liter/hr, and the pH of the buffer compartments checked at the end of the electrophoresis.

Temperature

Most transitions are temperature dependent, and in some cases (e.g., thermal melting) this is very important. To study helix–coil transitions in (A + T)-rich sequences in supercoiled DNA we use an apparatus for horizontal slab gel electrophoresis in which the gel is placed on top of a block through which water is circulated from a water bath. The temperature is measured by means of a platinum resistance element inserted into the gel itself.

Second Dimension

The conditions of the second dimension do not affect the structural transition. The main variable in the second dimension is the concentration of chloroquine chosen. A relatively low concentration of chloroquine is used to obtain optimal resolution of topoisomers around the relaxed position in the first dimension. Using an electrophoresis buffer of 90 m*M* Tris–borate (pH 8.3), 1 m*M* EDTA (TBE) we employ a concentration of 1–2 μg/ml chloroquine. These are good conditions for the study of transitions that occur at relatively low levels of supercoiling. For transitions that occur at higher values of $-\Delta Lk$, more chloroquine is required; we use 5–20 μg/ml chloroquine in TBE.

A further technique is possible with two-dimensional gels when a perturbed structure is either abnormally sensitive or resistant to a particular probe. For example a supercoil-dependent structure may be sensitive to cleavage by a particular nuclease, whereas the unperturbed molecule is not. If the distribution is incubated with the enzyme prior to electrophoresis, only the topoisomers that have undergone the transition should be cleaved, and thus these topoisomers will be selectively absent from the resulting gel. This may reveal the correspondence between topological and probing studies on a particular species.

Preparation of Topoisomer Distributions

A good distribution of topoisomers is an essential requirement for two-dimensional gel electrophoresis. This should cover the required range of topoisomers of $\Delta Lk/Lk^\circ$ from 0.01 to -0.08, and should have equal concentrations of all topoisomers. The distribution is generated by the action of a eukaryotic topoisomerase I on supercoiled DNA in the presence of a range of concentrations of ethidium bromide.

TABLE I Topoisomerase Incubations for Preparation of Topoisomer Distributions Required for Two-Dimensional Gel Electrophoresis[a]

Component	0	0.75	1.5	2.25	3.0	3.75	4.5	5.5	6.5	8.0
DNA	9	9	9	9	9	9	9	9	9	9
EB (5 μg/ml)	—	3	6	—	—	—	—	—	—	—
EB (10 μg/ml)	—	—	—	4.5	6	—	—	—	—	—
EB (20 μg/ml)	—	—	—	—	—	3.75	4.5	5.5	—	—
EB (40 μg/ml)	—	—	—	—	—	—	—	—	3.25	4
10× buffer	2	2	2	2	2	2	2	2	2	2
Water	8	5	2	3.5	2	4.25	3.5	2.5	4.75	4
Topoisomerase I	1	1	1	1	1	1	1	1	1	1

[a] Volumes of each component are given in microliters. A 300-μg/ml stock solution of supercoiled DNA is incubated with rat liver topoisomerase I, in the presence of a final concentration of ethidium bromide (EB) between 0 and 8 μg/ml. The 10× buffer stock used for the incubation is 100 mM Tris (pH 8.0), 1 M NaCl, 200mM EDTA. The concentrations of ethidium bromide may need to be changed if a different incubation buffer appropriate to an alternative topoisomerase is employed.

Ten 20-μl incubations with ethidium bromide concentrations of between 0 and 8 μg/ml are set up according to Table I. After the addition of enzyme, the samples are incubated at room temperature for 60 min. The DNA is then precipitated with ethanol, sodium acetate at $-70°$C, and the pellets redissolved in 20 μl 10 mM Tris (pH 7.5), 0.1 mM EDTA. These solutions are stored at $-20°$C.

Each incubation generates a Gaussian distribution of topoisomers about a mean, and the ethidium bromide concentration determines the position of the mean. We normally examine a new set of distributions by electrophoresis in 1% (w/v) agarose in TBE in one dimension. From the intensity and positions of the distributions, it can be judged how to mix the separate distributions to produce one complete distribution of topoisomers covering the complete range of ΔLk. A sample of the complete topoisomer distribution containing 2.5 μg of DNA in a volume of 8 μl is required for the electrophoresis. The DNA concentration is important, as the window between overloading and underloading is narrow. To this is added 2 μl of a solution of 0.35 mg/ml Ficoll 70, containing 1 mg/ml of xylene cyanol.

Procedure for Two-Dimensional Gel Electrophoresis

A 1% (w/v) agarose gel in the required running buffer is made by boiling the agarose in the appropriate buffer in a microwave oven until completely dissolved, followed by cooling to 60°C before pouring. Gels are 22 × 25 × 1 cm; the exact dimensions are not critical, but the gel should be approximately square, and the bigger the gel, the greater will be the resolution of topoisomers. A single circular well of about 2-mm diameter is placed at one corner (the top left as conventionally drawn). The sample is applied to this well, and electrophoresis carried out for 16–20 hr at 3.5 V/cm. The exact length of time will depend on the dimensions of the gel, the size of the plasmid, and the environmental conditions. The gel is soaked in TBE buffer containing the required concentration of chloroquine diphosphate for 6–10 hr in the dark, after which the gel is reassembled into a standard horizontal electrophoresis apparatus (temperature control is not important for the second dimension) such that the second dimension of electrophoresis is at 90° to the first. The running buffer for the second dimension is TBE containing the same chloroquine concentration that was used for the equilibration. Following electrophoresis in the second dimension the gel is stained in 1 μg/ml ethidium bromide in water for 1 hr, followed by extensive destaining in water at 4°C. The gel can then be photographed using a transilluminator at 300 nm. We photograph our gels with Kodak (Rochester, NY) Tri-X pan film, using red and green filters in front of the camera lens.

Probing Methods

When DNA structure becomes altered from the normal B-DNA geometry, this frequently results in an altered backbone conformation, or the exposure of bases, in such a way that the DNA is locally more sensitive to attack by an enzyme or chemical probe. For example, the N-7 atoms of guanine bases in the Z conformation are unusually reactive to carbethoxylation by diethyl pyrocarbonate (69), whereas the unpaired bases of a cruciform loop are reactive to a number of chemical agents (70–74). This may create a local hyperreactivity that may be the first indication of a perturbed DNA structure (17, 18, 70, 75–81). Such probes may then be used to gain further information about the local stereochemistry.

Enzyme and chemical probing experiments have generated important insight into a number of altered DNA structures, including cruciform structures (70–74), left-handed Z-DNA (69, 82–87), A-tract curved DNA (88, 89), H-triplexes (21, 22, 90–92), G-tetrad structures (24–26), and parallel stranded DNA (93), as well as exploring more subtle variations occurring in B-DNA (8, 94–98). Chemical probes have also been used in the study of DNA structure in the vicinity of bound drug molecules (99–101) or proteins (97, 102–104) and at base mismatches (105, 106). The latter is becoming important in the detection of single-base mutations by means of heteroduplexed molecules. A battery of probes is now available for the study of DNA structure.

Probing DNA with Enzymes

Nucleases are the most common enzymes used in the study of DNA structure, probing the accessibility of each phosphodiester bond. Some, such as DNase I and

S1 nuclease, may be relatively sequence neutral, whereas others, notably the restriction enzymes, may be highly sequence specific. A disadvantage of nuclease probes for probing supercoiled substrates is the relaxation brought about by cleavage anywhere within the circular molecule. The restriction methylases may be used to probe accessibility at particular locations. Enzyme probes have differing requirements for reaction conditions, but in general a given enzyme may have fairly strict requirements of buffer composition and conditions. Most enzymes will not tolerate extremes of ionic strength, pH, or temperature.

S1 Nuclease Cleavage

S1 nuclease (107) is a single strand-specific nuclease that has been extensively used for probing nucleic acid structure. Supercoiled DNA (0.1–1.0 μg) is incubated in 50 m*M* sodium acetate (pH 4.6), 50 m*M* NaCl, 1 m*M* $ZnCl_2$, on ice for 1–2 hr, in a volume of 20 μl containing 5 units of S1 nuclease. The DNA is then cleaved to completion with a restriction enzyme, radioactively labeled if required, and examined by gel electrophoresis.

Micrococcal Nuclease Cleavage

Micrococcal nuclease has been widely used to probe DNA structure. Incubations are typically carried out in 15 m*M* Tris-HCl (pH 7.4), 75 m*M* NaCl, 1 m*M* $MgCl_2$, 2 m*M* $CaCl_2$, 0.2 m*M* EDTA, and 15 m*M* 2-mercaptoethanol, on ice using 1.7×10^{-4} units of micrococcal nuclease in a 20-μl volume for 20–30 min.

Chemical Probes of DNA Structure

Chemical reagents will normally withstand a wider range of conditions than enzymes, and can thus be more useful for the study of structural transitions in response to changing conditions. For example, osmium tetroxide may be used over a wide range of conditions of cation chemistry and concentration, pH, temperature, and buffer composition. Table II summarizes the more commonly used chemical probes of DNA structure. Chemical probes can approach perfect sequence neutrality, such as hydroxyl radicals that attack the deoxyribose groups of the backbone (98), or can exhibit sequence selectivity. A further "chemical" probe that should not be forgotten is the photon, which can be used in a variety of photochemical experiments (108–110). In the following sections we present conditions for the use of the more commonly employed chemical probes.

TABLE II Chemical Probes Commonly Used to Study DNA Structure[a]

Probe	Comments	Refs.
Haloacetaldehydes	A > C base, etheno adduct	(70, 76, 122, 123)
Osmium tetroxide	T base, 5,6-diester adduct; cleaved with piperidine	(71, 112)
Permanganate	T base, diol product; cleaved with piperidine	(104)
DEPC	A > G base; carbethoxylation; cleaved with piperidine	(69, 119)
Glyoxal	G base, etheno adduct[b,c]	
Bisulfite	C base; deamination to dU[d]	(72)
Hydroxylamine	C base	(83, 120)
DMS	G base; N-7 methylation; cleaved with piperidine	(121)
Ethylnitrosourea	Phosphate ethylation[e]	
Carbodiimide	Single-strand-specific T and G base modification[f,g]	
EDTA-Fe	Deoxyribose cleavage by HO• generated uniformly	(98)
MPE-Fe	Binding of complex to DNA; strand scission[h]	
Cu(*o*-phen)	Attack on deoxyribose leads to backbone scission[i,j]	
UV light	Formation of cyclobutane and [6,4]pyrimidine dimers	(108)

[a] This list and the references given are not intended to be comprehensive, but are to be used as a guide to the probes available.
[b] From K. Nakaya, O. Takenaka, H. Horinishi, and K. Shibata, *Biochim. Biophys. Acta* **161,** 23 (1968).
[c] From D. M. J. Lilley, *in* "The Role of Cyclic Nucleic Acid Adducts in Carcinogenesis and Mutagenesis" (B. Singer and H. Bartsch, eds.), IARC Publ. 70, p. 85. IARC, Lyon, 1986.
[d] From R. Shapiro, B. Braverman, J. B. Louis, and R. E. Servis, *J. Biol. Chem.* **248,** 4060 (1973).
[e] From U. Siebenlist and W. Gilbert, *Proc. Natl. Acad. Sci. U.S.A.* **77,** 122 (1980).
[f] From P. Hale and J. Lebowitz, *J. Virol.* **25,** 305 (1978).
[g] From P. Hale, R. S. Woodward, and J. Lebowitz, *Nature* (London) **284,** 640 (1980).
[h] From P. B. Dervan, *Science* **232,** 464 (1986).
[i] From D. S. Sigman, D. R. Graham, V. D'Aurora, and A. M. Stern, *J. Biol. Chem.* **254,** 12269 (1979).
[j] From C.-H. B. Chen and D. S. Sigman, *Proc. Natl. Acad. Sci. U.S.A.* **237,** 1197 (1986).

Osmium Tetroxide Reactions

Osmium tetroxide (71, 111, 112) reacts with the 5,6-bond of thymine (and to a lesser extent with cytosine) to generate an alkali-labile *cis*-diester adduct. The electrophilic addition requires out-of-plane attack, which is hindered when an A · T base pair is stacked in the normal B-DNA helix. However, a variety of distortions can expose the thymine to attack, including (but by no means limited to) loss of base pairing. This probe, which has been developed for the detailed analysis of DNA structure by Palecek and colleagues, has been of enormous value in the last decade. It has been used in the study of cruciforms (71) and DNA junctions (47, 113), Z-DNA (69, 83, 84), curved DNA (114), H-triplex structures (22, 91, 115), melted regions (15, 116), underwound (117) and overwound (57) conformations of alternating adenine–thymine sequences, base mismatches (105, 106), and ligand-binding sites (100).

Supercoiled DNA (1–5 μg) is reacted with 1 m*M* osmium tetroxide (stored as 8 m*M* stock solution in water) and either 1–3% (w/v) pyridine or 1 m*M* 2,2-bipyridine as the basic ligand in 10 m*M* Tris-HCl (pH 8), 0.5 m*M* EDTA. Other buffers may be used; this is not critical. The reaction may be carried out over a wide range of temperatures, and the time of reaction will need to be adjusted as a function of reaction temperature. As a guide, single-hit conditions will probably be obtained with a 5-min incubation in the above buffer at a temperature of 20° C. The chemically modified DNA is cleaved with a restriction enzyme, and the resulting termini radioactively labeled. DNA may be 3′-^{32}P-labeled with Klenow DNA polymerase and [α-^{32}P]dATP, or 5′-^{32}P-labeled after dephosphorylation with calf intestinal alkaline phosphatase using polynucleotide kinase and [γ-^{32}P]ATP. After cleavage with a second restriction enzyme, the asymmetrically labeled fragment is purified by electrophoresis, and reacted with 1 *M* piperidine at 90° C for 30 min to cleave at the sites of modification. After extensive lyophilization, the DNA may be electrophoresed in a polyacrylamide sequencing gel containing 7 *M* urea in 90 m*M* Tris–borate (pH 8.3), 10 m*M* EDTA. Radioactive DNA fragments are observed by autoradiography of dried gels at −70° C with intensifier screens, or with storage phosphor screens and a 400S phosphor imager. The latter method is preferred where quantitation is required, because of the much higher dynamic range available.

Osmium tetroxide may be used with a variety of basic ligands (118). However some care may be necessary in the interpretation of results when pyridine is used, as this causes a significant destabilization of the helix in supercoiled DNA at the required concentration (15). By contrast there is no detectable destabilization caused by the presence of 1 m*M* 2,2′-bipyridine.

Diethyl Pyrocarbonate Reactions

Diethyl pyrocarbonate (DEPC) (69, 119) reacts with purine N-7 to generate an alkali-labile adduct. Unpaired adenine bases become strongly reactive, as do guanine bases in Z-DNA.

Diethyl pyrocarbonate (1–3 μl) is added to 1–5 μg of supercoiled DNA dissolved in 200 μl of 50 m*M* sodium cacodylate (pH 7.1), 1 m*M* EDTA. The reaction is carried out at 20° C for 15 min, with vigorous shaking at 2-min intervals, and stopped by ethanol precipitation. The DNA is asymmetrically radioactively labeled and cleaved with piperidine as above.

Hydroxylamine Reactions

Hydroxylamine hydrochloride (83, 120) is prepared in distilled water as a 4 *M* stock solution. Aliquots of this are freshly titrated with diethylamine to give a 3 *M* solution of pH 7. Reactions are performed by addition of 20 μl of reagent to 7 μl of DNA solution, and incubated at 20° C for 15 min. Reactions are terminated by ethanol precipitation of the DNA. The DNA is asymmetrically radioactively labeled and cleaved with piperidine as above. The high salt concentration required for hydroxylamine reactions may limit the applicability of this reagent.

Dimethyl Sulfate Reactions

Dimethyl sulfate (DMS) methylates predominantly the N-7 position of guanine bases, rendering the DNA sensitive to alkali cleavage (121). This position is normally exposed and reactive in double-stranded DNA, but becomes protected if N-7 is used to accept a hydrogen bond, as in Hoogsteen base pairing or the formation of guanine tetrads (24). In some structures N-3 of cytosine may also become reactive to DMS.

DNA is incubated with 0.5% (v/v) DMS in 50 m*M* sodium cacodylate (pH 8.0), 1 m*M* EDTA, at 0° C for 4 min. DNA is recovered by precipitation with ethanol and cleaved with 1 *M* piperidine at 90° C for 30 min as above.

Bromoacetaldehyde Reactions

The haloacetaldehydes (70, 76, 122, 123) react primarily with adenine and cytosine bases to generate etheno adducts. The adducts are not strongly alkali labile, and are normally identified using other strategies.

Bromoacetaldehyde (BAA) may be prepared by hydrolysis of the diethyl acetal according to the method of McLean *et al.* (124). The residual ether is evaporated under an air stream and aliquots stored at $-70°$ C. The stock solution contains BAA at a concentration of approximately 2.5 *M*. Supercoiled DNA (1 μg) is incubated with 2% (v/v) BAA, in a total volume of 100 μl in TBE/2 (other buffers may again be used). Reaction times vary between 120 min at 37° C to 210 min at 20° C. After reaction, BAA is removed by means of two extractions with diethyl ether, and the DNA is ethanol precipitated. Asymmetrically radioactively labeled DNA is prepared as outlined above. Sites of BAA modification may be determined at sequence resolution by treating the isolated fragment with either hydrazine or formic acid and reacting with piperidine (23). The enhanced reactivity of unpaired bases is determined by quantitative analysis of the gel tracks in the phosphor image.

Resolution of Different Probes

Some chemical probes (e.g., osmium tetroxide or diethyl pyrocarbonate) generate base adducts that are labile to alkali, so that they may be cleaved using piperidine just as in the Maxam–Gilbert (121) reactions. Thus their sites of modification are readily revealed at single-nucleotide resolution by electrophoresis on sequencing gels. Other probes, such as the haloacetaldehydes, generate base adducts that are not themselves alkali sensitive, but confer reactivity to a second modification reaction such that the secondary product may be cleaved with piperidine (125). An alternative strategy is to use primer extension methods for the detection of chemical adducts that stall a polymerase (22). These adducts may also be detected by virtue of the sensitivity of the linearized DNA to cleavage by S1 nuclease (70, 71, 76). The resolution of such an experiment is much lower. However, the double-stranded DNA may be electrophoresed in agarose, such that any modification within several kilobase pairs is revealed, and thus the sequence of almost an entire plasmid is probed in a single experiment. Thus the normal strategy for examining a new sequence would include an initial study of a wide region at low resolution, enabling any regions showing enhanced local reactivity to be subsequently studied at higher resolution.

An example of the chemical probing, using osmium tetroxide, of a section of Z-DNA contained within an (A + T)-rich sequence in supercoiled DNA is shown in Fig. 2.

In Situ Chemical Probing of DNA Inside Cells

It has become possible to probe DNA structure within cells. Thus we can ask questions about the structures adopted by DNA inside the cell. Moreover we can use structural transitions in cellular DNA as a probe of cellular conditions, such as the level of DNA superhelical stress.

There are two ways in which this can be achieved. Enzymes can be employed to cleave or modify DNA inside the cell. Because proteins cannot be introduced into the cell, they must be synthesized *in situ* using expression plasmids. In this way methylases (126) and resolving enzymes (127) have been used to reveal the presence of Z-DNA and cruciform structures, respectively, in living cells. Because these demonstrations have been confined to sequences that were introduced for the purpose, this still does not address a primary biological phenomenon. Some chemicals can be introduced into cells from the external medium in sufficient concentration to react with cellular DNA. Permanganate has been used to reveal open complex formation at promoters *in situ* (104), whereas osmium tetroxide has demonstrated the existence of Z-DNA (128, 129), cruciforms (130), and H-triplexes (131) inside bacterial cells. Another chemical probe that has been successfully used inside cells is psoralen (132).

In Situ Probing of Plasmid DNA in Escherichia coli Using Osmium Tetroxide–Bipyridine

The following is a protocol successfully used in this laboratory for the *in situ* modification of cruciform structures in plasmids.

1. Cells (30–50 ml) are grown to midlog phase by inoculation from an overnight culture. The cells may be amplified with 150 μg/ml chloramphenicol if required. Cells may be subjected to salt shock, or other treatments to be investigated.

2. Cells are collected by centrifugation at 10,000 rpm for 5 min at 4° C, washed in 100 m*M* potassium phosphate (pH 7.2), and after a second centrifugation they are resuspended in 8 ml 250 m*M* potassium phosphate (pH 7.2).

3. Two milliliters of 10 m*M* osmium tetroxide–bipyridine is added, to give a final concentration of 2 m*M*, and incubated at 37° C for 10 min, or at 20° C for 20 min. The reaction is stopped by addition of 10 vol of cold 200 m*M* potassium phosphate (pH 7.2), and the cells collected by centrifugation at 10,000 rpm for

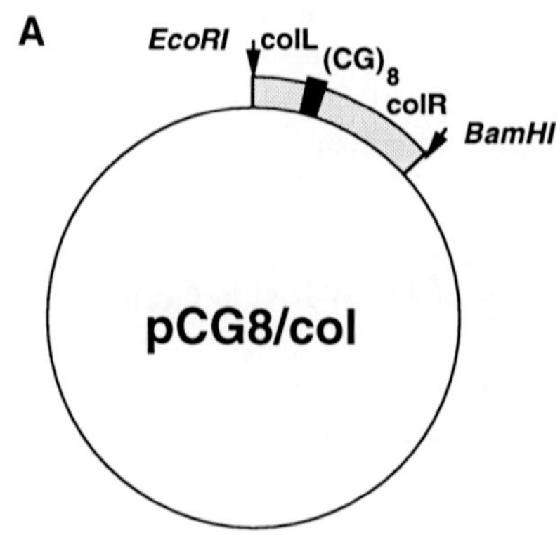

FIG. 2 Chemical probing of B–Z boundaries and helix opening in a supercoiled plasmid containing a $(CG)_8$ sequence embedded in an (A + T)-rich sequence (66). (A) Map of pCG8/col. The plasmid was constructed by insertion of oligonucleotides based on the sequence $(CG)_8$ into the unique *Xba*I sites of pColIRΔxba [K. M. Sullivan and D. M. J. Lilley, *Cell* **47,** 817 (1986)]. The *Xba*I site of pColIRΔxba is contained within the (A + T)-rich ColE1 sequence, shown stippled. These are subdivided into the colL and colR sequences. (B) Osmium tetroxide modification of pCG8/col as a function of temperature. Supercoiled pCG8/col of native superhelix density was reacted at the indicated temperature with 1 m*M* osmium tetroxide, 1% pyridine, in 45 m*M* Tris–borate (pH 8.3), 0.5 m*M* EDTA. The DNA was cleaved to completion by *Eco*RI and radioactively labeled at the 3′ termini. The 440-bp *Eco*RI–*Bam*HI fragment was isolated by electrophoresis, and osmium adducts cleaved by incubation in 1 *M* piperidine at 90° C. The DNA was electrophoresed on an 8% polyacrylamide sequencing gel in TBE containing 7 *M* urea. The leftmost tracks are G, A + G, T + C, and C reaction sequencing ladders generated by chemical modification of the DNA, and the track labeled "p" is a control sample that was incubated with piperidine without prior chemical modification. The remaining tracks contain pCG8/col modified with osmium tetroxide, and are labeled with their reaction temperature. The extent of the $(CG)_8$ tract is indicated by the stippled box shown on the right, together with the location of the colL and colR sequences. Note that the B–Z junctions are reactive to osmium tetroxide at all temperatures studied, and the colL (A + T)-rich sequences become uniformly reactive at 21° C.

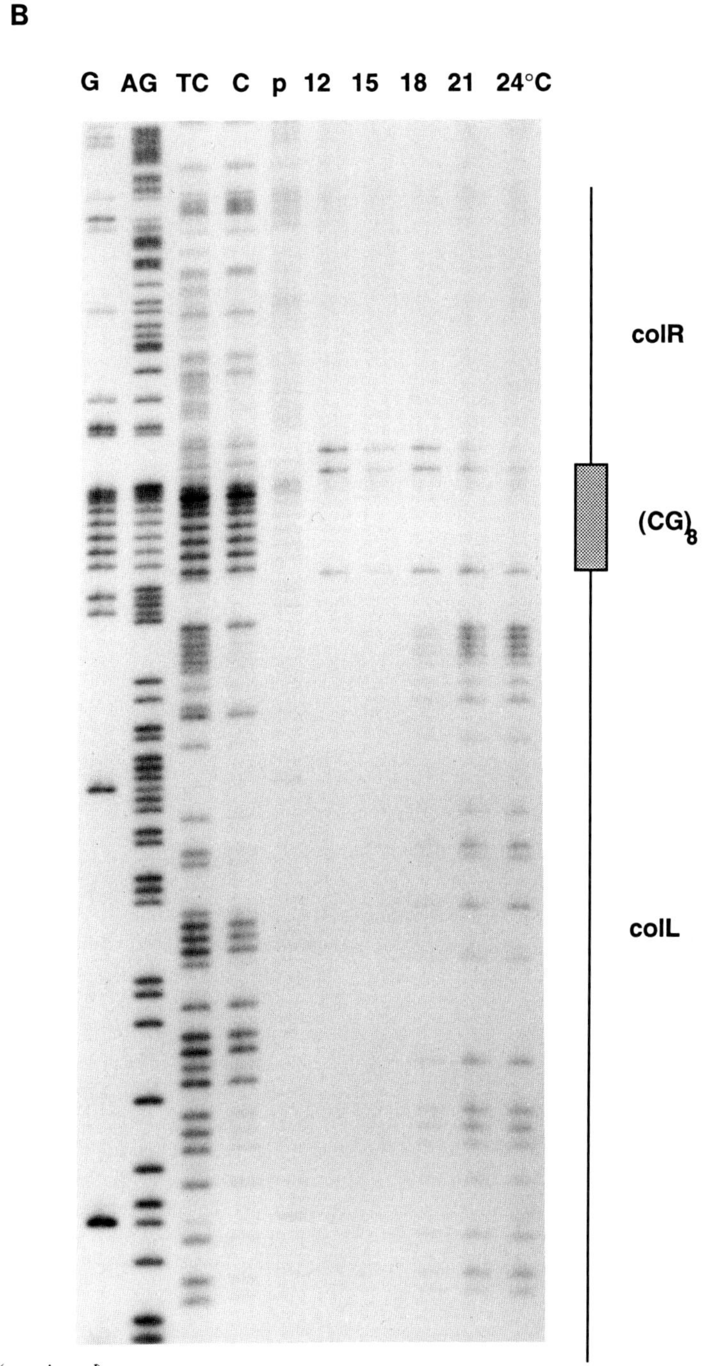

FIG. 2 (*continued*)

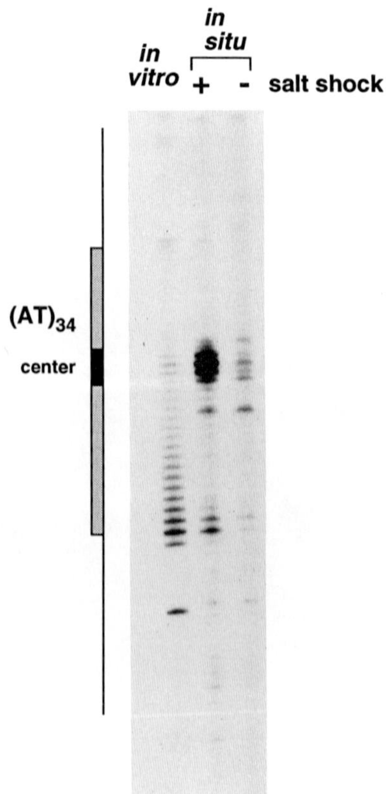

FIG. 3 *In situ* chemical probing of cruciform formation by an $(AT)_{34}$ sequence in *E. coli* (130): *In situ* osmium tetroxide modification of pXG540 with and without salt shock. The DNA was modified *in situ* using the procedure outlined in text. After isolation of plasmid DNA it was cleaved with *Eco*RI, and radioactively labeled at the resulting 3′ termini. The DNA in the leftmost track was obtained by *in vitro* modification of pXG540 with osmium tetroxide under conditions in which there was uniform reactivity of all thymines in the $(AT)_{34}$ tract [J. A. McClellan and D. M. J. Lilley, *J. Mol. Biol.* **197,** 707 (1987)], and can be used to identify the extent of the alternating sequence, indicated on the left-hand side (stippled box). In the *in situ* chemical modification experiments there was strong modification at the center of the $(AT)_{34}$ tract (corresponding to the unpaired loop of the extruded cruciform) by osmium tetroxide, provided that the cells had been exposed to osmotic shock. This experiment demonstrates both that cruciform formation was possible in *E. coli* under some circumstances, and that osmotic shock raises the effective superhelical stress in the cellular DNA.

5 min at 4° C. A further wash with 200 m*M* potassium phosphate (pH 7.2) removes any residual chemicals, and the cells are collected by centrifugation as before.

4. DNA is prepared from the cells by a method based on that of Holmes and Quigley (133). The cell pellets are resuspended in 0.7 ml 50 m*M* Tris-HCl (pH 7.6), 50 m*M* EDTA, 8% (w/v) sucrose, 5% (v/v) Triton X-100, and transferred to microcentrifuge tubes.

5. Ten microliters of 10 mg/ml lysozyme is added and, after mixing, the tube is transferred to a boiling water bath for 1 min. The mixture is centrifuged at 14,000 rpm for 15 min.

6. Three microliters of 10 mg/ml RNase A (DNase free) is added to the supernatant and incubated at 37° C for 15 min. Water-saturated phenol (500 μl) is added, and vortex mixed. The tube is briefly centrifuged to separate layers. The aqueous phase is transferred to a new tube, and 1 vol of phenol–chloroform (1 : 1) is added before vortexing and centrifugating again.

7. 2-Propanol (0.75 vol) is added to the aqueous phase, mixed, and left at room temperature for 10 min before centrifugation at 14,000 rpm for 15 min.

8. The pellet is washed with 1 ml 70% ethanol and recentrifuged. After aspiration of the supernatant the pellet is vacuum dried and dissolved in 100 μl of 10 m*M* Tris-HCl (pH 7.5), 1 m*M* EDTA. This is applied to a small-scale Sephadex G-50 column and centrifuged. DNA is recovered by addition of sodium acetate to 0.3 *M* and precipitation with ethanol.

9. Approximately 1-μg quantities of the *in situ*-modified DNA are digested with a restriction enzyme and radioactively labeled at the 5′ or 3′ termini, before restriction cleavage using a second enzyme. After purification of the asymmetrically labeled fragment by gel electrophoresis and electroelution, it is incubated with 1 *M* piperidine at 90° C for 30 min to cleave osmate adducts, and examined by electrophoresis on a polyacrylamide sequencing gel containing 7 *M* urea, and autoradiography. Chemical sequence reactions (121) on the equivalent unmodified DNA fragment are used to identify the positions of reactivity. An example of the *in situ* chemical probing of a cruciform extruded from an $(AT)_{34}$ sequence in *E. coli* is given in Fig. 3.

Strategies for Studying a New Structure in Supercoiled DNA

The twin approaches of two-dimensional gel electrophoresis and probing topological methods for the study of local structure in DNA both have their advantages and problems, and a combination of the two is the best policy.

A fundamental problem with probing experiments is that the modifying agent must react with the DNA in some way, and could therefore perturb the structure studied. This is of most concern when enzyme probes are employed, where specificity arises because of close protein–DNA interaction, but such a concern might also extend to chemical probes that require intimate interaction with their targets (e.g., intercalation). Such interactions might have the effect of catalyzing the formation of structures that are otherwise kinetically inaccessible. There is a more subtle concern if there is a dynamic equilibrium between two DNA structures; reaction of one form with a probe will drive the equilibrium into that conformation, and one might be led into overestimating the proportion of that structure.

For these reasons probing studies should be performed with care. The problems of structural perturbation can be largely overcome by working under "single-hit" conditions, using short times and low reagent concentrations. Use of a variety of different probes is also advisable. The targets for these would ideally be different; for example, if osmium tetroxide is used to probe thymine conformation, complementary information could be sought on adenine reactivity by use of bromoacetaldehyde or diethyl pyrocarbonate. The positive result with some probes might be protection against reaction (e.g., the inaccessibility of guanine N-7 in a Hoogsteen pairing) and thus, if the position of equilibrium is displaced, it will be driven in the opposite direction.

Two-dimensional gel electrophoresis does not suffer from these disadvantages, as it can be regarded as largely unperturbing, but there is a different problem. Whereas gel electrophoresis can reveal the existence of a topology-dependent structural transition in a circular DNA molecule, it gives no indication about which sequences are participating in the structural change unless mutants are available. However, this localization is the information that probing experiments can readily supply, and a combination of methods is highly recommended.

Using the kinds of methods outlined in this article, a new perspective on DNA structure has emerged. Many interesting new local structures have been described, and their properties analyzed. In many cases, however, the biological significance has not always been clearly established, and there are still a number of novel DNA conformations that remain structures in search of a function. For these reasons the new methods that allow us to probe the structure inside the cell are particularly significant, and are likely to grow in importance in the future.

Acknowledgments

We thank our co-workers and collaborators involved in the studies discussed here, including Emil Palecek, Anamitra Bhattacharyya, Dongrong Chen, Judy Furlong, and Derek Duckett, and we gratefully acknowledge the CRC, SERC, and MRC for financial support of work in our laboratory.

References

1. R. E. Dickerson and H. R. Drew, *J. Mol. Biol.* **149,** 761 (1981).
2. E. Palecek, *Crit. Rev. Biochem. Mol. Biol.* **26,** 151 (1991).

3. G. Yagil, *Crit. Rev. Biochem. Mol. Biol.* **26,** 475 (1991).
4. A. H.-J. Wang, G. J. Quigley, F. J. Kolpak, J. L. Crawford, J. H. van Boom, G. van der Marel, and A. Rich, *Nature (London)* **282,** 680 (1979).
5. H.-M. Wu and D. M. Crothers, *Nature (London)* **308,** 509 (1984).
6. S. Diekmann and J. C. Wang, *J. Mol. Biol.* **186,** 1 (1985).
7. P. J. Hagerman, *Biochemistry* **24,** 7033 (1985).
8. H. R. Drew and A. A. Travers, *J. Mol. Biol.* **186,** 773 (1985).
9. H.-S. Koo, H.-M. Wu, and D. M. Crothers, *Nature (London)* **320,** 501 (1986).
10. L. Ulanovsky, M. Bodner, E. N. Trifonov, and M. Choder, *Proc. Natl. Acad. Sci. U.S.A.* **83,** 862 (1986).
11. H. C. M. Nelson, J. T. Finch, B. F. Luisi, and A. Klug, *Nature (London)* **330,** 221 (1987).
12. M. Coll, C. A. Frederick, A. H.-J. Wang, and A. Rich, *Proc. Natl. Acad. Sci. U.S.A.* **84,** 8385 (1987).
13. F. S. Lee and W. R. Bauer, *Nucleic Acids Res.* **13,** 1665 (1985).
14. D. Kowalski, D. A. Natale, and M. J. Eddy, *Proc. Natl. Acad. Sci. U.S.A.* **85,** 9464 (1988).
15. R. Bowater, F. Aboul-ela, and D. M. J. Lilley, *Biochemistry* **30,** 11495 (1991).
16. M. Gellert, K. Mizuuchi, M. H. O'Dea, H. Ohmori, and J. Tomizawa, *Cold Spring Harbor Symp. Quant. Biol.* **43,** 35 (1979).
17. D. M. J. Lilley, *Proc. Natl. Acad. Sci. U.S.A.* **77,** 6468 (1980).
18. N. Panayotatos and R. D. Wells, *Nature (London)* **289,** 466 (1981).
19. V. I. Lyamichev, S. M. Mirkin, and M. D. Frank-Kamenetskii, *J. Biomol. Struct. Dyn.* **3,** 327 (1985).
20. V. Lyamichev, S. M. Mirkin, and M. D. Frank-Kamenetskii, *J. Biomol. Struct. Dyn.* **3,** 667 (1986).
21. J. C. Hanvey, M. Shimizu, and R. D. Wells, *Proc. Natl. Acad. Sci. U.S.A.* **85,** 6296 (1988).
22. H. Htun and J. E. Dahlberg, *Science* **241,** 1791 (1988).
23. Y. Kohwi and T. Kohwi-Shigematsu, *Proc. Natl. Acad. Sci. U.S.A.* **85,** 3781 (1988).
24. D. Sen and W. Gilbert, *Nature (London)* **334,** 364 (1988).
25. W. I. Sundquist and A. Klug, *Nature (London)* **342,** 825 (1989).
26. J. R. Williamson, M. K. Raghuraman, and T. R. Cech, *Cell* **59,** 871 (1989).
27. C. Kang, X. Zhang, R. Ratliff, R. Moyzis, and A. Rich, *Nature (London)* **356,** 126 (1992).
28. Z. Otwinowski, R. W. Schevitz, R.-G. Zhang, C. L. Lawson, A. Joachimiak, R. Q. Marmorstein, B. Luisi, and P. B. Sigler, *Nature (London)* **335,** 321 (1988).
29. S. C. Schultz, G. C. Shields, and T. A. Steitz, *Science* **253,** 1001 (1991).
30. G. B. Koudelka, P. Harbury, S. C. Harrison, and M. Ptashne, *Proc. Natl. Acad. Sci. U.S.A.* **85,** 4633 (1988).
31. L. Bossi and D. M. Smith, *Cell* **39,** 643 (1984).
32. K. Zahn and F. R. Blattner, *Nature (London)* **317,** 451 (1985).
33. A. A. Travers, *Cell* **60,** 177 1990).
34. R. Holliday, *Genet. Res.* **5,** 282 (1964).
35. T. R. Broker and I. R. Lehman, *J. Mol. Biol.* **60,** 131 (1971).
36. T. L. Orr-Weaver, J. W. Szostak, and R. J. Rothstein, *Proc. Natl. Acad. Sci. U.S.A.* **78,** 6354 (1981).

37. H. Potter and D. Dressler, *Proc. Natl. Acad. Sci. U.S.A.* **73,** 3000 (1976).
38. H. Potter and D. Dressler, *Proc. Natl. Acad. Sci. U.S.A.* **75,** 3698 (1978).
39. N. Sigal and B. Alberts, *J. Mol. Biol.* **78,** 789 (1972).
40. H. M. Sobell, *Proc. Natl. Acad. Sci. U.S.A.* **69,** 2483 (1972).
41. P. A. Kitts and H. A. Nash, *Nature (London)* **329,** 346 (1987).
42. S. E. Nunes-Düby, L. Matsomoto, and A. Landy, *Cell* **50,** 779 (1987).
43. R. Hoess, A. Wierzbicki, and K. Abremski, *Proc. Natl. Acad. Sci. U.S.A.* **84,** 6840 (1987).
44. M. Jayaram, K. L. Crain, R. L. Parsons, and R. M. Harshey, *Proc. Natl. Acad. Sci. U.S.A.* **85,** 7902 (1988).
45. G. W. Gough and D. M. J. Lilley, *Nature (London)* **313,** 154 (1985).
46. J. P. Cooper and P. J. Hagerman, *J. Mol. Biol.* **198,** 711 (1987).
47. D. R. Duckett, A. I. H. Murchie, S. Diekmann, E. von Kitzing, B. Kemper, and D. M. J. Lilley, *Cell* **55,** 79 (1988).
48. M. E. Churchill, T. D. Tullius, N. R. Kallenbach, and N. C. Seeman, *Proc. Natl. Acad. Sci. U.S.A.* **85,** 4653 (1988).
49. A. I. H. Murchie, R. M. Clegg, E. von Kitzing, D. R. Duckett, S. Diekmann, and D. M. J. Lilley, *Nature (London)* **341,** 763 (1989).
50. J. Vinograd, J. Lebowitz, R. Radloff, R. Watson, and P. Laipis, *Proc. Natl. Acad. Sci. U.S.A.* **53,** 1104 (1965).
51. F. B. Fuller, *Proc. Natl. Acad. Sci. U.S.A.* **68,** 815 (1971).
52. R. E. Depew and J. C. Wang, *Proc. Natl. Acad. Sci. U.S.A.* **72,** 4275 (1975).
53. D. E. Pulleyblank, M. Shure, D. Tang, J. Vinograd, and H.-P. Vosberg, *Proc. Natl. Acad. Sci. U.S.A.* **72,** 4280 (1975).
54. D. S. Horowitz and J. C. Wang, *J. Mol. Biol.* **173,** 75 (1984).
55. L. J. Peck, A. Nordheim, A. Rich, and J. C. Wang, *Proc. Natl. Acad. Sci. U.S.A.* **79,** 4560 (1982).
56. C. K. Singleton, J. Klysik, S. M. Stirdivant, and R. D. Wells, *Nature (London)* **299,** 312 (1982).
57. J. A. McClellan and D. M. J. Lilley, *J. Mol. Biol.* **219,** 145 (1991).
58. A. Bhattacharyya, A. I. H. Murchie, and D. M. J. Lilley, *Nature (London)* **343,** 484 (1990).
59. J. C. Wang, L. J. Peck, and K. Becherer, *Cold Spring Harbor Symp. Quant. Biol.* **47,** 85 (1983).
60. R. Bowater, F. Aboul-ela, and D. M. J. Lilley, *in* "DNA Structures" (D. M. J. Lilley and J. Dahlberg, eds: Methods in Enzymology, Vol. 212. Academic Press, San Diego, in press, 1992.
61. L. J. Peck and J. C. Wang, *Proc. Natl. Acad. Sci. U.S.A.* **80,** 6206 (1983).
62. M. J. Ellison, R. J. Kelleher, III, A. H.-J. Wang, J. F. Habener, and A. Rich, *Proc. Natl. Acad. Sci. U.S.A.* **82,** 8320 (1985).
63. T. R. O'Connor, D. S. Kang, and R. D. Wells, *J. Biol. Chem.* **261,** 13302 (1986).
64. R. J. Kelleher, III, M. J. Ellison, P. S. Ho, and A. Rich, *Proc. Natl. Acad. Sci. U.S.A.* **83,** 6342 (1986).
65. C. J. Benham, *J. Chem. Phys.* **92,** 6294 (1990).
66. F. Aboul-ela, R. Bowater, and D. M. J. Lilley, *J. Biol. Chem.* **267,** 1776 (1992).
67. L. F. Liu and J. C. Wang, *Proc. Natl. Acad. Sci. U.S.A.* **84,** 7024 (1987).

68. H.-Y. Wu, S. Shyy, J. C. Wang, and L. F. Liu, *Cell* **53,** 433 (1988).
69. W. Herr, *Proc. Natl. Acad. Sci. U.S.A.* **82,** 8009 (1985).
70. D. M. J. Lilley, *Nucleic Acids Res.* **11,** 3097 (1983).
71. D. M. J. Lilley and E. Palecek, *EMBO J.* **3,** 1187 (1984).
72. G. W. Gough, K. M. Sullivan, and D. M. J. Lilley, *EMBO J.* **5,** 191 (1986).
73. J. C. Furlong and D. M. J. Lilley, *Nucleic Acids Res.* **14,** 3995 (1986).
74. P. M. Scholten and A. Nordheim, *Nucleic Acids Res.* **14,** 3981 (1986).
75. C. C. Hentschel, *Nature (London)* **295,** 714 (1982).
76. T. Kohwi-Shigematsu, R. Gelinas, and H. Weintraub, *Proc. Natl. Acad. Sci. U.S.A.* **80,** 4389 (1983).
77. H. A. F. Mace, H. R. B. Pelham, and A. A. Travers, *Nature (London)* **304,** 555 (1983).
78. J. M. Nickol and G. Felsenfeld, *Cell* **35,** 467 (1983).
79. C. R. Cantor and A. Efstradiatis, *Nucleic Acids Res.* **12,** 8059 (1984).
80. H. Htun, E. Lund, and J. E. Dahlberg, *Proc. Natl. Acad. Sci. U.S.A.* **81,** 7288 (1984).
81. D. E. Pulleyblank, D. B. Haniford, and A. R. Morgan, *Cell* **42,** 271 (1985).
82. C. K. Singleton, M. W. Kilpatrick, and R. D. Wells, *J. Biol. Chem.* **259,** 1963 (1984).
83. B. H. Johnston and A. Rich, *Cell* **42,** 713 (1985).
84. K. Nejedly, M. Kwinkowski, G. Galazka, J. Klysik, and E. Palecek, *J. Biomol. Struct. Dyn.* **3,** 467 (1985).
85. J. K. Barton and A. L. Raphael, *Proc. Natl. Acad. Sci. U.S.A.* **82,** 6460 (1985).
86. G. Galazka, E. Palecek, R. D. Wells, and J. Klysik, *J. Biol. Chem.* **261,** 7093 (1986).
87. M. J. McLean, J. W. Lee, and R. D. Wells, *J. Biol. Chem.* **263,** 7378 (1988).
88. A. Millgram-Burkoff and T. D. Tullius, *Cell* **48,** 935 (1987).
89. J. G. McCarthy, L. D. Williams, and A. Rich, *Biochemistry* **29,** 6071 (1990).
90. T. Kohwi-Shigematsu and Y. Kohwi, *Cell* **43,** 199 (1985).
91. B. H. Johnston, *Science* **241,** 1800 (1988).
92. O. N. Voloshin, S. M. Mirkin, V. I. Lyamichev, B. P. Belotserkovskii, and M. D. Frank-Kamenetskii, *Nature (London)* **333,** 475 (1988).
93. J. Klysik, K. Rippe, and T. M. Jovin, *Biochemistry* **29,** 9831 (1990).
94. C. Dingwall, G. P. Lomonossoff, and R. A. Laskey, *Nucleic Acids Res.* **9,** 2659 (1981).
95. H. R. Drew, *J. Mol. Biol.* **176,** 535 (1984).
96. H. R. Drew and A. A. Travers, *Cell* **37,** 491 (1984).
97. A. Spassky and D. S. Sigman, *Biochemistry* **24,** 8050 (1985).
98. T. D. Tullius and B. A. Dombroski, *Science* **230,** 679 (1985).
99. K. R. Fox and G. W. Grigg, *Nucleic Acids Res.* **16,** 2063 (1988).
100. M. J. Mclean, F. Seela, and M. J. Waring, *Proc. Natl. Acad. Sci. U.S.A.* **86,** 9687 (1989).
101. B. M. G. Cons and K. R. Fox, *Nucleic Acids Res.* **17,** 5447 (1989).
102. U. Siebenlist, *Nature (London)* **279,** 651 (1979).
103. M. Buckle and H. Buc, *Biochemistry* **28,** 4388 (1989).
104. S. Sasse-Dwight and J. D. Gralla, *J. Biol. Chem.* **264,** 8074 (1989).
105. R. G. H. Cotton, N. R. Rodrigues, and R. D. Campbell, *Proc. Natl. Acad. Sci. U.S.A.* **85,** 4397 (1988).
106. A. Bhattacharyya and D. M. J. Lilley, *J. Mol. Biol.* **209,** 583 (1989).
107. V. M. Vogt, *Eur. J. Biochem.* **33,** 192 (1973).

108. M. M. Becker and J. C. Wang, *Nature* (*London*) **309,** 682 (1984).
109. M. M. Becker and J. C. Wang, *J. Mol. Biol.* **210,** 429 (1989).
110. V. I. Lyamichev, M. D. Frank-Kamenetskii, and V. N. Soyfer, *Nature* (*London*) **344,** 568 (1990).
111. K. Burton and W. T. Riley, *Biochem. J.* **98,** 70 (1965).
112. G. C. Glikin, M. Vojtiskova, L. Rena-Descalzi, and E. Palecek, *Nucleic Acids Res.* **12,** 1725 (1984).
113. D. R. Duckett, A. I. H. Murchie, and D. M. J. Lilley, *EMBO J.* **9,** 583 (1990).
114. E. Palecek, E. Makaturova-Rasovska, and S. Diekmann, *Gen. Physiol. Biophys.* **7,** 379 (1988).
115. M. Vojtiskova and E. Palecek, *J. Biomol. Struct. Dyn.* **5,** 283 (1987).
116. J. C. Furlong, K. M. Sullivan, A. I. H. Murchie, G. W. Gough, and D. M. J. Lilley, *Biochemistry* **28,** 2009 (1989).
117. J. A. McClellan, E. Palecek, and D. M. J. Lilley, *Nucleic Acids Res.* **14,** 9291 (1986).
118. E. Palecek, P. Boublikova, and K. Nejedly, *Biophys. Chem.* **34,** 63 (1989).
119. N. J. Leonard, J. J. McDonald, R. E. L. Henderson, and M. E. Reichmann, *Biochemistry* **10,** 3335 (1971).
120. C. M. Rubin and C. W. Schmid, *Nucleic Acids Res.* **8,** 4613 (1980).
121. A. M. Maxam and W. Gilbert, *in* "Nucleic Acids," Part I (L. Grossman and K. Moldave, eds.), Methods in Enzymology, Vol. 65, p. 499. Academic Press, New York, 1980.
122. K. Kayasuga-Mikado, T. Hashimoto, T. Negishi, K. Negishi, and H. Hayatsu, *Chem. Pharm. Bull.* **28,** 923 (1980).
123. J. A. Sechrist, III, J. R. Barrio, N. J. Leonard, C. Villar-Palasi, and A. G. Gilman, *Science* **177,** 279 (1972).
124. M. J. McLean, J. E. Larson, F. Wohlrab, and R. D. Wells, *Nucleic Acids Res.* **15,** 6917 (1987).
125. T. Kohwi-Shigematsu and Y. Kohwi, *Biochemistry* **29,** 9551 (1990).
126. A. Jaworski, W.-T. Hsieh, J. A. Blaho, J. E. Larson, and R. D. Wells, *Science* **238,** 773 (1988).
127. N. Panayotatos and A. Fontaine, *J. Biol. Chem.* **262,** 11364 (1987).
128. P. Boublikova and E. Palecek, *Gen. Physiol. Biophys.* **8,** 475 (1989).
129. A. R. Rahmouni and R. D. Wells, *Science* **246,** 358 (1989).
130. J. A. McClellan, P. Boublikova, E. Palecek, and D. M. J. Lilley, *Proc. Natl. Acad. Sci. U.S.A.* **87,** 8373 (1990).
131. P. Karlovsky, P. Pecinka, M. Vojtiskova, E. Makaturova, and E. Palecek, *FEBS Lett.* **274,** 39 (1990).
132. G. Zheng, T. Kochel, R. W. Hoepfner, S. E. Timmons, and R. R. Sinden, *J. Mol. Biol.* **221,** 107 (1991).
133. D. S. Holmes and M. Quigley, *Anal. Biochem.* **114,** 193 (1981).

[15] Glutathione *S*-Transferase Fusion Protein Affinity Chromatography to Assess Biochemical Interactions between DNA-Binding Proteins

Steven Artandi and Kathryn Calame

Introduction

Critical Importance of Protein–Protein Interactions for Regulated Eukaryotic Gene Transcription

Most biological processes involve multiple proteins that must interact in an ordered and specific fashion. After several components of a multiprotein system have been identified and cloned, understanding which of these proteins associate, how they associate, and what role their association plays in governing the biological process in question represents a significant challenge.

Eukaryotic gene transcription by RNA polymerase II is an example of a process in which many proteins, both DNA binding and non-DNA binding, must interact. These interactions determine the rate of transcription initiation, which is usually a key regulatory event governing overall expression of individual genes. A typical gene can be regulated by transcription factors that bind specific DNA sequences at the promoter, located within a few hundred base pairs upstream of the start site of transcription, and at enhancers, which may be located several kilobases away from the transcription start site. In addition, a complex of proteins termed the basal transcription machinery assembles just upstream of the start site. This complex includes TATA-binding protein (TBP), and several other proteins that assemble in an ordered fashion by interacting with TBP and the surrounding DNA (1, 2). Additional proteins, termed "coactivators" or "TAFs," bind TBP and may mediate the interaction of the basal transcription machinery with proteins bound at the promoter and enhancer (3, 4).

We have focused our attention on direct interactions between sequence-specific DNA-binding proteins that bind in promoter or enhancer regulatory regions. Most enhancer and promoter factors are not gene specific, that is, their cognate recognition sequences occur in many regulatory elements. However, regulatory elements in specific genes have combinations of factor binding sites that are unique with respect to both content and spacing, suggesting that particular combinations of transcription factors may be important for determining specific regulatory patterns of individual genes. One of the challenges for developing a complete understanding of the molecular events necessary for regulated transcription initiation is to understand

Methods in Molecular Genetics, Volume 1

how sequence-specific transcription factors bound at the enhancer and promoter of a gene interact with one another and how they interact with the components of the basal transcription machinery. A companion challenge is presented by the fact that for many factor binding sites, families of structurally related proteins have been identified that bind with nearly equal specificity and affinity to a given DNA sequence. Therefore it becomes important to determine which of these proteins are capable of appropriate protein–protein associations that are critical for carrying out regulation of transcription.

Transcription factors can be divided into several classes based on the structure of their DNA-binding and dimerization domains. These include basic leucine zipper (BZip) proteins, basic helix–loop–helix proteins (BHLH), helix–turn–helix homeobox or POU proteins, zinc finger proteins, ets family proteins, HMG-type proteins, and rel family proteins. Zip and HLH domains provide protein dimerization regions and factors containing them recognize DNA sequence dyads and bind DNA as dimers. The protein–protein interactions that govern dimer formation determine whether these proteins bind DNA as homodimers and what subset of proteins within a structural class can form heterodimers. When combined with different dimerization partners, a transcription factor can have altered capacity for binding DNA (5), altered DNA-binding specificity (6), or altered effector functions, that is, a different capacity for activation or repression of transcription (7, 8).

In addition to dimer formation required for DNA binding, transcription factors are capable of higher-order interactions in which monomers or dimers, competent for DNA binding, can associate with each other. Interactions of this type may be important for several aspects of promoter and enhancer function. First, because enhancer elements are usually located hundreds of base pairs of DNA from their target promoters, it is likely that protein–protein interactions between promoter-bound and enhancer-bound transcription factors mediate direct communication between the promoter and enhancer with the concomitant looping out of intervening DNA (9). Second, proteins bound proximally or distally within a promoter or enhancer may interact to bind DNA cooperatively (10). And third, proteins bound proximally or distally within a promoter or enhancer may interact to nucleate formation of a complex containing additional proteins that interact with the basal transcription machinery.

Assays to Detect Interactions between Transcription Factors

Assays for DNA binding, such as the gel electrophoretic mobility shift assay (GEMSA) and footprinting, can be utilized to reveal protein–protein interactions. Using GEMSA with a combination of long and truncated recombinant proteins has demonstrated homodimer and heterodimer formation (11) and higher-order interactions (12, 13). The GEMSA and footprinting have also been used to measure the off rates of proteins bound to DNA and thus demonstrate cooperative binding by two proteins to nearby sites on the same DNA molecule (14). Other assays that are independent of DNA binding have also been used to study protein–protein interactions. These include coimmunoprecipitation, chemical cross-linking, protein-probed Western blots ("far Westerns") (15), and affinity chromatography (16).

Affinity chromatography, in which one protein is attached to a solid support and a second protein in solution is exposed to the matrix, is a powerful technique because it can be used for both analytical and preparative purposes. However, classical affinity chromatography can be time consuming because it requires purification of the immobilized protein, usually by conventional chromatographic techniques, and subsequent conjugation to a solid support.

A fusion protein system that takes advantage of some of the properties of the enzyme glutathione *S*-transferase (GST) has been described (17). When expressed in bacteria, a large percentage of fusion proteins between GST and eukaryotic proteins are at least partly soluble and are therefore available for purification on a specific affinity matrix consisting of glutathione linked to agarose beads. Purified fusion proteins are therefore readily obtained from crude extracts of bacteria induced to express them. When bound to a glutathione-agarose matrix, these fusion proteins can themselves serve as affinity resins for other proteins that may interact with them and can be used directly for measuring these protein–protein interactions. In a short time, this technique has been used to characterize novel protein–protein interactions in many systems. These include interactions between the retinoblastoma gene product (Rb) and other cellular proteins (18), Rb and the transcription factor E2F (19, 20), products of the protooncogene c-*myc* (c-Myc) and its partner max (21), c-Myc and Rb (22), and between VP16 and TFIIB (23). This assay has become popular because it is rapid and the required reagents are relatively easy to generate.

Briefly, fusion proteins are expressed in bacteria and crude bacterial lysates are prepared. These lysates are incubated with glutathione–agarose and washed extensively to remove extraneous bacterial proteins. The protein whose interaction with the GST fusion protein is to be tested is synthesized *in vitro,* using a rabbit reticulocyte lysate system, and labeled with [^{35}S]methionine. The ^{35}S-labeled protein is incubated with the bound GST fusion protein matrix, using appropriate binding conditions. After thorough washing, the bound protein is quantitated by denaturation of the complex, analysis by sodium dodecyl sulfate-polyacrylamide gel electrophoresis (SDS-PAGE), and autoradiography. We have used the GST fusion protein affinity chromatography assay to analyze interactions between sequence-specific DNA-binding proteins: both dimer formation within a structural class and interactions between different structural classes.

Below we present a streamlined protocol for using the GST fusion protein method to analyze interactions among transcription factors. We also describe a simple way to ensure that potential artifacts caused by the presence of DNA in either the bacterial or reticulocyte lysates are eliminated.

Methodology

Expression of Glutathione S*-Transferase Fusion Proteins*

To measure the interaction between two transcription factors, at least one of them must be expressed as a translational fusion with GST. To accomplish this, we subclone the genes of interest into the vector pGEX-2T (Pharmacia, Piscataway, NJ), a plasmid that contains the *GST* gene under control of the isopropyl-β-D-thiogalactopyranoside (IPTG)-inducible *tac* promoter with a small polylinker 3′ of the *GST* coding sequences. Typically, the cDNA insert is cloned into pGEX-2T by blunt-ended cloning. To determine if the cDNA insert is cloned correctly in frame with the GST sequences and if the fusion protein is expressed properly, a small culture of *Escherichia coli* containing the plasmid is grown, induced with IPTG, lysed by boiling in SDS-PAGE loading buffer, and analyzed by SDS-PAGE, essentially as described in Sambrook *et al.* (24). Comparing uninduced bacteria with an induced aliquot allows visualization of the GST fusion protein band. From this analysis, one can determine if the fusion protein is the expected size and how efficiently the fusion protein is induced. It is helpful to use the pGEX-2T plasmid as a control for induction and for a comparison of induced protein sizes.

To make bacterial extracts on a preparative scale useful for the actual binding assay, a variety of induction conditions, lysis methods, and buffer conditions have been used. These conditions often need to be empirically adjusted for proteins that are difficult to solubilize, are poorly induced, or unusually labile. The following represent conditions that have routinely worked well in our hands.

1. A fresh overnight culture of *E. coli* harboring the recombinant pGEX-2T plasmid of interest is prepared. On the day of preparation of the extract, 4 ml of the overnight culture is added to 400 ml of LB medium containing 100 μg/ml ampicillin. The culture is grown with moderate shaking at 37° C to an optical density (OD) at 600 nm of approximately 0.5. IPTG is added to a final concentration of 0.2 m*M* and the culture is grown at 37° C for an additional 2–4 hr.

2. The bacterial culture is centrifuged at 5000 *g* for 15 min at 4° C. If the fusion protein is labile, it is helpful to quick-chill the culture by pouring it over an equal volume of frozen phosphate-buffered saline (PBS) prior to centrifugation. After centrifugation, the supernatant is discarded and the pellet is resuspended in 10 ml buffer A [20 mM *N*-2-hydroxyethylpiperazine-*N*′-2-ethanesulfonic acid (HEPES), (pH 7.6), 100 m*M* KCl, 0.2 m*M* ethylenediaminetetraacetic acid (EDTA), 1 m*M*

dithiothreitol (DTT), 20% (v/v) glycerol, 1% (v/v) Triton, 1 m*M* phenylmethylsulfonyl fluoride (PMSF), 2 μg/ml leupeptin, 2 μg/ml pepstatin, 5 m*M* benzamidine], and kept cold on ice. One milliliter of a solution containing 10 mg/ml lysozyme in buffer A is added to the bacterial suspension. After incubation for 15 min on ice, the cells are lysed by sonication on ice using three 20-sec bursts separated by 20-sec rest intervals to prevent excessive warming of the sample.

3. Bacterial membranes and unlysed bacteria are removed by centrifugation at 9000 *g* at 4° C for 10 min. This supernatant is the crude bacterial lysate and should contain GST fusion protein. This should be checked by SDS-PAGE to ensure that a sufficient percentage of the fusion protein is soluble and available for purification on the affinity resin.

In Vitro Transcription and Translation

In vitro transcription and *in vitro* translation are simple methods for generating labeled protein from cloned cDNAs. We have used these techniques to synthesize the labeled protein that is to be tested for interaction with the immobilized GST fusion protein. Because these techniques have been widely used, they will be described only briefly. The cDNA of interest is cloned into a vector containing a prokaryotic promoter such as T7 or T3. A translation initiation codon with a sufficiently good Kozak consensus (33) must be supplied either by the cDNA or the vector. RNA is synthesized from a linearized DNA template and this RNA is used to program reticulocyte lysates (Promega, Madison, WI) in the presence of [^{35}S]methionine, as described by the manufacturer. The labeled protein can be visualized by SDS-PAGE and autoradiography.

Potential Artifacts Due to DNA Contamination of Lysates

An important potential artifact in studying interactions between DNA-binding proteins has been noted. Genomic DNA is often present either in the crude bacterial lysate and/or the reticulocyte lysate. Because sequence-specific DNA-binding proteins also bind DNA, with lower affinity, in a non-sequence-specific manner, both the GST fusion protein and the labeled protein may bind nonspecifically to a single molecule of contaminating DNA. This bridging by DNA could result in the appearance of a protein–protein interaction when none actually occurred (25).

To eliminate this possibility, we have treated both our bacterial extracts and reticulocyte lysates with DEAE-Sephacel to remove contaminating DNA. This involves raising the salt concentration of each lysate to approximately 400 m*M*, conditions in which most transcription factors do not bind to DEAE.

To remove bacterial DNA from the crude bacterial lysate containing the GST fusion protein, we carry out the following steps.

1. Increase the salt concentration of 1.5 ml of extract by 300 m*M* to a total concentration of 400 m*M* by adding 90 μl 5 *M* NaCl.
2. Suspend DEAE-Sephacel (Pharmacia) in 20 m*M* HEPES (pH 7.6) and 400 m*M* NaCl.
3. Add 1 ml of this high-salt DEAE-Sephacel to 1.5 ml of high-salt bacterial lysate. Rock gently at 4°C for 1 hr.
4. Centrifuge at 100 *g* for 5 min at 4°C to remove DEAE-Sephacel.

Based on information from the manufacturer, we estimate the chloride concentration in reticulocyte lysates from Promega to be approximately 150 m*M*. We treat these lysates with DEAE as follows.

1. Increase the salt concentration of 100 μl of the lysate by 250 m*M* NaCl to a total of 400 m*M* NaCl by adding 5 μl of 5 *M* NaCl.
2. Add 100 μl of high-salt DEAE-Sephacel to the lysate and rock gently at 4°C for 1 hr.
3. Tap spin in microfuge (10 sec) at 4°C and 16,000 g to remove DEAE-Sephacel.

To determine what percentage of contaminating DNA is removed by this procedure, we add a trace amount of radiolabeled DNA to a GST bacterial extract and a reticulocyte lysate preparation. After addition of the radiolabeled tracer DNA to these samples, a small aliquot is removed as "pre-DEAE treatment." The remainder is treated with DEAE-Sephacel as above, and a second aliquot is removed from the sample as "post-DEAE treatment." Both samples are phenol extracted and ethanol precipitated in the presence of 2 μg of carrier tRNA. Nucleic acids are resolved on an 8% (w/v) acrylamide gel and analyzed by autoradiography. Figure 1 shows that DNA is quantitatively removed by DEAE-Sephacel treatment of both the bacterial lysates and the reticulocyte lysates. We conclude that treatment of lysates with DEAE-Sephacel is rapid, gentle, and quantitatively removes contaminating DNA.

Binding Glutathione S-*Transferase Fusion Proteins to Glutathione-Agarose*

The first step in the affinity chromatography assay is to make an affinity resin with the GST fusion protein as follows.

1. Prepare reduced glutathione–agarose (Sigma, St. Louis, MO) by swelling beads in PBS for 2 hr. Wash the beads that will be used in assay twice with water.
2. In a 1.5-ml Eppendorf tube, add 25 μl DEAE-treated GST lysate, 30 μl of a

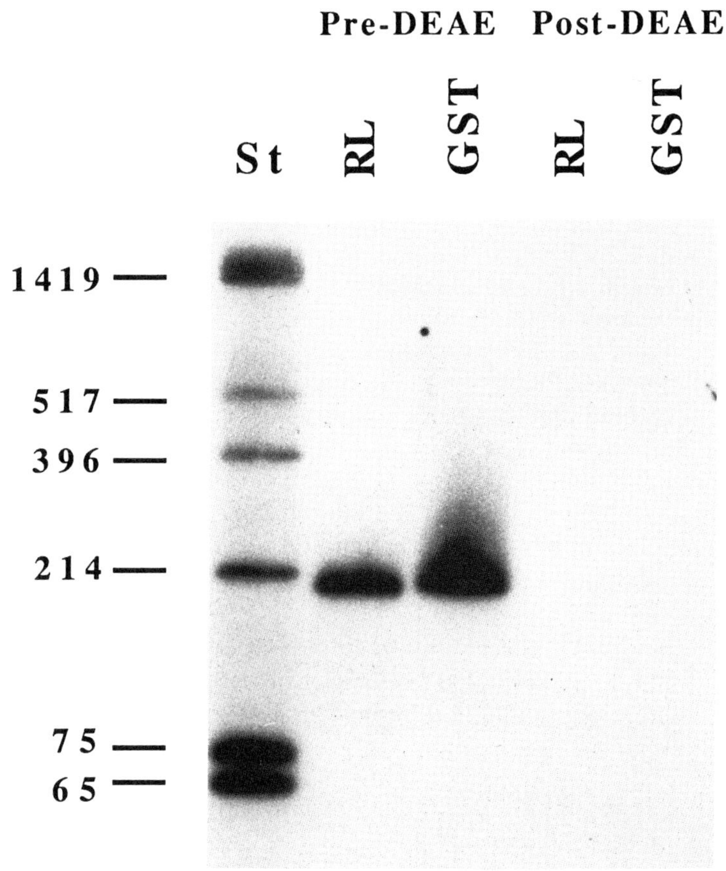

FIG. 1 A trace amount of a radiolabeled DNA fragment was added to a reticulocyte lysate (RL) and a GST bacterial extract (GST). Small aliquots of these samples were removed and labeled "Pre-DEAE." The extracts were then treated with DEAE-Sephacel and an equal amount of extract was removed and labeled "Post-DEAE." These samples were phenol extracted, ethanol precipitated, and run on an 8% (w/v) acrylamide gel with radiolabeled *Hin*fI-cut pUC19 standards (St).

50% slurry of glutathione–agarose in water, and 45 μl of water to reduce the salt concentration to 100 m*M* and add 1 *M* HEPES, pH 7.6, to make the final concentration 20 m*M*. Mix gently by rocking at 4°C for 1–2 hr.

3. Wash the beads by gentle mixing followed by brief centrifugation in a microfuge as follows: three times with PBS, then twice with buffer B [20 m*M* HEPES (pH 7.6), 50 m*M* NaCl, 200 μg/ml bovine serum albumin (BSA), 5 m*M* DTT] to remove bacterial proteins not specifically binding to glutathione.

(*Note:* This binding step can be scaled up if multiple-labeled proteins are to be tested with the bound GST fusion protein.)

Comments

Glutathione *S*-transferase fusion proteins vary significantly in terms of the amount of protein induced and the efficiency of binding to the glutathione matrix. Therefore, if one wishes to compare the ability of a labeled protein to interact with two different matrices, it is imperative to determine if similar amounts of GST fusion protein are bound to these matrices. This can be accomplished by denaturing the immobilized GST fusion proteins, followed by SDS-PAGE, and visualizing the bound proteins by Coomassie staining. If there is a significant discrepancy in the amounts of the two proteins, one can include less bacterial extract in the binding reaction with glutathione–agarose, thereby adjusting the amount of bound fusion protein empirically.

Testing Interaction of Labeled Protein with Immobilized Glutathione S-Transferase Fusion Protein

The DEAE-treated, reticulocyte lysate-expressed protein is incubated with the immobilized GST fusion protein. Note that the high salt concentration in the lysate is taken into account in setting up the binding reaction to achieve the desired final salt concentration.

1. To 30 μl of 50% slurry of bound glutathione–agarose, add 35 μl buffer B minus salt (buffer B containing no NaCl) and 5 μl DEAE-treated reticulocyte lysate protein. This will reduce the salt concentration to 50 m*M* for the binding reaction. Rock at 4°C for 1.5 hr.
2. Wash four times with 250 μl PBS containing 0.1% (v/v) Nonidet P-40 (NP-40) at room temperature.
3. Add 20 μl 2× SDS sample buffer. Heat at 90°C for 3 min.
4. Centrifuge at 12,000 *g* for 1 min at room temperature. Load the supernatant on the SDS-polyacrylamide gel. Analyze the gel by autoradiography or flurography.

Applications of Glutathione *S*-Transferase Column Technique

Glutathione *S*-transferase fusion protein affinity columns can be used either analytically or preparatively. Furthermore, either cloned proteins expressed *in vitro* or proteins from whole cell extracts can be tested for interaction with immobilized GST fusion proteins (18, 20). Here we briefly describe a few analytical applications of the method, using cloned proteins produced *in vitro*.

Interaction of Dimerization Partners

Two distinct structural classes of transcription factors that we study are basic region leucine zipper (BZip) proteins and basic region helix–loop–helix leucine zipper proteins (BHLHZip). Both of these classes contain a basic region of amino acids, proposed to contact DNA directly, and an adjacent domain that facilitates dimer formation. The leucine zipper is an amphipathic helix containing a heptad repeat of leucines that dimerizes to form a coiled-coil structure (26, 27). BHLHZip proteins also contain a leucine zipper, carboxy terminal to two other amphipathic helices separated by a loop of variable length. This HLHZip domain serves a function similar to that of the leucine zipper in mediating dimer formation and allowing the basic regions from two molecules to interact with DNA (28–30).

Because some, but not all, members within a structural family can form heterodimers and because many of these proteins are expressed within the same cell, it is important to determine which members of the family can heterodimerize. Immunoglobulin enhancer binding protein (Ig/EBP) (31) and liver-enriched transcriptional activator protein (LAP) (32) (also called NF-IL6 or c/EBPβ) are members of the CAAT/enhancer binding protein (C/EBP) family of BZip proteins that bind DNA as homodimers but have also been shown to form heterodimers by GEMSA analysis (Roman and Calame, unpublished results, 1991). Using an [^{35}S]methionine-labeled LAP protein, expressed in reticulocyte lysates, we showed that LAP can interact with immobilized GST-LAP and immobilized GST-Ig/EBP but not with GST itself (Fig. 2).

TFE3 (30, c-Myc, and Max (21) are all proteins in the BHLHZip class of proteins. TFE3 is a transcriptional activator that binds as a homodimer to a site in several genes, including the immunoglobulin heavy chain enhancer, the V1 heavy chain promoter, and the T cell receptor β chain enhancer. c-*myc* is a protooncogene whose expression is altered in many different tumors and *max* is its heterodimer partner. Together, c-Myc and Max form a sequence-specific DNA-binding heterodimer. Figure 2 shows that TFE3 can associate with itself in the form of a GST–TFE3 fusion protein but not with GST alone. As shown by Blackwood and Eisenman (21), c-Myc associates with GST–Max but does not interact with GST–TFE3, GST–LAP, or GST itself (Fig. 2).

Although the GEMSA for heterodimer formation is a simple and elegant assay, it requires having a DNA-binding site that a heterodimer can bind. In cases in which a binding site for a heterodimer cannot be predicted, the GST fusion protein affinity chromatography assay should be quite useful in evaluating heterodimer formation.

Interaction between Structural Classes

Interactions between proteins in different structural classes may serve many potentially important functions, as stated above. To determine if proteins in the BZip class could interact with proteins in the BHLHZip class, we tested the interaction

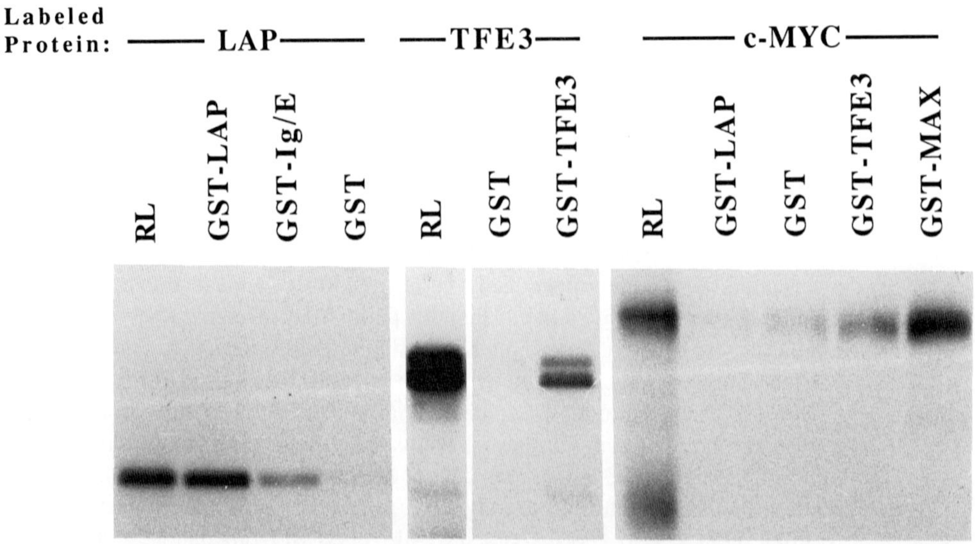

FIG. 2 Interaction of ^{35}S-labeled proteins with immobilized GST fusion proteins demonstrates specific interactions within a structural class. Labeled proteins were incubated with the indicated GST fusion protein matrices, and the bound protein was denatured and run on a 10% (w/v) SDS-acrylamide gel. In each gel, the first lane represents 1 μl of protein directly from the reticulocyte lysate (RL).

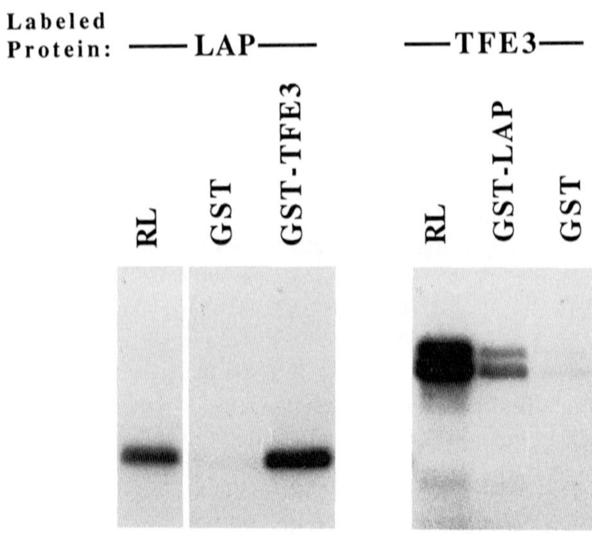

FIG. 3 Interaction of ^{35}S-labeled proteins with immobilized GST fusion proteins demonstrates interactions between BZip and BHLHZip structural classes. Assays were performed and presented as in Fig. 2.

of LAP and TFE3. Figure 3 shows that labeled TFE3 associates with the GST-LAP matrix and not the GST matrix. The interaction of TFE3 with GST-LAP is dependent on the integrity of the HLH domain because mutant proteins altered in helix 1 or helix 2 do not interact with LAP (data not shown). Conversely, labeled LAP interacts with TFE3 and not GST alone. One limitation of this assay is that the stoichiometry of the proteins in a complex on the matrix is unknown. For example, we cannot distinguish between interactions of a monomer of TFE3 and a monomer of LAP and interactions between dimers of each.

Salt Sensitivity of Complexes Bound to Immobilized Glutathione S-Transferase Proteins

Salt sensitivity is often informative in terms of the type of protein-protein interactions (hydrophobic or ionic) and the strength of these interactions. To measure the salt sensitivity of these interactions, instead of washing with PBS-NP-40 after binding the labeled protein to the immobilized fusion proteins, we wash with buffer B containing 50, 100, or 150 m*M* NaCl/ Figure 4 shows that the amount of bound LAP protein on the GST-TFE3 fusion protein matrix decreased with increasing salt,

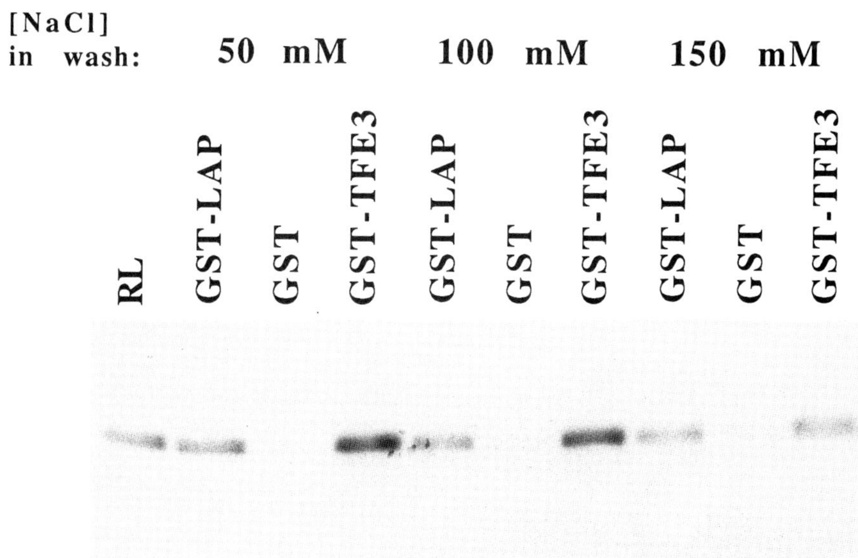

FIG. 4 Salt sensitivity of ^{35}S-labeled LAP protein interaction with various immobilized GST fusion proteins. Labeled LAP was incubated with GST-LAP, GST, or GST-TFE3 and then washed with buffer B containing 50, 100, or 150 m*M* NaCl.

indicating that over this range of salt concentration thses interactions are salt sensitive. LAP interaction with immobilized GST-LAP also decreased over this salt range but to a lesser degreee than the LAP-TFE3 interaction.

Acknowledgments

We are grateful to our colleagues at Columbia University for helpful discussions. This work was supported by U.S. Public Health Grants GM29361 and CA38571 to K.C.S.A. was supported by an NIGMS Medical Scientist Training Program training grant.

References

1. S. Buratowski, S. Hahn, L. Guarente, and P. A. Sharp, *Cell* **56,** 549 (1989).
2. D. Reinberg and R. G. Roeder, *J. Biol. Chem.* **262,** 3310 (1987).
3. B. D. Dynlacht, T. Hoey, and R. Tjian, *Cell* **66,** 563 (1991).
4. B. F. Pugh and R. Tjian, *Genes Dev.* **5,** 1935 (1991).
5. X.-H. Sun and D. Baltimore, *Cell* **64,** 459 (1991).
6. T. K. Blackwell and H. Weintraub, *Science* **250,** 1104 (1990).
7. N. S. Foulkes and P. Sassone-Corsi, *Cell* **68,** 411 (1992).
8. C. Roman, L. Cohn, and K. Calame, *Science* **254,** 94 (1991).
9. R. Li, J. D. Knight, S. P. Jackson, R. Tjian, and M. R. Botchan, *Cell* **65,** 493 (1991).
10. L. Poellinger, B. K. Yoza, and R. G. Roeder, *Nature* (*London*) **337,** 573 (1989).
11. I. A. Hope and K. Struhl, *EMBO J.* **6,** 2781 (1987).
12. P. E. Shaw, H. Schroter, and A. Nordheim, *Cell* **56,** 563 (1989).
13. T. M. Kristie, J. H. LeBowitz, and P. A. Sharp, *EMBO J.* **8,** 4229 (1989).
14. M. Sawadogo and R. G. Roeder, *Cell* **43,** 165 (1985).
15. P. M. Lieberman and A. J. Berk, *Genes Dev.* **5,** 2441 (1991).
16. K. F. Stringer, C. J. Infles, and J. Greenblatt, *Nature* (*London*) **345,** 783 (1990).
17. D. B. Smith and K. S. Johnson, *Gene* **67,** 31 (1988).
18. W. G. Kaelin, D. C. Pallas, J. A. DeCaprio, F. J. Kaye, and D. M. Livingston, *Cell* **64,** 521 (1991).
19. S. Bagchi, R. Weinmann, and P. Raychaudhuri, *Cell* **65,** 1063 (1991).
20. T. Chittenden, D. M. Livingston, and J. W. G. Kaelin, *Cell* **65,** 1073 (1991).
21. E. M. Blackwood and R. N. Eisenman, *Science* **251,** 1211 (1991).
22. A. K. Rustgi, N. Dyson, and R. Bernards, *Nature* (*London*) **352,** 541 (1991).
23. Y.-S. Lin and M. R. Green, *Cell* **64,** 971 (1991).
24. J. Sambrook, E. F. Fritsch, and T. Maniatis, "Molecular Cloning: A Laboratory Manual," 2nd Ed. Cold Spring Harbor Lab. Press, Cold Spring Harbor, New York, 1989.
25. J.-S. Lai and W. Herr, *PNAS* **89,** 6958 (1992).
26. W. H. Landschultz, P. F. Johnson, and S. L. McKnight, *Science* **240,** 1759 (1988).

27. E. K. O'Shea, J. D. Klemm, P. S. Kim, and T. Alber, *Science* **254,** 539 (1991).
28. P. D. Gregor, M. Sawadogo, and R. G. Roeder, *Genes Dev.* **4,** 1730 (1990).
29. H. Beckmann and T. Kadesch, *Genes Dev.* **5,** 1057 (1991).
30. C. Roman, A. G. Matera, C. Cooper, S. Artandi, S. Blain, D. C. Ward, and K. Calame, *Mol. Cell. Biol.* **12,** 817 (1992).
31. C. Roman, J. S. Platero, J. Shuman, and K. Calame, *Genes Dev.* **4,** 1404 (1990).
32. P. Descombes, M. Chojkier, S. Lichtsteiner, E. Falvey, and U. Schibler, *Genes Dev.* **4,** 1541 (1990).
33. M. Kozak, *Nucleic Acids Res.* **12,** 857 (1984).

[16] Gel Mobility Shift Assay to Study Nuclear Hormone Receptor–DNA Interactions

Leonard P. Freedman and Iris Alroy

Introduction

Gene transcription is a central biological control point. By modulating the level of mRNA synthesis, cells can effectively regulate the expression of the vast number of genes encoding proteins that contribute to the structure, growth, and differentiation of a particular cell type. Thus a detailed, mechanistic description of the regulatory processes that occur during transcription forms a foundation for our understanding of cellular regulation and development.

Steroid/nuclear receptors comprise an important group of transcriptional regulatory proteins. Like many eukaryotic transcription factors, nuclear hormone receptors are organized into relatively discrete functional domains (1), and it is this common domain organization that groups these receptors into a superfamily of functionally and most likely structurally related proteins (2, 3). The superfamily includes receptors for steroid hormones, such as glucocorticoids, progesterone, estrogen, aldosterone, and androgens, as well as hormonal forms of vitamins A and D, thyroid hormone, and peroxisomal activators (4–7). Several other receptor-like proteins whose ligands have yet to be identified have been isolated utilizing molecular cloning techniques. In the same way, several putative nuclear receptors from *Drosophila* that appear to act as regulators of early development have been identified, including a receptor for the insect steroid ecdysone (8).

On association with a particular ligand, nuclear receptors, as soluble intracellular proteins, bind to specific DNA sites generally located upstream of responsive genes, and up or down regulate the transcription of those genes, presumably by functionally interacting with other components of the transcriptional apparatus. Domains have been delineated for ligand binding (located in the carboxy-terminal region), DNA binding (located toward the central or amino terminus), nuclear localization (within both the DNA and ligand-binding domains), and transcriptional modulation (localized to more variable regions of the receptors, including the N terminus). In addition, a putative dimerization region has been described that overlaps part of the ligand-binding domain for some receptors (7).

Because of this modular organization of nuclear hormone receptors, the functional domains can confer their particular activities when linked to unrelated, nonreceptor proteins (9–14). This has led to the efficacy of examining in detail the properties of a particular receptor functional domain, such as the DNA-binding

Methods in Molecular Genetics, Volume 1

domain, and avoiding, at least initially, the potential physical/chemical complexities in studying a full-length, native receptor.

Along these lines, this article will focus exclusively on approaches we have utilized to study the functional properties of two nuclear hormone receptor DNA-binding domains, those of the glucocorticoid and vitamin D_3 receptors. We use these receptor derivatives here as examples of some of the techniques and manipulations applied to study protein–DNA interactions, utilizing the gel mobility shift to assay these interactions qualitatively and quantitatively.

Materials and Methods

The basic strategy we have used to study the structure and function of nuclear receptor DNA-binding domains has been to overexpress and purify the domains to high quantities, and then to study the function of these protein derivatives *in vitro* using the gel mobility shift assay. High yields of these purified derivatives have also produced detailed three-dimensional structures, as solved by both two-dimensional nuclear magnetic resonance (2-D NMR) and protein crystallographic analyses (15–17).

Vectors and Bacterial Strain

Receptor sequences encoding amino acids 440–525 of the rat glucocorticoid receptor (GR) (18) and amino acids 15–114 of the vitamin D_3 receptor (called VDRF; 19) (Fig. 1A and B, respectively) are cloned into the T7 RNA polymerase-dependent expression vector pT7X556 (20), using either a unique *Nde*I site at the initiator ATG (for the GR construct) or a unique *Bam*HI site that creates an in-frame fusion with the first 11 amino acids of T7 gene 10 (for VDRF). The vector also contains a polylinker with multiple restriction sites and translation stop codons in all three reading frames immediately downstream of these sites. The plasmids are used to transform BL21(DE3)/pLysS, a strain carrying a stable integrant of T7 gene 1 (T7 RNA polymerase) under the control of the *lacUV5* promoter (21).

Oligonucleotides

Oligonucleotides are synthesized using an Applied Biosystems (Foster City, CA) DNA/RNA synthesizer (model 392). This synthesizer automates all steps of single-stranded oligonucleotide synthesis. It uses the phosphoramidite method of oligonucleotide synthesis. The synthesis begins with the 3′ terminal nucleoside attached

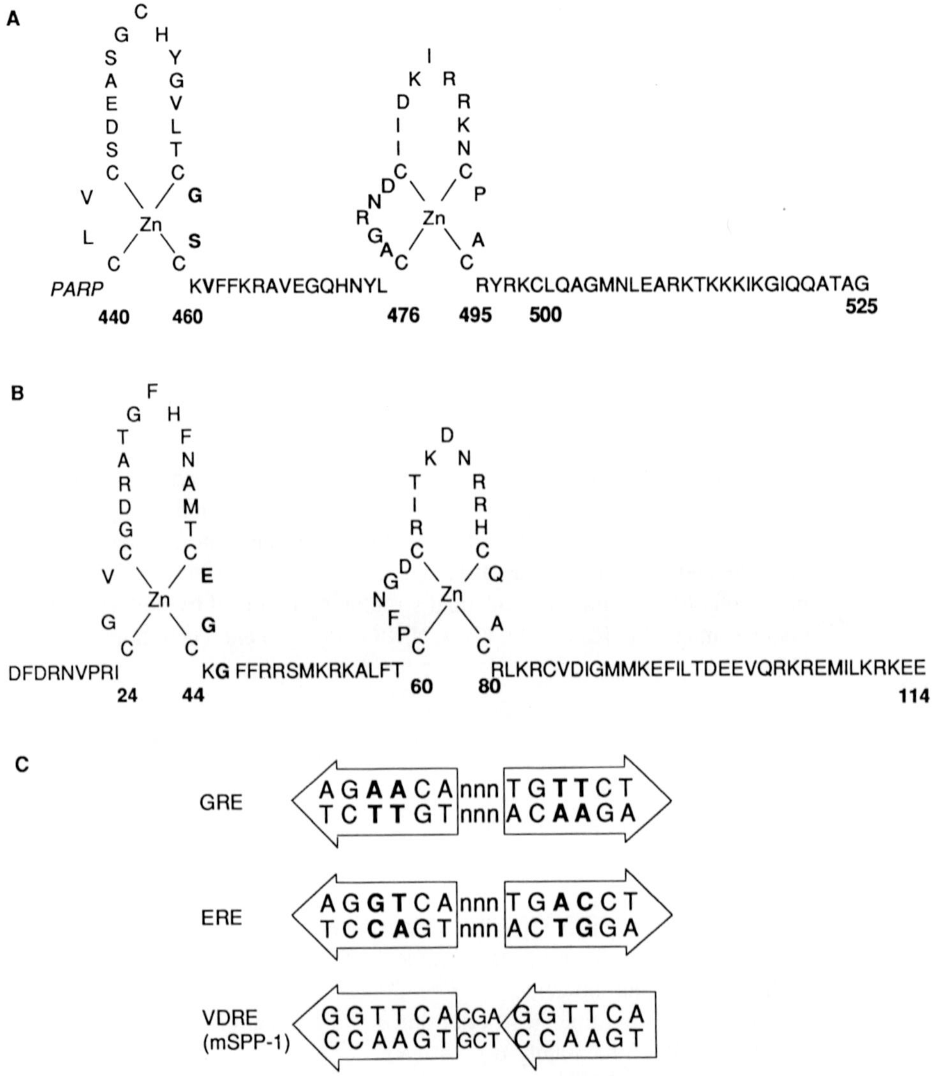

FIG. 1 Amino acid sequence of (A) the rat glucocorticoid receptor and (B) human vitamin D_3 receptor DNA-binding domains. Numbering corresponds to the translated full-length receptor cDNAs [(18); A. R. Baker, D. P. McDonnel, M. Hughes, T. M. Crisp, D. J. Mangelsdorf, M. R. Haussler, J. W. Pike, J. Shine, and B. W. O'Malley, *Proc. Natl. Acad. Sci. U.S.A.* **85,** 3294 (1988)]. (C) DNA sequence of consensus glucocorticoid and estrogen response elements (GRE and ERE) and a mouse *Spp-1* vitamin D response element (VDRE) [M. Noda, R. L. Vogel, A. M. Craig, J. Prahl, H. F. DeLuca, and D. T. Denhardt, *Proc. Natl. Acad. Sci. U.S.A.* **87,** 9995 (1990)]. Functional GREs or EREs consist of two hexameric half-sites oriented as inverted repeats (enclosed by arrows) spaced by 3 bp. Boldfaced bases in each half-site indicate the only differences between consensus GRE and ERE. The *Spp-1* VDRE consists of two direct repeats (enclosed by the arrows) with a spacing of 3 bp.

to a solid support, which is contained within a disposable column. Nucleotide bases are added one at a time to the support-bound DNA chain until the sequence is fully synthesized. The phosphate groups and exocyclic amines of the bases are protected to prevent side reactions during synthesis. After the synthesis is complete, the oligonucleotide is cleaved from the support using concentrated ammonium hydroxide at room temperature. During the cleavage, the β-cyanoethyl phosphate-protecting groups are also removed. Benzoyl(dA), benzoyl(dC), and isobutyryl(dG) base-protecting groups are completely removed by addition of concentrated ammonium hydroxide and heating for 8 to 15 hr at 55° C. The oligonucleotide solution is then dried in a Speed-Vac (Savant, Hicksville, NY) and resuspended in water. For gel shifts, either two single-stranded complementary oligonucleotides, or a self-annealing oligonucleotide can be used. When using two complementary oligonucleotides, equal amounts are annealed (see below). A recurrent problem is that not all oligonucleotides anneal completely, resulting in a lower effective concentration of double-stranded target site than believed. An alternative is to synthesize a long, self-annealing oligonucleotide that contains two complementary sequences with an intervening five-polynucleotide stretch that connects them. Here, essentially all of the oligonucleotides will self-anneal.

In Vitro Mutagenesis

Mutagenesis of the GR DNA-binding domain is carried out by subcloning the domain as an *Xba*I fragment from the pT7-GR 440–525 overexpression plasmid (17, 22) into the ampicillin repair phagemid vector pSELECT-1 (Promega, Madison, WI). Single-stranded pSELECT-GR 440–525 is generated by superinfection with helper phage. The pSELECT-1 vector contains a functional tetracycline resistance gene and a nonfunctional ampicillin resistance gene that carries a point mutation in the coding sequence. An ampicillin repair oligonucleotide that can revert the gene to resistance and a 5′ phosphorylated mutagenic oligonucleotide are both annealed to the single-stranded DNA template, and subsequent synthesis (T4 DNA polymerase) and ligation (T4 DNA ligase) of the mutant strand links the two. A repair-defective strain of *Escherichia coli* (BMH 71-18 *mutS*) is transformed with this DNA and the cells are grown in the presence of ampicillin; mutagenic clones are selected for their ability to grow with the antibiotic. A second round of transformation in JM109 ensures proper segregation of mutant and wild-type clones. Seventy to 80% of the ampicillin-resistant clones also carry the desired mutation. We note that it is also possible to use any phagemid vector that carries a functional ampicillin gene and to simply perform double-oligonucleotide mutagenesis with the ampicillin repair oligonucleotide and a mutagenic oligonucleotide. We have found that even without the selection for ampicillin repair, 40–50% of the colonies contain an introduced mutation. The advantage to not using the pSELECT-1 vector is that no recloning is required if an overexpression plasmid carries a single-stranded phage (f1 or M13) origin of replication.

The mutagenic oligonucleotide should have at least 12 bases on each side of the mismatch. It is possible to introduce several closely linked mutations with the same oligonucleotide, but in such cases the homology on each side should be greater than 12 bases. The mutant constructions described here are done in three stages. First, GR Val-462 is mutated to alanine using a "V → A" oligonucleotide (5′ GAA AGC TTT GCA GCC TTC ACA TGT CAG CAC CCC 3′). The GR-V → A single-stranded DNA is then used as template for a second round of mutagenesis in which Gly-458 and Ser-459 are simultaneously mutated to glutamine and glycine, respectively, using an "EGA" oligonucleotide (5′ GCT CTT TTA AAG AAA GCT TTG CAG CTT CCA C 3′). The GR-**EGA** DNA template is used to generate a third mutant, GR-**EG**V, by converting the alanine back to Val-462, with an "EGV" oligonucleotide (5′ GCT CTT TTA AAG AAT ACT TTG CAG CCT TCA C 3′). Following mutagenesis, clones are selected for ampicillin resistance, and screened for the appropriate mutations by dideoxy-DNA sequencing. Confirmed mutated DNA is subcloned back into the original T7 overexpression vector, and pT7 GR 440–525 mutants reconfirmed by sequencing following subcloning.

Overexpression and Purification of Receptor Derivatives in Escherichia coli

A culture of BL21(DE3)/pLysS carrying pT7-GR 440–525 or pT7-VDRF is grown overnight at 37°C in LB medium (10 g tryptone, 5 g yeast extract, 5 g NaCl, per liter) containing 50 μg/ml ampicillin and 34 μg/ml chloramphenicol and then diluted 1:100 in fresh LB containing only ampicillin. The culture is induced at mid-logarithmic growth ($OD_{600} \approx 0.6$) by addition of isopropyl-β-D-thiogalactoside to a final concentration of 0.5 m*M*. After 3 hr of induction, cells are collected by centrifugation [4000 rpm in a Sorvall (Norwalk, CT) RC3C], and overexpressed proteins purified in one of two ways, depending on their relative solubilities in *Escherichia coli*. GR 440–525 is soluble even at millimolar concentrations, whereas VDRF is extremely insoluble. Although these two nuclear receptor derivatives are highly homologous in amino acid sequence (≈50%) (see Fig. 1) and presumably fold into similar three-dimensional structures based on the conservation of critical residues (23), it is difficult if not impossible to predict *a priori* whether a bacterially overexpressed protein or protein fragment will turn out to be stable, soluble, and easy to purify. We provide purification protocols for both proteins below as relatively simple strategies to isolate proteins depending on such parameters.

Purification of GR 440–525

The cell pellet is resuspended in 3 vol (w/v) of lysis buffer [50 m*M* Tris-HCl (pH 7.5), 1 m*M* ethylenediaminetetraacetic acid (EDTA), 10% (v/v) glycerol, 500 m*M* NaCl, 4 m*M* $CaCl_2$, 40 m*M* $MgCl_2$, and, added fresh, 5 m*M* dithiothreitol

(DTT), 0.5 m*M* phenylmethylsulfonyl fluoride (PMSF), lysozyme to 0.13 mg/ml, and DNase I to 20 μg/ml] and incubated on ice for 10 min. Sodium deoxycholate is added slowly over a period of 10 min to 0.05%, and the mixture mixed for 15 min at 4° C. The lysate is centrifuged at 60,000 *g* for 30 min, poly(ethyleneimine) (Aldrich Chemical Co., Milwaukee, WI) [10% (v/v) stock, pH 7.5] is added slowly over a period of 10 min to the supernatant to a final concentration of 0.2% (v/v), and the solution mixed for an additional 10 min at 4° C. This supension is centrifuged at 15,000 rpm for 30 min in an SS-34 rotor, and the pellet discarded. The supernatant is precipitated with ammonium sulfate to 50% (w/v) stirred gently for 10 min at 4° C, and centrifuged at 10,000 rpm. The pellet is resuspended in $TGED_{50}$ buffer [50 m*M* Tris-HCl (pH 7.5), 0.5 m*M* EDTA, 10% (v/v) glycerol, 5 m*M* DTT, 50 m*M* NaCl] and dialyzed against $TGED_{50}$. The dialysate is applied to a Bio-Rex 70 (Bio-Rad, Richmond, CA) column equilibrated with 5 vol of $TGEDZ_{50}$ buffer ($TGED_{50}$ plus 0.05 m*M* $ZnSO_4$), the column is then washed with 3 vol of $TGEDZ_{50}$ and the protein eluted with a 50–600 m*M* NaCl gradient, and 0.5-ml fractions are collected. The fractions are run on a 15% (w/v) sodium dodecyl sulfate polyacrylamide gel electrophoresis (SDS-PAGE) gel to determine which fractions contain the purified protein. The pooled eluant is 90–95% pure protein, as assessed by SDS-PAGE.

Purification of VDRF

The cell pellet is resuspended in 3 vol of buffer 1 [phosphate-buffered saline (PBS) containing 1 m*M* DTT, 1 m*M* PMSF, and 0.1 m*M* $ZnCl_2$], and sonicated on ice for three 20-sec intervals. The lysate is centrifuged at 14,000 rpm for 20 min at 4°C, and the pellet is resuspended in 3 vol buffer 1 plus 1 *M* NaCl. The mixture is sonicated on ice three times for 20 sec and the pellet spun down at 14,000 rpm for 20 min at 4°C, resuspended in 3 vol of buffer 1 plus 1 *M* NaCl, and sonicated and centrifuged as before. The protein pellet is resuspended in 10 ml of $TGEDZ_{50}$ plus 5 *M* urea and denatured overnight by inversion at 4° C. After centrifugation for 10 min at 14,000 rpm, the supernatant containing denatured, solubilized VDRF is retained and subjected to stepwise renaturation by dialysis for 2.5 hr each against $TGEDZ_{50}$ plus 0.1% (v/v) Triton X-100 and then against $TGEDZ_{50}$. The dialysate is centrifuged at 14,000 rpm for 10 min and then applied to a CM-Sepharose CL-6B (Pharmacia, Piscataway, NJ) column equilibrated with $TGEDZ_{50}$, and eluted with 50–800 m*M* NaCl gradient. The pooled eluant is 80% pure VDRF, as assessed by SDS-PAGE.

Gel Mobility Shifts Using Purified Nuclear Receptor Derivatives

DNA binding is assessed by gel mobility shift electrophoresis. Complementary strands of synthetic oligonucleotides are resuspended in 5 m*M* sodium cacodylate, 50 m*M* NaCl, and annealed by heating for 2 min to 65° C, followed by slow cooling to room temperature. Duplex oligonucleotide (50 ng) is end labeled with 20 μCi

[γ-^{32}P]ATP (3000 Ci/mmol) using T4 polynucleotide kinase. The labeled oligonucleotide is separated from unincorporated nucleotides using a Chroma Spin-10 column (Clontech, Palo Alto, CA). Purified GR 440–525 or VDRF is incubated with 0.5 ng (10,000 cpm) of the labeled, double-stranded oligonucleotide at 33°C for 30 min, together with 500 ng poly[d(I-C)], and a binding buffer to give a final concentration of 20 m*M* Tris-HCl (pH 7.9), 1 m*M* EDTA (pH 8), 0.05% (v/v) Nonidet P-40 (NP-40), 10% (v/v) glycerol, 50 m*M* KCl, and 1 m*M* DTT. At the end of the incubation, 3 μl of 6× gel loading dye [0.025 g bromphenol blue, 0.025 g xylene cyanol, in a total of 10 ml water plus 0.5 m*M* EDTA (pH 8.0)] is added and protein–DNA complexes are then resolved by electrophoresis on 10% non-denaturing acrylamide gels (75:1 (v/v), acrylamide–bisacrylamide) run at 26 V/cm at 4°C in 0.5× TBE [45 m*M* Tris, 45 m*M* boric acid, 0.1 m*M* EDTA (pH 8)]. Gels were dried and subjected to autoradiography with an intensifying screen for 3 hr at −70°C.

Results

Among the superfamily of nuclear hormone receptors, the DNA-binding domain is by far the most highly conserved. Several amino acids in this domain are invariant throughout the family, including eight cysteines that were shown, in the case of the glucocorticoid receptor (GR), to tetrahedrally coordinate two zinc ions (20) in an arrangement reminiscent of the "zinc finger" coordination scheme originally proposed for transcription factor IIIA of *Xenopus* (24, 25). Mutagenesis experiments indicate that specific residues within the zinc finger region of the glucocorticoid receptor are critical for DNA-binding specificity and DNA-dependent dimerization (22, 23, 26–29). These results have been confirmed and expanded by three-dimensional structural analysis (15–17). Thus, the DNA-binding domain, a small yet self-contained region within nuclear receptors, is itself subdivided into substructures of distinct conformation and function (reviewed in Ref. 23).

Purification of Nuclear Hormone Receptor DNA-Binding Domains

The scheme presented above for protein purification can serve as an outline. GR 440–525 and VDRF proteins both belong to the nuclear receptor family and contain highly homologous DNA-binding domains (Fig. 1). The purification profiles of these proteins, however, are different due to the fact that GR 440–525 is soluble in *E. coli,* whereas VDRF is not. Thus, a protein such as VDRF must be extracted out of inclusion bodies with high concentrations of a denaturant such as urea or guanidine hydrochloride, and then renatured by dilution or step dialysis (see Materials and Methods). Insolubility is a rather common problem facing investigators performing overexpression in heterologous systems, and must be balanced with the advantage of high yields that these systems can produce. For soluble proteins, ammonium sulfate precipitation can be an effective step during purification, because it provides the dual advantage of concentration (and stabilization) as well as fractionation prior to column chromatography. Optimal ammonium sulfate concentrations required for concentration and/or enrichment need to be empirically tested as part of the overall optimization of the purification strategy. The same is true for the column resins that will ultimately yield the best fractionation. For relatively basic DNA-binding domains that utilize, in part, ionic interactions with the net negatively charged phosphate backbone of DNA, cationic exchangers such as Bio-Rex 70 (Bio-Rad) or CM-Sepharose (Pharmacia) generally work well as a first column, because they selectively retain many nucleic acid-binding proteins, and have a relatively high protein capacity. For further purification, affinity resins, such as Cibacron blue or heparin-Sepharose, retain many DNA-binding proteins. Finally, if a high-affinity DNA-binding site is known for the protein of interest, oligonucleotide affinity chromatography is often an excellent last step in purification, not only resulting in a highly purified pool but one that essentially contains exclusively active protein, because this purification step demands specific protein–DNA interaction.

Gel Mobility Shift Assay: General Considerations

The gel mobility shift serves as an extremely convenient assay for protein–DNA interaction, specificity, and affinity determination. This assay is the simplest and, in one sense, the most sensitive, of all *in vitro* protein–DNA binding methods. It is based on the fact that DNA–protein complexes migrate through low ionic strength, native polyacrylamide gels at a slower rate than free, unbound DNA (30–33). Because of its ease of use, it can be utilized as an assay for the purification of specific DNA-binding proteins. In addition, the gel mobility shift assay can be used quantitatively. Methods will be described that allow the determination of a dissociation constant for a given DNA sequence of a DNA-binding protein, the stoichiometry and cooperativity of binding, and relative specificity a protein may have for a given DNA element.

To perform the assay, a DNA fragment carrying a sequence of interest is radioactively labeled with ^{32}P. Small amounts of this DNA (10 pmol or less) are then incubated with a crude nuclear extract, or a partially purified or purified protein, and a carrier DNA to reduce nonspecific binding. The synthetic copolymer poly[(d(I-C)] is commonly used. The binding reaction is adjusted with a buffer that allows for variations in salt and Mg^{2+}, as well as pH and nonionic detergents. This needs to be empirically determined depending on the properties of the particular DNA-binding protein being studied (see below). These components are allowed to incubate, often at room temperature, to equilibrium (the time may vary from a few minutes to

1–2 hr), and then loaded directly on a native polyacrylamide gel that has been prerun for a short time. We load the gels while they are running, in order for the samples to enter the gel as rapidly as possible. Because the gels are electrophoresed at relatively high voltage (26 V/cm of gel), they are run at 4°C to avoid excess heating. The gels are generally electrophoresed in buffers at low ionic strengths, in order not to disrupt the essentially ionic interactions of the protein and DNA. The gel matrix itself provides a "caging effect," such that dissociated protein and DNA do not diffuse away from each, effectively raising the local concentration of the two components and thereby resulting in an increase in a bimolecular reassociation reaction between the protein and DNA. After the gel is run, it is exposed to X-ray film. The typical pattern is that of a series of bands: the fastest migrating band represents unbound, or free, DNA. Retarded, or slower migrating bands within the same lane correspond to protein–DNA complexes; each shifted band represents a distinct complex. The size of the protein, or the number of proteins bound to a given DNA site, further slows the mobility of the DNA, resulting in a higher shift. As increasing amounts of protein are added to the reaction, a greater intensity of the shifted bands should be observed as the free band diminishes in intensity, due to the proportion of protein-bound DNA relative to unbound DNA.

The conditions we have described for gel mobility shift assays, that is, salt and detergent concentrations, as well as temperature, should be adjusted for a given protein. For example, a mutant GR DNA-binding domain derivative we made in which three amino acids were changed toward the estrogen receptor (ER) (22), called GR-**EGA,** preferred an estrogen response element (ERE) over a glucocorticoid response element (GRE) (Fig. 1C). Our initial binding buffer, however, contained no salt and thus GR-**EGA** retained some affinity for a GRE as assessed using the gel shift assay. We tried different salt concentrations, temperatures, and NP-40 concentrations to render the binding more discriminating (i.e., a clear preference for an ERE over a GRE). KCl (50 m*M*) and incubation at 33°C proved to be optimal in this case. The composition of the binding buffer should also be tested. We have noticed that *N*-2-hydroxyethylpiperazine-*N*′-2-ethanesulfonic acid (HEPES)-based, Mg^{2+}-containing buffers can be inhibitory for some receptor proteins over time. Incubation time can also be critical, depending on the off-and-on rate of the protein for DNA.

When using nuclear extracts, it is advantageous to preincubate the extract with a nonspecific competitor such as poly[d(I-C)] in the absence of the labeled probe. Preincubation time should be between 30 min to an hour, depending on the extract and the probe. Because there are a variety of nucleic acid-binding proteins in nuclear extracts, it is of course necessary to prove specific DNA–protein interactions. In Fig. 2, a nuclear extract derived from cultured T cells was incubated with a VDRE oligonucleotide (DR + 3′) either with (lanes 4–9, Fig. 2) or without (lane 3, Fig. 2) overexpressed vitamin D_3 receptor (VDR) purified from *E. coli*. Only when both the extract and 20 ng of purified VDR are present is binding detected (two arrowheads, Fig. 2). In the presence of 10-fold excess cold DR + 3′

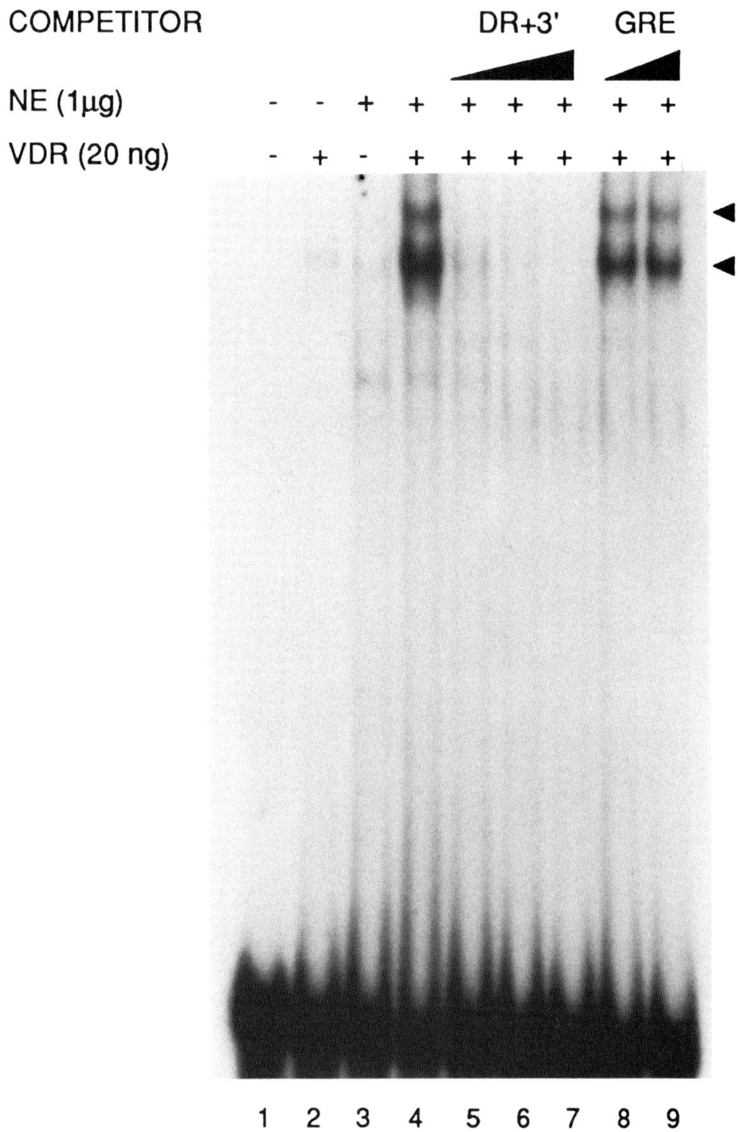

FIG. 2 *In vitro* binding of purified human vitamin D_3 receptor (hVDR) and T cell nuclear extract to a VDRE. Gel mobility shift assay of *E. coli* purified hVDR without (lane 2) or with (lanes 4–9) nuclear extract prepared from a cultured T cell line, 22F6. In each lane, 20 ng of hVDR and 1 μg of nuclear extract were incubated with 25,000 cpm (170 pg) of a VDRE probe (called DR + 3′; top strand, 5′ AGCTTCGC**AGTTCA**AGG**AGTTCA**GAGCG 3′). In lanes 5–7, cold DR + 3′ was added to each reaction at 10-, 50-, and 100-fold excess, respectively. As a nonspecific competitor, a GRE oligonucleotide (Fig. 1C) was used in lanes 8 and 9 (50- and 100-fold, respectively). The arrowheads indicate hVDR-specific shifts.

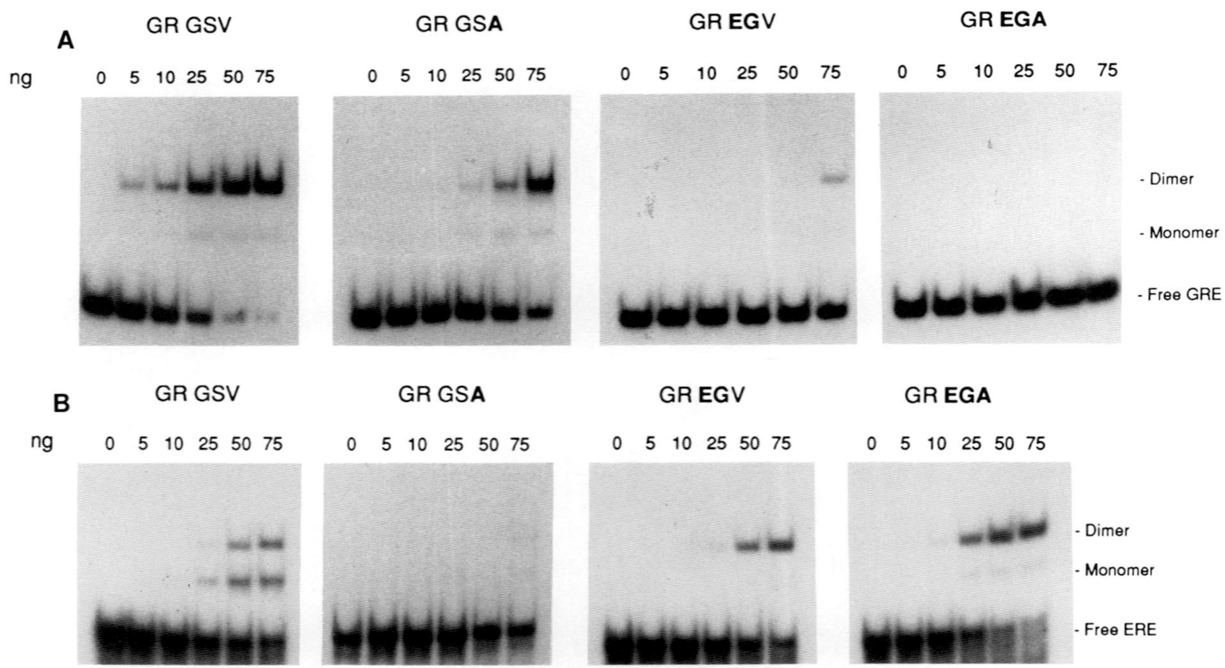

FIG. 3 *In vitro* binding of purified GR 440–525, GR-GS**A**, GR-**EG**V, and GR-**EGA** proteins to a GRE and ERE. (A) Gel mobility shift assay of the indicated purified proteins to a double-stranded GRE duplex oligonucleotide (see Fig. 1C). The total amount of protein (ng) in each binding reaction is indicated above each lane. Bands representing the free DNA probe, and DNA bound by a monomer or a dimer, are indicated. (B) Gel mobility shift assays of indicated purified proteins to an ERE duplex oligonucleotide (see Fig. 1C).

oligonucleotide (lane 5, Fig. 2), the two upper bands disappear, whereas a GRE oligonucleotide does not compete, even at a 100-fold excess (lane 9, Fig. 2). The results indicate that the shifted species bind specifically to a VDRE. To confirm that it is the vitamin D_3 receptor that is part of the complex, an antibody directed specifically against it can be included in the reaction. If the receptor is in fact present in the protein–DNA complex, the band will be supershifted, that is, the complex will migrate even more slowly through the gel.

Specificity of Receptor–DNA Interactions

Gel mobility shifts can be used to delineate which amino acids within a given protein are important for contact with DNA, and which base pairs are critical for specifying contacts. As an example, we consider the DNA-binding domains of the estrogen and glucocorticoid receptors (ER and GR, respectively). This highly conserved, short (<70 amino acids) region is the sole determinant for specific target gene recognition by these receptors. Functional consensus GREs and EREs, which consist of two half-sites organized as imperfect palindromes, can differ by only 2 bp in each half-site; T(5)T(6) of a GRE becomes A(5)C(6) in an ERE (AGAACAnnnTG**TT**CT and AGGTCAnnnTG**AC**CT, respectively; see Fig. 1C). Thus it is here that binding discrimination must take place within the response element. Combining the results of transient chloramphenicol acetyltranserase (CAT) trans-activating assays from three laboratories (26–28), it is apparent that three amino acids (Fig. 1A, residues in bold), located at the C-terminal side of the first finger, play a key role in distinguishing between a GRE and ERE. The three-dimensional structure of the GR DNA-binding domain complexed to a GRE has been solved (17). Based on this structure, it appears that of the three discriminating amino acids, only the valine at position 462 is making a direct side-chain contact with DNA. To understand the role of the other residues in mediating target site discrimination, the structures of the ER DNA-binding domain, as well as GR substitution mutants, complexed to an ERE, will have to be solved. Toward that end, we have generated specific mutations in this subregion of the GR DNA-binding domain utilizing oligonucleotide-directed *in vitro* mutagenesis, overexpressed the mutant proteins in *E. coli,* and purified them to homogeneity. Three mutants were made in the context of the GR zinc finger region (residues 440–525): GR-GS**A,** in which Val-462 of GR was substituted to the corresponding alanine of ER; GR-**EG**V, where Gly-458 was changed to the corresponding glutamic acid of ER, and Ser-459 to the ER glycine; and GR-**EGA,** which combines all three amino acid changes to the corresponding residues of GR. Replacing Gly-458 and Ser-459 by the corresponding amino acids Glu and Gly, in ER is enough to change binding specificity to an ERE, even though the specifying GR valine is still present (GR-**EG**V; Fig. 3A and B). On the other hand, making a single substitution for the valine with the corresponding residue from ER, alanine (GR-GS**A**), does not convert the protein to an ERE-binding species (Fig. 3B).

Thus, the conclusion we draw from these mutants is that although residues within the first α helix of both the GR and ER DNA-binding domains are clearly critical for discrimination, their positions are not equivalent. That is to say, whereas valine at position 462 within GR is necessary for GRE recognition, the corresponding alanine in ER is not the determining residue for ERE recognition. Rather, based on these results, it is probable that the ER-specific glutamic acid, five amino acids more N terminal than the alanine, is what specifies protein binding to an ERE (22).

Changing specific base pairs in an oligonucleotide used in gel shifts can give additional information on particular bases important for recognition. For example, a battery of oligonucleotides with systematic base pair changes can be used as specific competitors of protein–DNA binding using the wild-type sequence as the probe. The degree of competition from the various oligonucleotides used can be quantitated to assess the relative contribution a given base may make in determining affinity and/or specificity.

Protein–Protein Interactions on DNA

The gel shift assay can also be used for assessing DNA-dependent protein–protein interactions that are, for example, mediated by homo- or heterodimerization. Proteins that bind to DNA as dimers often do so by recognizing half-site sequences that are oriented symmetrically (i.e., head to head, as in GREs or EREs), or asymmetrically (i.e., head to tail, as in VDREs; see Fig. 1C). The crystal structure of the GR DNA-binding domain (17) clearly shows that the protein, monomeric in solution (15, 34), symmetrically dimerizes on DNA binding by forming reciprocal salt bridges and hydrogen bonds between residues in each monomer that reside exclusively in the second finger. Among the important contacts are a symmetrical pair of salt bridges between Arg-479 and Asp-481 of each monomer. These residues lie within the so-called "D box" of the second finger (28). The strength of the dimer interface that results from these symmetrical contacts explains why GRE half-sites are always in an inverted orientation with a strict spacing of 3 bp. This is manifested in the gel shift assay as a strongly cooperative, DNA-induced binding by GR 440–525 to the GRE probe (a quick shift to the higher shifted species even at low protein concentrations; Fig. 3A). When that spacing is changed to 4 bp, this cooperativity is abolished (19), because the spacing between the two GRE half-sites and the two protein monomers is no longer in complementary register (i.e., 3 bp) (17).

Homo- or heterodimerization of proteins on DNA can also be demonstrated using two different-sized protein derivatives that can recognize the same target sequence. As shown in Fig. 4, we used an ER DNA-binding domain (ER DBD) derivative that encompasses amino acids 171–281 [110 amino acids (35); apparent molecular mass, 20 kDa] and the GR DNA-binding domain (GR DBD) (85 amino acids; apparent molecular mass, 13.5 kDa) to test the interaction of these proteins on a chimeric GRE/ERE oligonucleotide, in which one half-site derives from an ERE

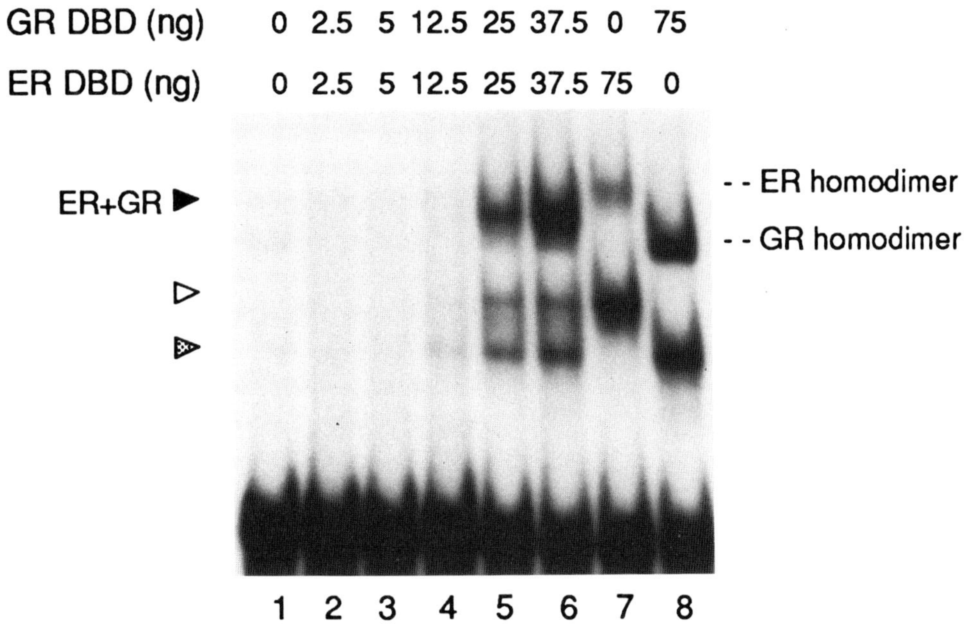

FIG. 4 *In vitro* binding of GR and ER DNA-binding domains to a hybrid GRE/ERE oligonucleotide ("EGRE"; top strand, 5′ GATCCGACCG**AGGTCA**AGA**TGTTCT**GTCGAG 3′). The amount (ng) of each protein in each binding reaction is indicated above the lanes. Bands representing an ER/GR heterodimer (lanes 5 and 6; black arrowhead) are indicated. The open arrowhead indicates a GR monomer and the stippled arrowhead indicates an ER monomer.

and the other from a GRE (AGGTCAnnnTGTTCT). The wild-type ER and GR typically bind with high affinity as apparent homodimers to the two half-sites that comprise a functional ERE or GRE, respectively (Fig. 4, lanes 7 and 8). When the two proteins are mixed (lanes 5 and 6, Fig. 4), three bands can be seen, corresponding to a GR DBD monomer (stippled arrowhead, Fig. 4), an ER DBD monomer (open arrowhead, Fig. 4), and a band (solid arrowhead, Fig. 4) that has a mobility intermediate of the GR and ER homodimers, consistent with the notion that a GR DBD/ER DBD heterodimer can form on a chimeric GRE/ERE DNA probe.

Quantitation of Protein Affinity

The relative affinity of a protein for a particular DNA sequence can be determined in a variety of ways from a mobility shift. For example, the dissociation constant (K_D) can be calculated by determining the concentration of competitor oligonucleotide required to reduce the percentage conversion of free to bound probe to 50%.

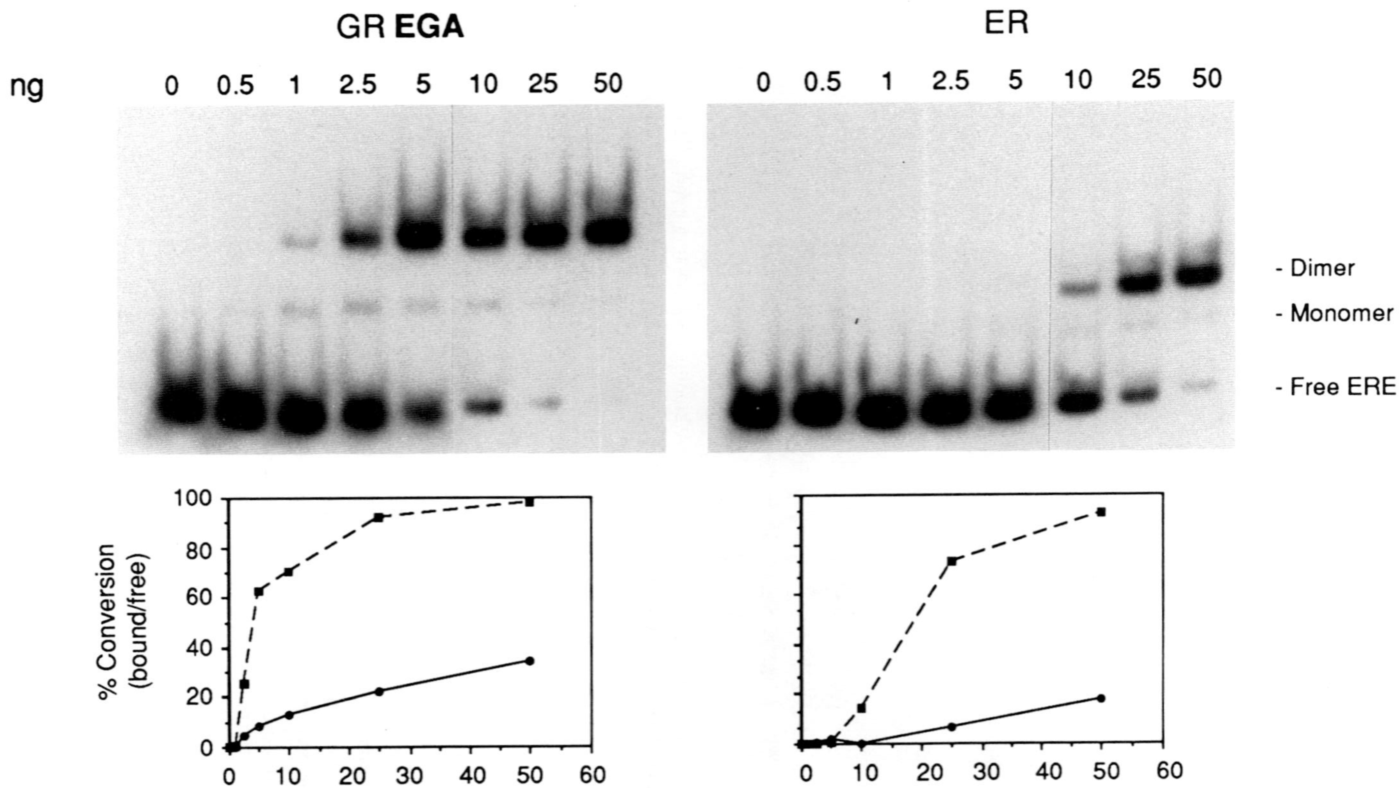

FIG. 5 Determination of relative affinities (K_D) of GR-**EGA** and ER DNA-binding domains to an ERE probe, using the gel mobility shift assay. The probe used here is a 51-bp restriction fragment from the plasmid pATC1 (35). Note that the ER DNA-binding domain retards the probe to a lesser extent than GR-EGA due to its slightly smaller size (13 vs 13.5 kDa, respectively). The graphs show the proportion of the probe bound by a monomer (—) or a dimer (– – –) as a percentage of total input probe. Annotation as in Fig. 3.

The vitamin D receptor DNA-binding domain (VDRF) has a low affinity for a response element from the human osteocalcin gene promoter (hOC VDRE), but a relatively high affinity for a VDRE from the mouse *Spp-1* gene, because an excess of the latter oligonucleotide can effectively compete for binding at a much lower concentration than the hOC VDRE (19). The K_D of VDRF for the *Spp-1* VDRE was determined from this experiment to be 50 n*M* (19). With pure protein in hand, perhaps a more accurate method of determining the K_D is to keep the probe and competitor concentration constant while increasing the concentration of protein, and then determining what concentration results in half occupancy of the specific probe.

For both approaches, a rough estimate of the fraction of free DNA converted to bound can be made visually. As an example, in Fig. 5 we compare the relative affinities of the GR DNA-binding domain specificity mutant EGA and the ER DNA-binding domain itself for an ERE recognition site. To quantitate, bands can be excised directly from the gel and counted, or the autoradiogram subjected to densitometry tracing. Another method for quantitation that is extremely convenient and accurate is the use of a phosphoimager. In this method, the dried gel is placed in a phosphoimager cassette and individual bands are recorded on a phosphor screen and scanned using a Phosphoimager 400-E (Molecular Dynamics, Sunnyvale, CA) set to detect ^{32}P radioactive emissions; the signals of each band are then digitized and quantified with Image Quant Software (Molecular Dynamics) and corrected for background. The fraction of DNA converted from free to bound for each protein concentration is plotted (Fig. 5) and the K_D is calculated by determining the amount of protein at which half the DNA is bound, and then using the formula:

$$K_D\ (\mathrm{n}M) = \frac{\text{ng of protein at 50\% conversion}}{[\text{reaction volume (ml)}][\text{molecular mass protein (Da)}]} \times 10^3$$

In Fig. 5 the calculated K_D dimer for GR-EGA is $9.3 \times 10^{-9}\,M$ and for ER DBD it is $5 \times 10^{-8}\,M$.

At least one problem with determining the K_D using the gel shift assay is that it is unclear to what degree the fraction of free and bound complexes is perturbed in the process of performing the assay (i.e., running the gel). It has been suggested (36) that utilizing a method that does not affect this equilibrium is a more accurate and quantitative way to calculate a binding constant. DNase I footprinting is a method that fits this criterion, because the enzyme apparently acts without disrupting the equilibrium distribution of protein bound to DNA. Thus, regions of DNA carrying multiple protein-binding sites can be resolved into individual K_D values simultaneously, and can be used to analyze multiple, cooperatively interacting sites.

One potential drawback of DNase I footprinting, however, is that the enzyme does not always cut uniformly, presumably because it has some preference for certain sequences. It is therefore often not unusual to observe some regions of the cleaved DNA ladder that are more enhanced than other areas. To overcome this, other footprinting methods have been described that utilize small molecules, rather than enzymes, to cleave the DNA. For example, both the hydroxyl radical (37) and 1,10-phenanthroline copper (38), because of their small size relative to DNase I, uniformly nick DNA, resulting in protection patterns that are often of higher resolution than that generated by DNase I. However, because the DNase I technique is easier to perform than other methods, it is usually the first footprinting method to attempt in beginning to define regions of protein–DNA interactions.

Kinetic Determinations of Protein–DNA Interactions

Kinetic analysis of a protein–DNA complex can be determined using the gel shift assay. Rate measurements can also be used to confirm a previously determined dissociation constant, because $K_D = k_{off}/k_{on}$. On rates can be generated by mixing protein and DNA and loading the sample on the gel at a specific time interval, because any free DNA is separated from protein–DNA complexes once the sample is loaded on a electrophoresing gel. When determining off rates, complexes are allowed to come to equilibrium, and competed at varying time intervals with a given amount of specific competitor DNA at a concentration previously determined to compete with the protein effectively. For example, at 44 n*M*, the vitamin D_3 receptor DNA-binding domain (VDRF) is completely competed by a 500-fold molar excess of an *Spp-1* VDRE-containing oligonucleotide (19). The reaction is then challenged with this amount of competitor for a specific time interval, and then loaded on a gel (Fig. 6). To load all of the samples simultaneously, one should

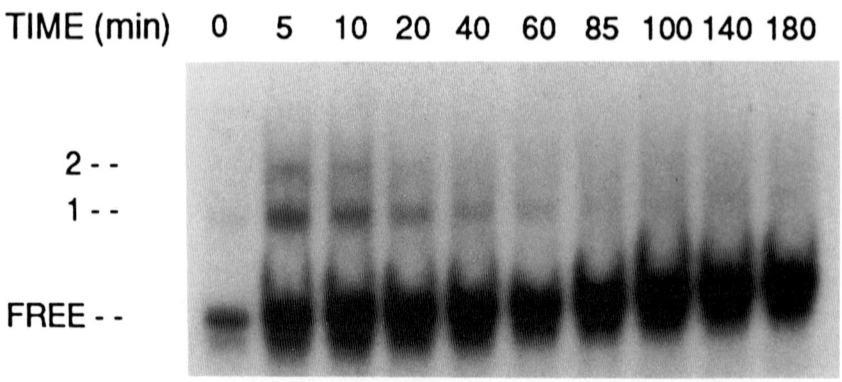

FIG. 6 Off-rate determination of hVDR DNA-binding domain (VDRF) to an *Spp-1* VDRE. Purified VDRF (280 ng) was incubated for 90 min with 5 ng labeled *Spp-1* probe (see Fig. 1C) in a 100-μl reaction. Starting with the longest time point (180 min, lane 10), 10 μl (containing 28 ng of protein, 0.5 ng of probe) was removed from the total reaction and added to a 500× molar excess of cold *Spp-1* oligonucleotide; 20 min later 10 μl was removed again and added to a second tube for the 140-min incubation, and so on. At time 0, all samples were loaded together and subjected to a gel shift assay.

begin with the longest time point and end with the shortest. The gel is then run as any shift assay is, exposed to film, and quantitated. The dissociation rate constant, k_{off}, can be calculated from the equation: $-k_{off}$ = ln[protein–DNA]/[protein–DNA]$_0$, where [protein–DNA]$_0$ is the amount of complex present immediately after adding the excess competitor (39). Our calculated k_{off} of VDRF for *Spp-1* VDRE was 2.5×10^{-4} sec^{-1}, and the half-life for the complex was approximately 30 min (Fig. 6).

Stoichiometry of Binding

Because the gel shift assays resolves protein–DNA complexes into discrete bands that represent the same probe with varying numbers of a protein or proteins bound to it, it can be used as an effective assay to determine the stoichiometry of a given shifted species. This can be accomplished by performing a standard double-label experiment, in which the DNA probe is labeled with ^{32}P, and the protein with ^{3}H, ^{14}C, ^{35}S, or ^{125}I. Spillover and counting efficiencies of the isotopes used should be taken into account, and the precise DNA concentration determined to calculate the specific radioactivity of the probe. It is also important to use enough of both the protein and DNA so that accurate quantitations of each shifted species are possible. Haran *et al.* (40) described a convenient procedure for determining the stoichiometry of the *E. coli trp* repressor to its operator using iodinated *trp* repressor. When a constant amount of ^{32}P-labeled probe (0.1 μM) was mixed with increasing amounts of ^{125}I-labeled *trp* repressor, this complex was found to contain 2495 counts per minute (cpm) for ^{32}P (background and quenching corrected), and 19,069 cpm for ^{125}I (background, spillover, and quenching corrected) at 0.1 μM repressor. Because the specific activity for ^{32}P-labeled probe DNA was 22,200 cpm/pmol and for ^{125}I-labeled *trp* repressor was 88,970 cpm/pmol, the molar ratio of repressor to operator DNA is 1.91, consistent with the notion that this protein occupies its operator binding site as a dimer.

Conclusion

Many, but not all, factors involved in transcriptional control are DNA-binding proteins. Over the past several years, powerful techniques have been developed to study DNA binding *in vitro,* ranging from the gel mobility shift assay to the more complicated DNase I footprinting and methylation protection/interference assays. The gel mobility shift assay is a technically simple assay that nevertheless allows the investigator to (1) identify key DNA sequences that are recognized specifically by a trans-acting factor, (2) assess whether one or more proteins are binding to a given sequence and, if they are, whether or not this association is cooperative, (3) determine the affinity and rate constants a protein has for a given sequence, and

(4) identify new proteins that recognize a specific target sequence. Some of these experiments do not require the acquisition of pure proteins, but rather will work quite well with a relatively crude nuclear extract. Thus, the combination of its ease of use and its ability to give both qualitative and quantitative results usually makes it the first assay of choice when studying protein–DNA interactions.

References

1. A. D. Frankel and P. S. Kim, *Cell* **65,** 717 (1991).
2. S. A. Amero, R. H. Kretsinger, N. D. Moncrief, K. R. Yamamoto, and W. R. Pearson, *Mol. Endocrinol.* **6,** 3 (1992).
3. V. Laudet, C. Hanni, J. Coll, F. Catzeflis, and D. Stehélin, *EMBO J.* **11,** 1003 (1992).
4. R. M. Evans, *Science* **240,** 889 (1988).
5. M. Beato, *Cell* **56,** 335 (1989).
6. W. Wahli and E. Martinez, *FASEB J.* **5,** 2243 (1991).
7. M. G. Parker, "Nuclear Hormone Receptors: Molecular Mechanisms, Cellular Functions, Clinical Abnormalities." Academic Press, London, 1991.
8. W. A. Seagraves, *Cell* **67,** 225 (1991).
9. D. Picard, S. J. Salser, and K. R. Yamamoto, *Cell* **54,** 1073 (1988).
10. P. J. Godowski, D. Picard, and K. R. Yamamoto, *Science* **241,** 812 (1988).
11. N. J. G. Webster, S. Green, J. R. Jin, and P. Chambon, *Cell* **54,** 199 (1988).
12. S. M. Hollenberg and R. M. Evans, *Cell* **55,** 899 (1988).
13. M. Eilers, D. Picard, K. R. Yamamoto, and J. M. Bishop, *Nature (London)* **340,** 66 (1989).
14. R. M. Umek, A. D. Friedman, and S. J. McKnight, *Science* **251,** 288 (1991).
15. T. Hard, E. Kellenbach, R. Boelens, B. A. Maler, K. Dahlman, L. P. Freedman, J. Carlstedt-Duke, K. R. Yamamoto, J.-A. Gustafsson, and R. Kaptein, *Science* **249,** 157 (1990).
16. J. W. R. Schwabe, D. Neuhaus, and D. Rhodes, *Nature (London)* **348,** 458 (1990).
17. B. F. Luisi, W. X. Xu, Z. Otwinowski, L. P. Freedman, K. R. Yamamoto, and P. B. Sigler, *Nature (London)* **352,** 497 (1991).
18. R. Miesfeld, S. Rusconi, P. J. Godowski, B. A. Maler, S. Okret, A. C. Wikstrom, J. A. Gustafsson, and K. R. Yamamoto, *Cell* **46,** 389 (1986).
19. L. P. Freedman and T. Towers, *Mol. Endocrinol.* **5,** 1815 (1991).
20. L. P. Freedman, B. F. Luisi, Z. R. Korszun, R. Basavappa, P. J. Sigler, and K. R. Yamamoto, *Nature (London)* **334,** 543 (1988).
21. F. W. Studier, A. H. Rosenberg, J. J. Dunn, and J. W. Dubendorff, *in* "Gene Expression Technology" (D. Goeddel, ed.), Methods in Enzymology, Vol. 185, p. 60. Academic Press, San Diego, 1990.
22. I. Alroy and L. P. Freedman, *Nucleic Acids Res.* **20,** 1045 (1992).
23. L. P. Freedman, *Endocrinol. Rev.* **13,** 129 (1992).
24. J. Miller, A. D. McClachlin, and A. Klug, *EMBO J.* **4,** 1609 (1985).
25. R. S. Brown, C. Sander, and P. Argos, *FEBS Lett.* **186,** 271 (1985).

26. S. Mader, V. Kumar, H. deVereneuil, and P. Chambon, *Nature* (*London*) **338,** 271 (1989).
27. M. Danielson, L. Hinck, and G. M. Ringold, *Cell* **57,** 1131 (1989).
28. K. Umesono and R. M. Evans, *Cell* **57,** 1139 (1989).
29. K. Dahlman-Wright, A. Wright, J.-A. Gustafsson, and J. Carstedt-Duke, *J. Biol. Chem.* **266,** 3107 (1991).
30. M. Freed and D. M. Crothers, *Nucleic Acids Res.* **9,** 6505 (1981).
31. M. M. Garner and A. Revzin, *Nucleic Acids Res.* **9,** 3047 (1981).
32. W. Hendrickson and R. F. Schleif, *J. Mol. Biol.* **174,** 611 (1984).
33. J. Carey, *in* "Protein–DNA Interactions" (Robert T. Sauer, ed.), Methods in Enzymology, Vol. 208, p. 103. Academic Press, San Diego, 1991.
34. L. P. Freedman, K. R. Yamamoto, B. F. Luisi, and P. J. Sigler, *Cell* **54,** 444 (1988).
35. A. M. Nardulli, D. Lew, L. Erijman, and D. J. Shapiro, *J. Biol. Chem.* **266,** 24070 (1991).
36. M. Brenowitz, D. F. Snear, M. A. Shea, and G. K. Ackers, *in* "Enzyme Structure," Part K (C. Hirs and S. Timasheff, eds.), Methods in Enzymology, Vol. 130, p. 132. Academic Press, San Diego, 1986.
37. W. J. Dixon, J. J. Hayes, J. R. Levin, M. F. Weidner, B. A. Dombroski, and T. D. Tullius, *in* "Protein–DNA Interactions" (Robert T. Sauer, ed.), Methods in Enzymology, Vol. 208, p. 380. Academic Press, San Diego, 1991.
38. D. S. Sigman, M. D. Kuwabara, C.-H. B. Chen, and T. W. Brucie, *in* "Protein–DNA Interactions" (Robert T. Sauer, ed.), Methods in Enzymology, Vol. 208, p. 415. Academic Press, San Diego, 1991.
39. M. Affolter, A. Percival-Smith, M. Muller, W. Leupin, and W. Gehring, *Proc. Natl. Acad. Sci. U.S.A.* **87,** 4093 (1990).
40. T. Haran, A. Joachimiak, and P. B. Sigler, *EMBO J.* **11,** 3021 (1992).

Section V

Gene Expression

[17] Tissue-Specific Gene Regulation: Emphasis on Heart and Skeletal Muscle

Howard Prentice, Vittorio Sartorelli, and Larry Kedes

Introduction: Muscle-Specific Gene Regulation

The purification and molecular cloning of tissue-specific and ubiquitously expressed transcription factors (1) have illuminated transcription mechanisms based on cis-acting DNA elements and proteins or protein complexes that bind to such elements. The mechanisms underlying such transcriptional regulation may be investigated (1) by measurement of mRNA synthesis, accumulation, or turnover, (2) by identification of DNA–protein interactions with binding or footprinting analysis either *in vitro* or *in vivo,* or (3) by examination of the effects on gene expression of mutating promoter regions or altering the structure of the DNA (e.g., by methylation). The role of transcriptional regulatory sequences may be examined by their ability to direct transcription *in vitro,* or after transfection into cells in culture, or following introduction of exogenous DNA *in vivo* to create transgenic animals. This article will focus on promoter analyses by *in vitro* transfection and direct DNA injection techniques, measurements of specific mRNA levels, and the investigation of DNA–protein interactions in myogenic cells and cardiomyocytes. The approaches are not organ specific and are generally applicable but we will focus our discussion on heart and skeletal muscle.

Reporter Genes for Analysis of Promoter Activity

The introduction into target cells or tissues of promoter sequences linked to reporter genes is the principal method for the assessment of promoter function. These chimeric constructs typically involve plasmid vectors that have cloning sites upstream and downstream of the reporter gene that allow the assessment of regulatory activity for a sequence irrespective of location or orientation.

The bacterial chloramphenicol acetyltransferase (CAT) gene is the most commonly used reporter gene. Its product catalyzes the transfer of acetyl groups from acetyl coenzyme A (acetyl-CoA) to chloramphenicol. Chloramphenicol acetyltransferase activity is measured as the percentage conversion of chloramphenicol to its acetylated forms. After incubation with cell extracts and acetyl-CoA, [^{14}C]chloramphenicol and its acetylated forms are separated by thin-layer chromatography (TLC). It is important to ensure that the acetylated forms of chloramphenicol constitute no more than 30% of the total or the measurements of percentage

CAT conversion will be out of the linear range (2). A less expensive extraction method using [^{3}H]acetyl-CoA involves extraction of acetylated and unacetylated chloramphenicol into the same benzene phase (3). Alternatively, butyryl-CoA and [^{3}H]chloramphenicol may be used with a differential extraction of butyrylchloramphenicol from chloramphenicol. This method can detect 5×10^7 molecules of CAT enzyme, comparable with the sensitivity of the standard acetyl-CoA technique (4, 5). A CAT–ELISA (enzyme-linked immunosorbent assay) technique (Boehringer Mannheim, Indianapolis, IN) can also measure activity from promoter–*CAT* constructs. CAT protein is quantitated with an anti-CAT antibody conjugated to digoxigenin and an anti-digoxigenin antibody conjugated to peroxidase. Finally, cleavage of a peroxidase substrate is quantitated spectrophotometrically. This method is as sensitive as methods that employ TLC but is faster (approximately 4 hr).

Firefly luciferase as reporter gene has the advantage over the *CAT* gene of up to 1000-fold greater sensitivity (3, 5). Luciferase catalyzes the reaction:

$$\text{Luciferin} + \text{ATP} + \text{O}_2 \xrightarrow[\text{luciferase}]{\text{Mg}^{2+}} \text{oxyluciferin} + \text{AMP} + \text{PP}_\text{i} + \text{CO}_2 + \text{light}$$

A luminometer measures activity by detection of light emission in the presence of an excess of ATP. Peak light emission and integrated light output values are proportional to the concentration of luciferase in the reaction.

Human growth hormone (hGH) operates as a reporter protein in most mammalian cell types because its endogenous expression is restricted to the anterior pituitary. Human growth hormone has an advantage over other reporter proteins because it is secreted from the cell and may be measured from the culture medium (6). This leaves the cells intact for further analysis. Human placental alkaline phosphatase is relatively stable and can also function as a reporter protein (7). This protein is relatively cell type specific and may be distinguished from endogenously expressed alkaline phosphatase after thermal denaturation and 1-homoarginine treatment. Not all nonplacental alkaline phosphatases can be inactivated in this way, and before using alkaline phosphatase as a reporter protein the endogenous expression of the enzyme should be measured (5, 8).

The frequent use of the bacterial β-galactosidase gene as a reporter has the advantage that its activity can be both localized histochemically and measured enzymatically. Quantitation of enzyme activity in unheated cell extracts measures hydrolysis of the substrate *O*-nitrophenyl-β-D-galactopyranoside (ONPG). For histochemical localization, the substrate X-Gal (5-bromo-4-chloro-3-indolyl-β-D-galactopyranoside) is incubated with cells or tissue with the catalyst ferricyanide/ferrocyanide (9).

β-Galactosidase can also be detected in living cells by its catalytic release of fluorescence from the fluorescein di-β-D-galactopyranoside (FDG) substrate (10).

This method can identify stably transfected cells that can then be selectively enriched by fluorescence-activated cell sorting (FACS) (11).

Stable and Transient Transfections in Models of Myogenic Differentiation

Methods of Transfection

Transfections into cultured cells often yield variability due to differences in cell number or DNA uptake between dishes. To circumvent this problem an internal control construct such as Rous sarcoma virus (RSV)-luciferase may be cotransfected with a promoter–*CAT* test construct. This enables a normalization of CAT values to luciferase values. The luciferase assay is very sensitive: only one-twentieth or less of RSV-luciferase DNA relative to *CAT* construct DNA need be used, which is sufficiently low to prevent the phenomenon of promoter competition (12). Growth hormone constructs are also often used as an internal control for other reporter genes such as those for CAT or luciferase.

Commonly used techniques for DNA transfection include those based on the addition of DEAE-dextran (13), or the formation of a calcium phosphate–DNA precipitate (14), or methods that employ a liposome carrier. The DEAE-dextran method for plasmid transfection involves addition of 1.5 ml of Dulbecco's modified Eagle's medium (DMEM) containing 8–16 μg/ml DNA plus 200 μg/ml DEAE-dextran to 60-mm culture dishes containing 6×10^5 cells/dish. The transfection goes on for 4–20 hr. If cells are also to be subjected to dimethyl sulfoxide (DMSO) or glycerol shock the transfection solution is removed and replaced with 1 ml of HBS [137 m*M* NaCl, 5 *M* KCl, 0.7 m*M* Na_2HPO_4, 6 m*M* dextrose, and 21 m*M* *N*-2-hydroxyethylpiperazine-*N'*-2-ethanesulfonic acid (HEPES) (pH 7.1)] containing 10% DMSO or 15% glycerol. The DMSO shock is for 2 min, or alternatively glycerol shock is for 90 sec. Cells are then washed with phosphate-buffered saline and incubated at 37°C for up to 72 hr before harvesting.

For optimal transfection by the calcium phosphate precipitation method, DNA is diluted in 0.5 ml of 2× HEPES-buffered saline (2× HeBS) (5 g/liter HEPES, 8 g/liter NaCl, 0.37 g/liter KCl, 0.125 g/liter $Na_2HPO_4 \cdot 2H_2O$, and 1.0 g/liter dextrose). One-half milliliter 250 m*M* $CaCl_2$ is added under a gentle stream of nitrogen or while vortexing to allow a fine precipitate to form. Precipitates are allowed to stand 30–60 min before being added to the culture dishes. The quality of the calcium phosphate–DNA precipitate is important for effective transfection and is dependent on the pH of the 2× HeBS, which should be 7.05 for maximal transfection. No transfection is observed above pH 7.4 or below pH 6.9.

Transfectability depends on DNA concentration. Addition of nonspecific carrier DNA (12.5 μg) and test DNA (5 μg) per 60-mm dish resulted in maximal transfectability with a decrease in transfectability at carrier concentrations greater than 12.5 μg/dish (14).

A glycerol shock step can increase transfection efficiency of the calcium phosphate precipitation method. With C2 and L8 myogenic cells we have found no increase in transfection efficiency with glycerol shock and so we do not include this part of the procedure. Because calcium is toxic to cells by varying degrees, we change the medium on C2 cells after 20 hr and on L8 cells after 8 hr of exposure to the precipitate.

DNA carried by liposomes can efficiently transfect C2 myoblasts. Because the reagent lipofectin (Bethesda Research Laboratories, Gaithersburg, MD)* is sensitive to serum, we reduce the serum levels in growth medium from 20 to 1% (v/v) fetal calf serum (FCS) during transfection. DNA (400 ng/μl in 12.5 μl H_2O), added to an equal volume of lipofectin, is mixed gently and allowed to stand for 15 min at room temperature. Cells are incubated for 16 hr at 37°C. For subconfluent cultures the medium is then changed to 20% (v/v) FCS. Alternatively, 2% (v/v) horse serum can be used to enable fusion of 100% confluent cultures. Transfection efficiencies with lipofectin are at least 5–10 times higher than with $CaPO_4$-mediated transfections. The cost of this reagent is relatively high and for this reason experiments are often carried out in six-well plates with 6.25 μl lipofectin per well as opposed to 100-mm dishes with 50 μl lipofectin reagent (Fig. 1).

A newer liposome reagent, DOTAP (Boehringer Mannheim), has also been successfully used in transfection of C2 myoblasts. This reagent is reported to transfect at efficiencies from 5 to 100 times higher than either calcium phosphate or DEAE-dextran. We routinely transfect 60-mm dishes of C2 cells at 60% confluency by using 30 μl DOTAP plus 70 μl HBS [20 m*M* HEPES, 150 m*M* NaCl (pH 7.4)] with 5 μg DNA in 100 μl HBS. The mixture is left to stand for 10 min and then added directly to the culture medium. Compared with other liposome-based transfection reagents, DOTAP has the advantage that it is effective in high-serum conditions such as those used in myoblast growth medium [20% (v/v) FCS]. Unlike DOTMA (a principal component of lipofectin), DOTAP also has the advantage that it can be decomposed by nonspecific esterases of the cells and so is relatively nontoxic.

The highly efficient DNA transfer technique of electroporation is effective for a range of cell types including mammalian cells, plant protoplasts, and bacteria (15–17). Promoter constructs have been transfected by electroporation into newborn and fetal cardiac myocytes (18, 19). The application of a high-voltage electric field temporarily damages cell membranes and forms pores that later close. Whereas DNA transfected by the calcium phosphate procedure often integrates as many as hundreds of copies of the transfected sequence, electroporation can be adjusted to

* When specific manufacturers or supplies are mentioned by name, we do not intend to imply that similar products from other manufacturers are not equally useful. We mention specific companies whose products have been used satisfactorily in our laboratory.

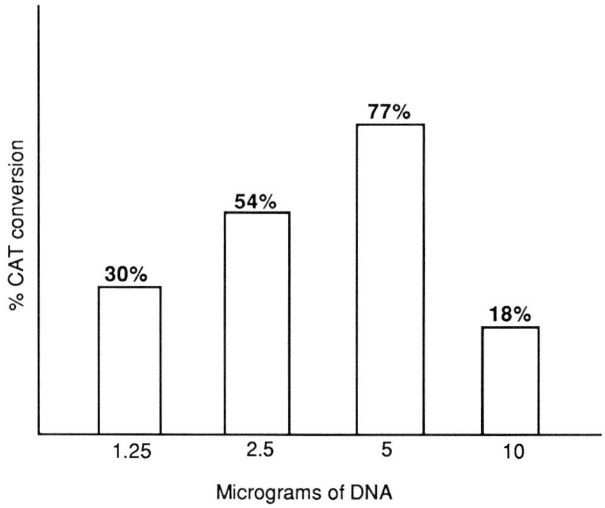

FIG. 1 β-Actin-CAT activities in C2 cells following lipofectin-mediated transfection. DNA (1.25, 2.5, 5, or 10 μg) was transfected into six-well plates with a DNA:lipofectin ratio of 1 μg:2.5 μg. Optimal transfection efficiencies were obtained with 5 μg of DNA.

yield 1–20 copies of an inserted gene (15). Electroporation therefore is preferable for gene expression studies that require low copy numbers of transferred genes.

Electroporation of Plasmid DNA into Newborn Rat Cardiac Myocytes

1. Resuspend 7×10^6 cardiomyocytes in 700 μl of 140 m*M* NaCl, 15 m*M* HEPES (pH 7.2), with 20 μg of plasmid DNA in a 0.4-cm cuvette.
2. Subject cells to a 240-V, 960-μF pulse in a Bio-Rad (Richmond, CA) electroporation unit. Although the optimal electric pulse delivered differs depending on the apparatus, the conditions described here have been used successfully (18).
3. Surviving cells are plated at 10^5 cells/cm^2 on tissue culture dishes in modified Eagle's medium (MEM; GIBCO, Grand Island, NY) supplemented with 5% (v/v) fetal calf serum, 2 m*M* glutamine, 100 U/ml streptomycin and penicillin. This procedure usually leads to an approximately 50% cell recovery.

Transient Transfections into Primary Skeletal Myoblasts and Myogenic Cell Lines

Cell culture models of myogenesis allow the comparison of predifferentiation myoblasts with postdifferentiation multinucleated myotubes. Chick, rodent, and human primary skeletal muscle cells, isolated as satellite cells, can grow in culture

(20–22). These cells undergo the transition from myoblasts to myotubes on withdrawal of mitogens. Immortalized muscle cell lines display different capacities for differentiation. Table I describes a number of widely used myogenic cell lines. Mouse C2 myoblasts (23) differentiate more rapidly than rat L8 cells (24) and in this respect they resemble primary cultures of rat muscle. Other similarities between the C2 cell line and primary cultures are the spontaneous contraction of its myotubes and the synthesis of acetylcholinesterase and acetylcholine receptor clusters.

TABLE I Properties and Derivations of Myogenic Cell Lines

Cell line	Property	Source	Refs.
L6	Fuse to form multinucleated muscle fibers. Ultimately cross-striations and contractility are observed	Derived from rat skeletal muscle cell cultures prepared from newborn noninbred Wistar rats	(20)
BC3H1[a]	Nonfusing; differentiated quiescent nonfused myoblasts can reenter the cell cycle in response to mitogens	May be of smooth muscle origin; line was derived from a chemically induced mouse brain tumor	
H9[b]	Produce action potentials in response to electrical stimulation or iontophoretic application of acetylcholine. Myotubes contain a branched tubular system and bundles of thick filaments with distinct M bands.	Cell line was derived by selective serial passaging of embryonic BDIX rat heart cells	
C2	Fuse to form multinucleated myofibrils in 24–36 hr with high endogenous cardiac α-actin mRNA (equivalent to that of adult mouse heart)	Derived from thigh muscle of 2-month-old normal mouse by selective serial passage of myoblasts 72 hr after crush injury	(24)
L8	Display low levels of transfected cardiac α-actin gene expression in myoblasts with greatly increased levels after differentiation	Cell line was derived by serial passage of myoblasts from the skeletal muscle of newborn noninbred Wistar rats	(24)
L6E9[c]	Displays induction of endogenous skeletal but not cardiac α-actin mRNA on differentiation, implying a defect in its myogenic program	Subclone of L6	
MM14[d]	Cells proliferate in the presence of serum and FGF. Terminal differentiation is repressed by FGF	This permanent skeletal muscle line was derived from the leg muscle of a 60-day-old BALB/c mouse	
Sol 8[e]	Expresses β-MHC; induction of α-MHC promoter–CAT expression in the presence of rat TRα1 receptors and thyroid hormone	This cell line was cloned by limiting dilution from primary cultures of slow twitch soleus muscle prepared from 4-week-old C3H mice	

[a]From D. A. Schubert, A. J. Harris, C. E. Devine, and S. Heinemann, *J. Cell. Biol.* **61,** 389 (1974).
[b]From B. W. Kimes and B. L. Brandt, *Exp. Cell Res.* **98,** 367 (1976).
[c]From B. Nadal-Ginard, *Cell* **15,** 855 (1978).
[d]From T. A. Linkhart, C. H. Clegg, and S. D. Hauschka, *Dev. Biol.* **86,** 19 (1981).
[e]From C. Mulle, P. Benoit, C. Pinset, M. Roa, and J.-P. Changeux, *Proc. Natl. Acad. Sci. U.S.A.* **85,** 5728 (1988).

Whereas the transfected cardiac α-actin gene is highly expressed in both C2 myoblasts and myotubes, it is induced and only expressed after L8 cell differentiation. The discrepancy in cardiac α-actin gene regulation between these two cell lines might be accounted for by an increase in transcriptional modulating factors during the transition from L8 myoblasts to myotubes (25).

The rat L6E9 cell up-regulates expression of skeletal α-actin during differentiation, but does not express cardiac α-actin mRNA in myoblasts or myotubes (26). There is no gross abnormality in the endogenous cardiac actin gene and these cells are capable of up-regulating a stably transfected human cardiac α-actin. The abnor-abnormal regulation of the endogenous cardiac α-actin gene in L6E9 cells could result either from a mutation in the regulatory region or from the failure of the cell to make a necessary transcription factor.

For analysis of promoter constructs, transfections are typically performed at the myoblast stage while cells are rapidly dividing. The cells can be maintained in growth medium or transferred to low-serum fusion medium after 24 hr. Cells are harvested 24–48 hr later either as myoblasts or as myotubes. We have also successfully transfected C2 cells as 1-day-old myotubes that are harvested 48 hr later.

Isolation of Primary Neonatal Cardiomyocytes

The availability of cell lines derived from different tissues and organs, including skeletal muscle, has provided a valuable tool for investigating the transcriptional regulation of tissue-restricted genes. The analysis of cardiac regulated genes has suffered in this respect because of the lack of a reliable and well-characterized cardiac cell line in which to study the expression of transfected genes. In addition, unlike primary myoblasts isolated from skeletal muscle, primary cardiomyocytes cannot be expanded in culture. The primary rat cardiocyte cell culture model developed by Simpson and Savion (27) offers a unique opportunity for transient transfection of heart cells and is useful in conjunction with the application of short-term stimuli such as α-adrenergic stimulation resulting in hypertrophy (28). The following protocol enables the isolation of fetal rat cardiomyocytes suited for gene transfer and nuclear extract preparation.

Materials

Sprague-Dawley neonatal rats (0–4 days postpartum)
Dry ice and ethanol
Phosphate-buffered saline (PBS): 137 m*M* NaCl, 2.7 m*M* KCl, 4.3 m*M* $Na_2HPO_4 \cdot 7H_2O$, 1.4 m*M* KH_2PO_4
Calcium- and bicarbonate-free Hanks' with HEPES (CBFHH) buffer: 137 m*M*

NaCl, 5.4 m*M* KCl, 0.81 m*M* $MgSO_4 \cdot 7H_2O$, 5.55 m*M* dextrose, 0.44 m*M* KH_2PO_4, 0.34 m*M* $Na_2HPO_4 \cdot 7H_2O$, 20 m*M* HEPES (pH 7.4). Filter sterilize and adjust pH to 7.5 with NaOH. The buffer is supplemented with streptomycin and penicillin (100 U/ml of each)

Trypsin (1:250, w/v) (Difco, Detroit, MI) (50×): Trypsin is dissolved at 5 mg/50 ml in CBFHH buffer and stored overnight at 4° C. The solution is filter sterilized and frozen

BrdU (Cat. No. B-5002; Sigma, St. Louis, MO) (100×): 30.74 mg in 10 ml DMEM. Filter sterilize and store frozen

DMEM (GIBCO): Supplemented with 5% (v/v) fetal calf serum, 2 m*M* glutamine, 100 U/ml streptomycin and penicillin

DNase II type V, bovine spleen (Cat. No. D8764; Sigma, St. Louis, MO): 20 mg/10 ml sterile solution in 0.15 *M* NaCl. Filter sterilize and store frozen

Sterile operating scissors (Cat. No. 13-804-33; Fisher, Pittsburgh, PA)

Sterile Braun microforceps (Cat. No. 08-953F; Fisher)

Sieve (85 ml), pan, and key (Cat. No. 1985-18500; Bellco, Vineland, NJ)

Mesh 80 (Cat. No. 1985-00080; Bellco)

Methods

Thaw DNase, BrdU, and trypsin solutions at room temperature. Prepare the following solutions:

Trypsin solution (TS): 1 ml stock trypsin solution (50×), 0.7 ml stock DNase solution, 0.5 ml penicillin/streptomycin (P/S) solution (100×) (10,000 U of each antibiotic); adjust to a final volume of 50 ml with CBFHH buffer in a 50-ml Falcon (Becton Dickinson, Oxnard, CA) tube

DNase solution (DS): 1 ml fetal calf serum, 0.7 ml stock DNase solution, 0.5 ml penicillin/streptomycin (P/S) solution (100×); adjust to a final volume of 50 ml with CBFHH buffer in a 50-ml Falcon tube

Collecting solution (CS): 2 ml of fetal calf serum in a 50-ml Falcon tube

Prepare an adequate number of tubes of TS, DS, and CS. For a typical experiment involving 50–60 neonatal rats, 5 TS tubes, 4 DS tubes, and 6 CS tubes will be sufficient.

After humane killing of 20 neonatal animals, expose and isolate the hearts using sterile surgical tools (curved miniforceps and scissors).

1. Place about 20 hearts in each 100-mm tissue culture dish containing 20 ml of cold PBS.

2. At this stage, carry out all work with cardiac tissue under a laminar flow

hood under sterile conditions. Carefully remove noncardiac tissues from the hearts (mainly pericardium and vessels), using sterile surgical tools.

3. Transfer the hearts to a second 100-mm tissue-culture dish containing 10 ml of cold CBFHH and cut the hearts in half. Remove the CBFHH buffer.

4. Wash the hearts twice with 10 ml of cold CBFHH.

5. Add 5 ml of CBFHH and carefully mince the hearts into pieces less than 2 mm in diameter.

6. Transfer the tissues with a wide-tip 10-ml pipette to a 50-ml conical polystyrene tube. Rinse the original dish twice with 5 ml CBFHH and pool the solutions in the same tube. Be careful to use a wide-tip pipette to preserve the integrity of the tissue.

7. Let the tissues settle at the bottom of the tube and then gently remove the CBFHH.

8. Add 10 ml TS to the tube containing the cells and gently shake on a rocker table for 20 min at ~50 rpm at room temperature.

9. Remove the TS using a wide-tip Falcon pipette and transfer it to a 50-ml collecting tube (containing CS) without disturbing the cells. Return the collecting tube to 4° C.

10. Add 10 ml of TS to the cells and rock for 10 min. Remove the TS and transfer it to a collecting tube.

11. Add 9 ml of DS to the cells. Pipette up and down 10 times with a wide-tip 10-ml Falcon pipette. Transfer the DS to the collecting tube.

12. Add 10 ml of TS to the cells and rock for 5 min; transfer the TS to the collecting tube. Add 9 ml of DS, pipette up and down 10 times, and transfer the DS to the collecting tube.

13. Repeat step 12 until the cell pellet becomes white; Generally cell isolation will take 3–5 hr. The remaining material is connective tissue. During the process slowly reduce the volume of the TS added to the cells by 1-ml increments from 10 to 6 ml. The volume of DS used should be 1 ml less than that of TS.

14. When the collecting tubes are filled, spin them in a refrigerated clinical centrifuge for 5 min at 1000 rpm. The cell pellet will be loose. Aspirate the supernatant with a pipette, leaving 10–15 ml, and continue to aspirate the remainder with a 10-ml pipette. Discard the supernatant.

15. Add 2 ml of supplemented DMEM to each collecting tube. Gently resuspend the cell pellets, using a wide-tip 10-ml pipette.

16. Rinse the original tubes with 2 ml supplemented DMEM. The final volume in which the cells are resuspended should be now approximately 20 ml.

17. Add 200 μl of stock DNase solution to the 20 ml DMEM containing the cells (if the volume is larger, add more DNase solution).

18. Triturate the cells by pipetting up and down with a fine-tip 10-ml Falcon pipette until all clumps are broken up.

19. Spin the cells 5 min at 1000 rpm in a clinical centrifuge at 4° C and carefully discard the supernatant.

20. Prewet a sterile Cellector (80 mesh) with DMEM and prepare four 100-mm tissue culture dishes.

21. Resuspend the cells with 32 ml of supplemented DMEM.

22. Using a 10-ml Falcon pipette, pour 8 ml of the cell suspension through the sieve into a tissue culture dish. Repeat this procedure for the other three tissue culture dishes. Rinse the sieve immediately with alcohol to avoid blocking the mesh with tissue fragments.

23. Incubate these for 30 min at 37° C, 5% CO_2.

24. Remove the dishes from the incubator, gently swirl them, and transfer the medium to a fresh 50-ml tube. Add 2 ml of DMEM to each dish, tap them two or three times, and transfer the medium to the 50-ml tube.

25. Wash the dishes with a further 5 ml of DMEM, and transfer this to the Falcon tube. The adherent cells (mainly fibroblasts and endothelial cells) may be cultured with 10 ml of supplemented DMEM.

26. The volume in which the cells are now resuspended should be 45–50 ml. After resuspending, count the cells with a hemacytometer.

27. Plate the cardiomyocytes, at a concentration of $3-4 \times 10^6$/60-mm tissue culture dish, in 5 ml of supplemented DMEM plus 1% (v/v) BrdU. The BrdU prevents the division of nonmyocardial cells that may contaminate the preparation.

28. Culture the cardiomyocytes at 37° C, 5% CO_2.

29. After 24 hr the cardiomyocytes should start beating synchronously.

If the calcium phosphate precipitation method is used, transfect cells before they start to beat (at 20–22 hr after isolation).

Optimally, the isolation procedures described here can provide 4×10^6 cells per neonatal heart with less than 10% cardiac fibroblast or smooth muscle cell contamination.

Trans-Activation

Characterized eukaryotic trans-acting factors fall into a variety of categories based on their DNA-binding or activation domains. Zinc-dependent DNA-binding domains known as zinc fingers are found in transcription factor Sp1 (29). A second type of DNA-binding domain, the homeodomain (HD) (30), is similar to the helix–turn–helix binding structure of prokaryotic repressors and is found in the octamer binding factors (31, 32) and the pituitary-specific factor Pit 1 (33). A third type of DNA-binding domain contains a stretch of 30 amino acids with a net basic charge followed by a region of four leucine residues located at intervals of seven amino acids. The leucine-rich segment, termed the leucine zipper, is required for dimerization and DNA binding and is found in transcription factors C/EBP, Jun, Fos, and CREB (34).

A single transcription factor such as yeast GAL4 can activate transcription because of its DNA-binding site and an acidic activating region that interacts with a target DNA-binding protein causing the intervening DNA to loop out (35, 36). The herpesvirus protein VP16, however, contains an acidic activating region but no binding domain, but it can activate transcription when attached to an Oct-1 DNA-binding region. Activation again involves looping of the DNA (35, 36).

The MyoD family of muscle-specific transcriptional regulators contains a motif that can form two amphipathic α helices separated by a nonhelical loop (the HLH motif) and a basic region amino terminal to this (37). Mutational analysis indicates that the HLH domain mediates protein–protein interactions, whereas the basic region interacts with the DNA. Heterodimerization of MyoD with an E2A protein [a ubiquitous basic HLH (bHLH) protein] appears to be crucial for functional activity.

The transcriptional activation of the human cardiac α-actin gene (*HCA*) is dependent on both ubiquitously expressed transcription factors (SRF and Sp1) and the myogenic determination factor MyoD. Using the nonmuscle cell lines CV1 and HeLa, it has been possible to activate the exogenous cardiac α-actin gene by cotransfection with the long terminal repeat (LTR)-*MyoD* expression vector. In CV-1 cells, a mutation in -177 cardiac α-actin-*CAT* at the CArG box, the MyoD site, or the Sp1 site abolished trans-activation (38).

The *Drosophila* Schneider cell line is devoid of many transcriptional activator homologs, including Sp1. The *HCA* promoter was assayed for transcriptional activity when cotransfected with either or both the LTR-*MyoD* expression vector and the plasmid pPAcSp1 (39), which carries the *Drosophila* nonmuscle actin gene promoter driving human Sp1. Neither the MyoD vector nor the Sp1 plasmid alone affected *HCA* promoter activity, but both plasmids in combination induced the *HCA* promoter activity synergistically (24-fold). When constructs containing mutations in either the CArG box, the Sp1 site, or the MyoD site were cotransfected with the MyoD and Sp1 expression vectors, no trans-activation occurred. These results indicate that the CArG-binding factor, Sp1, and MyoD are all simultaneously required for expression of the *HCA* gene.

Retroviral-Mediated DNA Transfer into Primary Cardiac Fibroblasts: Conversion of Cellular Phenotype by MyoD

Prepare rat neonatal primary cardiac fibroblasts as described above.

1. Primary cardiac fibroblasts are grown in three 60-mm dishes at 5×10^5 cells per dish with DMEM containing 20% (v/v) fetal calf serum and 2 m*M* glutamine. Add Polybrene (8 μg/ml) to each dish. Polybrene supplies a negatively charged anion that facilitates the attachment of the retroviral particles to the cell membrane. Culture cells overnight in 5% CO_2 at 37° C.

2. Infect the cells with 100 μl of a supernatant containing the retroviral vector bearing the gene of interest. As a control, infect one plate with a supernatant con-

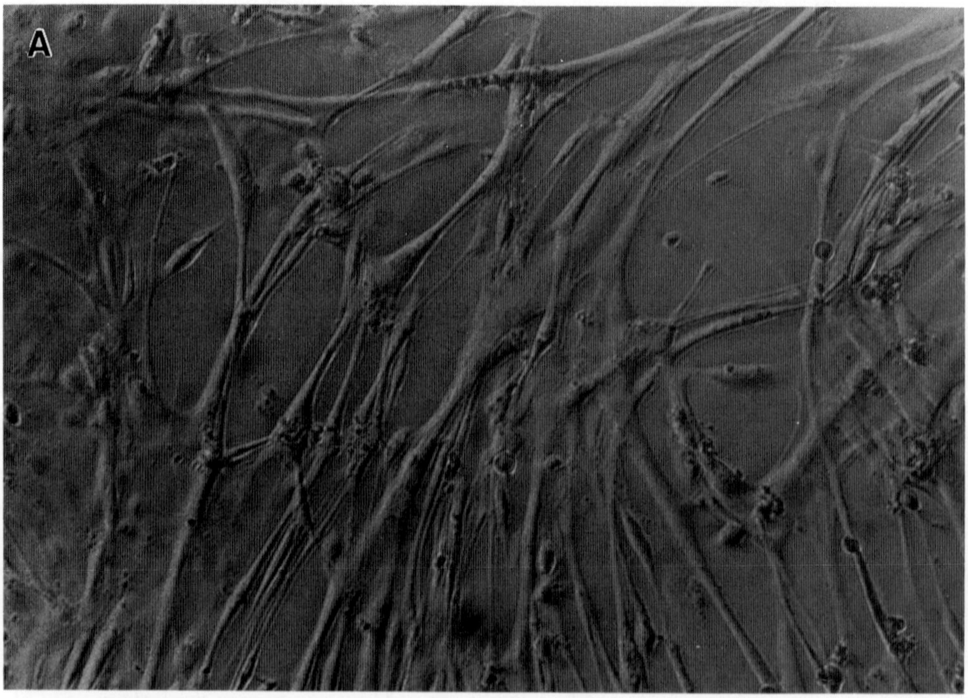

FIG. 2 Cardiac fibroblasts transduced with LMDSN (A) or LXSN (B).

taining the retroviral vector with no insert and another plate with DMEM medium alone (mock-infected cells).

3. After 2 days split the cells to give densities of 300, 600, 3000, 6000, 30,000, and 60,000 cells per dish. Culture fibroblasts in selection medium (DMEM and G418 for neomycin selection).

4. Change the medium every 4 days and after 12 days reduce the fetal calf serum concentration from 20 to 2%.

Experiments were conducted using primary cardiac fibroblasts infected with a retroviral vector bearing the MyoD cDNA (LMDSN) kindly provided by A. D. Miller of the Fred Hutchinson Cancer Research Center. Primary cardiac fibroblasts isolated from neonatal rats were infected with the retroviral construct LMDSN, LXSN, or mock-infected with DMEM. No G418-resistant colonies were formed by the mock-infected cells, whereas numerous colonies resulted with the retrovirus infected cells. As shown in Fig. 2, the morphology of the LMDSN-infected fibroblasts strongly differed from those transduced with LXSN. Whereas the latter cells conserved the typical rounded fibroblast morphology and were separate cells, the LMDSN-infected fibroblasts appeared elongated and frequently contained several

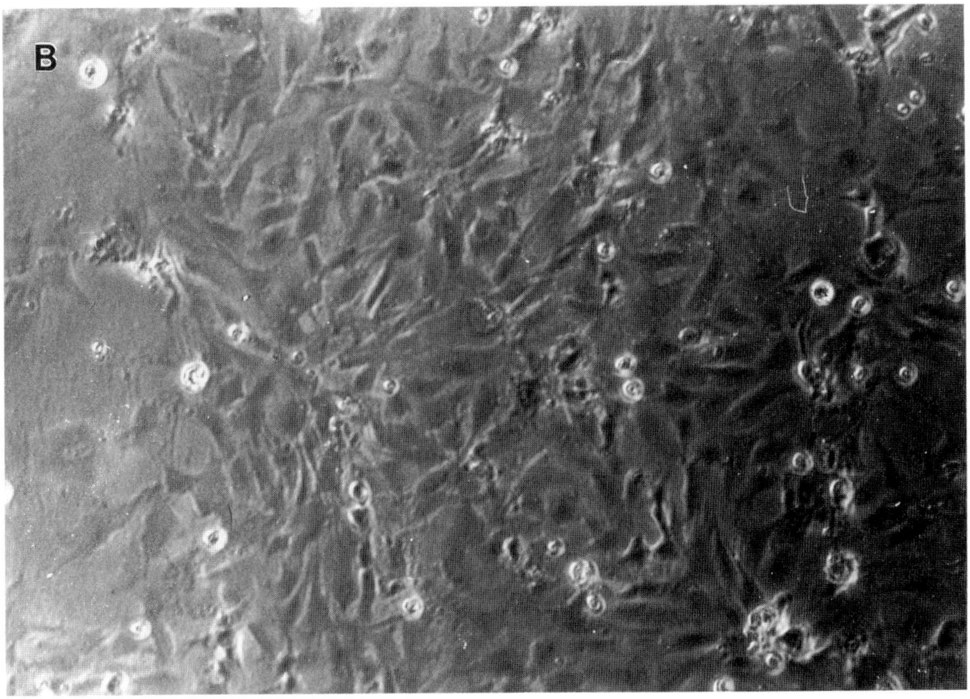

FIG. 2 (*continued*)

nuclei. These morphological changes are consistent with a conversion to the myogenic phenotype. To confirm the myogenic conversion resulting from overexpression of MyoD, the LMDSN- and LXSN-infected cells were immunostained with a monoclonal antibody recognizing the skeletal isoform of MHC. Only cardiac fibroblasts that had been transduced with the retrovirus containing the MyoD coding sequence were reactive with the MHC antibody.

These results demonstrate that cardiac fibroblasts, like many other nonmyogenic cell types, can be converted to myogenic cells by the overexpression of MyoD protein (V. Sartorelli and L. Kedes, unpublished observations, 1992).

Strategies for Deletion and Mutation Analyses

*Bal*31 deletions, site-directed mutations, and linker scanner mutations have helped pinpoint key sites of action for transcription factors in the promoter regions of a number of muscle-specific genes (40–42).

For unidirectional deletion analysis, digestion of promoter regions with *Bal*31 yields regulatory regions that are successively truncated from either the 5′ end or

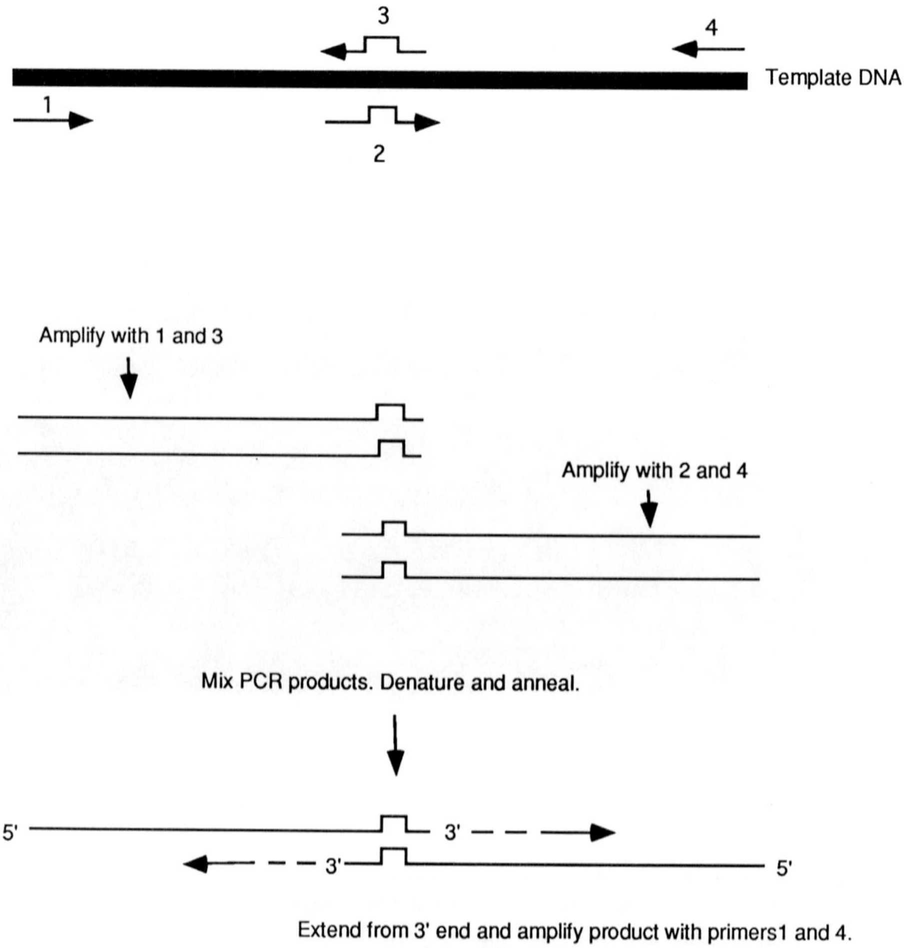

FIG. 3 Strategy for introducing mutations in the absence of a convenient unique restriction site for cloning: left and right PCR products are amplified using primers 1 and 3, and 2 and 4, respectively. PCR products are mixed, denatured, and annealed, after which 3′ ends are extended with *Taq* polymerase. The final mutated product is amplified with primers 1 and 4.

the 3′ end. To analyze the role of regulatory factors in an enhancer domain of the troponin C fast gene (*TnCf*), linker scanner mutations were created in the wild-type −1625 *TnCf* promoter construct. Initially a series of constructs was created with 3′ deletions from −1500 into an enhancer element known to be located between −1625 and −1500. *Bgl*II linkers were added to the 3′ end of these constructs that were then recircularized. DNA primers were synthesized that contained the wild-type *TnCf* sequences located immediately downstream of the 3′ end of each deletion construct (primer 1). These primers were used to amplify the remainder of the promoter from the wild-type template with the use of a second, downstream primer (primer 2). Besides *TnCf* sequences, primer 1 contained a *Bgl*II site corresponding to the same location as the *Bgl*II site in the 3′ deletion construct. Each polymerase chain reaction (PCR) product was digested with *Bgl*II and ligated to the corresponding *Bgl*II-digested 3′ deletion construct. The promoter region containing a *Bgl*II site at a particular location was then reamplified by PCR with an upstream primer and a downstream primer. The product was digested with *Xba*I and *Sac*I and cloned into a −1625 *TnCf* promoter-*CAT* construct also digested with *Xba*I and *Sac*I. The resulting constructs contained promoter sequences from −1625 with *Bgl*II site substitutions at various locations.

The most convenient approach to site-directed mutational analysis involves a single-step PCR with the first primer both spanning a unique restriction site in the DNA and containing the desired mutation. Typically this primer may be 25 to 50 bp long. A PCR product is generated with this primer and a second primer corresponding to a sequence that is near or as much as 1 or 2 kb away. Finally the PCR product is cleaved at two restrictions sites that are unique to the original DNA and the mutated sequence is cloned in place of the wild-type sequence. A second, more complex method for introducing mutations is necessary when the desired mutation site is too far from a unique restriction site for a single-step PCR strategy (Fig. 3). This technique requires two sets of primers. The first set involves an upstream primer (primer 1) that forms a PCR product in conjunction with another primer (primer 3) that contains the desired mutation. A second PCR product is made with a complementary mutated primer (primer 2) and a downstream primer (primer 4). Finally, a third PCR product is amplified from a mixture of the first two products using primers 1 and 4. As the first two PCR products are complementary only at the region containing the mutation, the final PCR product will span the entire sequence from primer 1 to primer 4 and will contain the desired mutation.

Studies of DNA–Protein Interactions

Gel Electrophoresis Mobility Shift Assay and Antibody-Mediated Complex Perturbation

The goal of the gel electrophoretic mobility shift assay (GEMSA) is to define DNA regulatory regions essential for gene expression and to clarify the nature of regulatory proteins that interact with these regions. DNA that forms stable complexes with sequence-specific binding proteins displays altered mobility when subjected to gel electrophoresis (43). DNA fragments or synthetic oligonucleotides that interact with known transcription factors can also be used as competitors to help define the nature of the DNA–protein interaction. Furthermore, to reduce non-sequence-specific

DNA-binding activity, different DNA competitors are included in the reaction. Genomic mouse or salmon sperm DNA, plasmid DNA, and polynucleotides are the most frequently used. Genomic DNA is most effective but can mask labile interactions of some sequence-specific binding proteins. Polynucleotides are more gentle competitors. For each experimental condition the type and quantity of DNA competitor to be used should be assessed in pilot experiments. The identity of interacting proteins may be further investigated with a battery of mono- or polyspecific antibodies. The addition of an antibody that is reactive with a peptide from the protein–DNA complex will either block the interaction or perturb the electrophoretic migration of the complex. In this section we will describe the GEMSA and antibody-mediated complex perturbation as applied to transcription factors that regulate human cardiac α-actin gene expression in myogenic skeletal and cardiac cells.

DNA Dephosphorylation and Labeling

Reagents

DNA oligomers/fragments
Alkaline phosphatase, calf intestinal
T4 polynucleotide kinase
T4 polynucleotide kinase buffer (10×): 400 m*M* Tris (pH 7.5), 100 m*M* $MgCl_2$, 50 m*M* dithiothreitol (DTT)
Alkaline phosphatase buffer (10×): 10 m*M* $ZnCl_2$, 10 m*M* $MgCl_2$, 100 m*M* Tris (pH 8.3)
Low-salt buffer: 0.2 *M* NaCl, 20 m*M* Tris (pH 7.4), 1 m*M* ethylenediamine tetraacetic acid (EDTA)
High-salt buffer: 1 *M* NaCl, 20 m*M* Tris (pH 7.4), 1 m*M* EDTA
TEN buffer: 150 m*M* NaCl, 10 m*M* Tris (pH 7.4), 1 m*M* EDTA
[γ-^{32}P]ATP: >5000 Ci/mmol
Elutip-D columns (Scheicher & Schuell, Keene, NH)

Methods

1. Combine the following as the Labeling reaction mixture:

DNA oligomer	100ng
T4 polynucleotide kinase	5–10U
[γ-^{32}P]ATP	100μCi
T4 polynucleotide kinase buffer (10×)	2μl
doubly distilled H_2O	20μl (final volume)

2. Incubate at 37°C for 90 min. Heat inactivate (90°C for 5 min).

3. Add 1 ml low-salt buffer to the labeling reaction tube.

4. Load the reaction on a preequilibrated (low-salt buffer) Elutip-D column. Wash the column once with low-salt buffer (2–3 ml) and elute the labeled DNA with 500 μl of high-salt buffer into a 1.5-ml microcentrifuge tube. Add 1 μl of glycogen (Boehringer Mannheim), 1 ml cold 100% ethanol and precipitate. Wash with 70% ethanol and dry the probe.

If double-stranded DNA fragments are used, the 5′-phosphate must be removed before labeling:

DNA fragment	1μg
Alkaline phosphatase, calf intestinal	1U
Alkaline phosphatase buffer (10×)	5μl
H_2O	50μl (final volume)

Incubate at 37°C for 30 min; add 1 U of alkaline phosphatase and continue the incubation for an additional 60 min. If the DNA fragment has 3′ overhanging ends, the first incubation should be performed at 37°C for 15 min and the second at 55°C for 60 min.

Heat inactivate the enzyme (90°C for 5 min); extract with phenol–chloroform twice and ethanol precipitate the labeled DNA fragment.

Nuclear Extract Preparation

Nuclear extract from isolated cardiomyocytes and nonmuscle cardiac cells were prepared as reported (44) with minor modifications. Other cell types such as CV-1, NIH 3T3, and HeLa can also be processed with this method to obtain nuclear extracts.

Reagents

Primary rat cardiomyocytes and nonmuscle cardiac cells

PBS

Buffer A: 10 m*M* HEPES (pH 7.9), 10 m*M* KCl, 0.1 m*M* EDTA, 0.1 m*M* ethylene glycol-bis(β-aminoethyl ether)-*N,N,N′,N′*-tetraacetic acid (EGTA), 1 m*M* DTT, 0.5 m*M* phenylmethylsulfonyl fluoride (PMSF)

Buffer C: 20 m*M* HEPES (pH 7.9), 0.4 *M* NaCl, 1 m*M* EDTA, 1 m*M* EGTA, 1 m*M* DTT, 1 m*M* PMSF, 25% (v/v) glycerol

Nonidet NP-40 (NP-40): 10% (v/v) solution

Methods

Wash 4 × 10^6 adherent cells twice, briefly, with cold PBS. Transfer to a 1.5-ml microcentrifuge tube and spin at 2000 rpm for 1 min at room temperature. Remove the PBS and resuspend the cells in 400 μl of buffer A by repipetting three or four times. Leave the cells on ice for 15 min, then add 25 μl of 10% NP-40 solution. Vortex the tubes briefly, then centrifuge for 1 min at 12,000 rpm. Remove the supernatant and resuspend the nuclear pellet in 50 μl of buffer C and shake for 15 min at 4°C. Recover the supernatant after a 15-min centrifugation at 12,000 rpm at 4°C and determine the protein concentration. Store aliquots of the nuclear extract at −80°C.

Binding Reaction

Reagents

- Labeled specific DNA
- Unlabeled specific DNA
- Poly(dI–dC) · poly(dI–dC), genomic mouse DNA, salmon sperm DNA, or plasmid DNA
- Nuclear extract or purified recombinant proteins
- Binding buffer: 20 m*M* HEPES (pH 7.6), 5% (v/v) glycerol, 50 m*M* NaCl, 1.5 m*M* $MgCl_2$
- Mono- or polyclonal antibodies

Methods

If DNA oligomers are used, anneal them to the complementary DNA sequences. Combine:

Labeled DNA	100 ng
Unlabeled complementary DNA	300–500 ng
TEN buffer	to 40 μl

Incubate the reaction at 90°C for 5 min and gradually cool until the temperature reaches 37°C. Combine:

Labeled DNA	0.1–0.5 ng
Unlabeled specific DNA (competitor)	10–50 ng
Nuclear extract	5–10 μg (purified or recombinant protein 100 ng)
Poly(dI–dC) · poly(dI–dC)	0.5 μg (100 ng with purified recombinant proteins)
Binding buffer	20–30 μl (final volume)

1. Incubate the reaction mixture at room temperature for 15 min. Mono/poly-specific antibodies may be added to the reaction either before or after addition of the

labeled DNA. Addition of antiserum before the DNA may result in complex disruption, whereas superretardation may occur when antiserum is added after the DNA.

2. Load the reaction mixture onto a 5% (w/v) polyacrylamide gel containing 50 m*M* Tris, 50 m*M* boric acid, and 1 m*M* EDTA. Electrophorese for 5 hr (200 V, 4°C). Include an additional lane with binding buffer plus bromphenol blue/xylene cyanol as a marker.

3. After electrophoresis, soak the gel in 10% acetic acid, dry, and expose for autoradiography. Reaction conditions must be optimized for a particular protein in terms of salt concentration (especially $MgCl_2$) and concentrations of nonspecific competitor and nuclear extract. Mouse genomic DNA can be used on the order of nanogram to microgram quantities. Plasmid DNA is usually employed on the order of micrograms. The quantity of specific DNA necessary for competition is dependent on the affinity of the protein for the sequence.

Results

Activation of the human cardiac α-actin (*HCA*) promoter in primary cardiomyocytes depends on an interaction with three regulatory proteins (45). We performed GEMSA and antibody-mediated complex perturbation assays on *HCA* promoter sequences with nuclear extracts derived from primary rat cardiomyocytes and from nonmuscle cardiac cells (mainly cardiac fibroblasts and endothelial cells). Figure 4 shows the results of a typical experiment. In this study we determined that a factor that is antigenically related to the *E2A* gene products and present in cardiomyocytes but not nonmuscle cardiac cells interacts with the *HCA* proximal E-box.

Methylation Interference Assay

The methylation interference assay is a powerful technique that enables one to analyze the guanine residues of the DNA contacted by a protein. DNA–protein interaction is prevented or weakened if guanine residues in the DNA target site are methylated. Extensive DNA methylation may result in completely abolishing the accessibility of the protein. It is therefore critical to aim for partial methylation of the DNA, ideally one guanine residue per DNA molecule. This can be achieved by varying the temperature and time of the reaction or the concentration of the methylating agent. As a result, a GEMSA performed with a methylated DNA probe yields shifted complexes depleted of DNA molecules whose protein-binding sites contain methylated guanines. Treatment of shifted and unbound DNA with piperidine results in cleavage at methylated guanine residues (and to a lesser extent adenine residues). A comparison of bound and free DNA enables precise identification of guanines contacted by the protein. Protein contacts with adenine residues may be investigated using DNA that is modified with diethyl pyrocarbonate (DEPC).

The methylation interference assay offers a number of advantages over other tech-

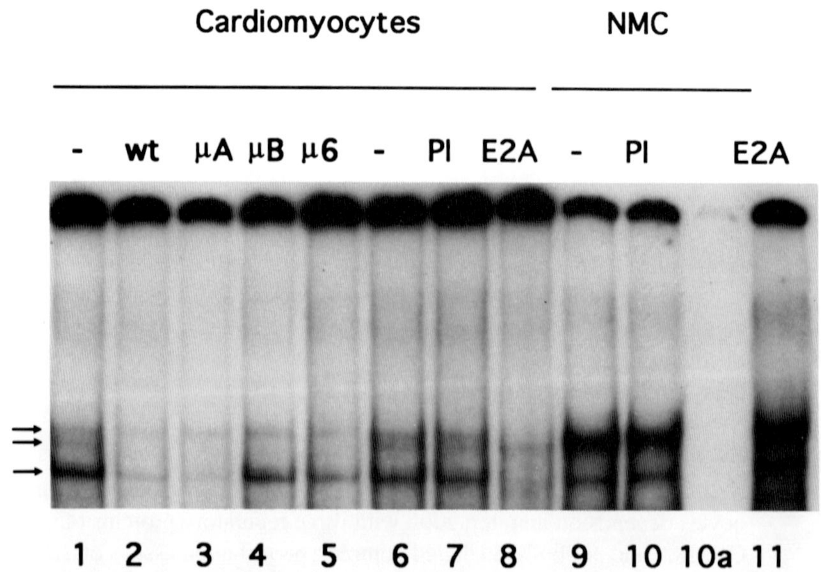

FIG. 4 Gel mobility shift assay indicating that a factor that is antigenically related to the *E2A* gene products and present in cardiomyocytes but not nonmuscle cardiac cells interacts with the *HCA* proximal E-box.

niques employed to study DNA–protein interaction. These include (1) the localization of precise nucleotide contacts made by the protein compared to extended regions of protection in DNase footprinting and phenanthroline-copper ion methods, (2) the isolation of a pure population of nonmethylated DNA molecules in the shifted complex compared to a mixed population of digested and undigested DNA molecules as obtained by DNase footprinting, and (3) the absence of sequence-dependent preferential digestion within the specific site resulting from the dynamic and transient protein interactions associated with DNase footprinting. Methylation interference, DNase footprinting, and phenanthroline-copper ion methods are complementary techniques that may be applied in combination to obtain detailed information on a specific protein–DNA interaction.

Reagents

Labeled DNA fragment or oligomer
Dimethyl sulfate (DMS) (Aldrich Chemical Company, Madison, WI)
Piperidine (Sigma)

Nuclear extract
DMS modification buffer: 50 mM sodium cacodylate (pH 8.0), 1 mM EDTA
DMS stop buffer: 1 M 2-mercaptoethanol, 1.5 M sodium acetate
Binding buffer
Glycogen (Boehringer Mannheim)
Dialysis tubing
TBE (5X): Trisbase 54 g/l, boric acid 27.5 g/l, 0.01 M EDTA
Elutip-D columns (Schleicher & Schuell)

Methods

DNA Labeling

1. Radiolabel DNA probes at one end on either the top or bottom strand.
2. Radiolabel DNA oligomers or fragments as described for the GEMSA.

Alternatively, use Klenow enzyme and [α-^{32}P]dNTP.

DNA Methylation

1. Combine and incubate in a 1.5-ml microcentrifuge tube for 5 min at room temperature:

Labeled DNA	100 ng (oligomer)–1000 ng (DNA fragment)
DMS modification buffer 100–1000 ng (fragment)	200 μl (final volume)
	1 μl

(Remember that DMS is extremely toxic. All manipulations should be done with double gloves in a fume hood. Liquid waste must be stored separately.) The incubation time should be optimized for each DNA-binding protein under investigation.

2. Add 50 μl of DMS stop buffer and 1 μl of glycogen. Precipitate with 600 μl 100% ethanol.
3. Centrifuge (12,000 rpm, 15 min, 4°C).
4. Precipitate with 250 μl of 0.3 M sodium acetate and 750 μl of 100% ethanol.
5. Centrifuge (12,000 rpm, 15 min, 4°C).
6. Wash with 70% ethanol and dry the methylated probe.

Gel Electrophoresis Mobility Shift Assay and Isolation of Free and Bound DNA Complexes

1. Perform a GEMSA as described above. The quantity of protein and DNA should be increased 10-fold.

2. After electrophoresis, remove one glass plate and cover the wet gel with plastic wrap. Expose for autoradiography at 4°C.

3. Cut out the bands corresponding to bound and free probes from the X-ray film with a razor blade and use this film as a template for excising the appropriate fragments from the gel.

4. Monitor the polyacrylamide strips containing free and bound DNA with a hand-held Geiger counter.

Electroelution of Free and Bound DNA

Free and bound DNA are eluted separately according to the same procedure.

1. Place the acrylamide strip containing the DNA probe in a dialysis bag. Close one end of the bag with a clip and fill with sufficient 0.5× TBE to cover the strip. Close the other end of the bag.

2. The dialysis bag is placed in a horizontal gel apparatus and subjected to 150 V for 90 min in 0.5× TBE.

3. Remove the TBE from the dialysis bag and transfer the solution to a plastic tube. Carefully rinse the remaining DNA from the walls of the dialysis bag with 1 ml of low-salt buffer (see DNA Dephosphorylation and Labeling, above) and add this solution to the same tube. The polyacrylamide strips may be monitored with a Geiger counter to check for residual activity. If the radiolabeled probe is not yet eluted then repeat steps 1–3.

4. Pass the DNA solution through a plastic syringe with a preequilibrated Elutip-D column. Wash the column with an additional 2–3 ml of low-salt buffer and elute the sample with 500 μl of high-salt buffer into a 1.5-ml siliconized microcentrifuge tube. Add 1 μl of glycogen and precipitate with 1 ml of 100% ethanol. Recover the pellet by centrifugation (15 min, 4°C, 12,000 rpm), wash with 500 μl of 70% ethanol, and dry.

Piperidine Treatment

1. Resuspend unbound and bound probes in 100 μl of 10% piperidine and incubate for 30 min in a 90°C water bath. A weight should be placed on the tubes to prevent them from opening.

2. Allow reactions to cool and centrifuge (12,000 rpm, 30 sec) and lyophilize the samples. Add 100 μl of distilled water to the dried pellet and lyophilize the samples. Repeat the washing procedure one additional time.

Analysis of DNA Fragments

1. Resuspend the dried probes in 3–4 μl of formamide sequencing buffer.

2. Measure radioactivity from 1 μl of unbound and bound probes in a scintilla-

tion counter. To use the same number of counts per minute (cpm) for each probe in the methylation interference footprinting, dilute the samples proportionally with formamide sequencing buffer.

3. Heat denature the probes in a boiling water bath for 3 min and load 2 to 3-μl samples (ideally 500–1000 cpm) on a prerun (6–15%, w/v) polyacrylamide, 8 *M* urea sequencing gel in the following order: unbound, bound, unbound.

4. After electrophoresis, soak the gel in 10% acetic acid, 10% methanol, dry, and expose for autoradiography with intensifying screens at $-80°$C.

Interpretation of Methylation Interference Assay

The order of sample loading proposed in step 3 allows one to easily compare the intensity of individual bands obtained after autoradiography. High-affinity binding of a protein to DNA is indicated by complete interference. In this instance virtually every DNA molecule present in the reaction is occupied by the protein, preventing the guanine residues contacted from chemical cleavage by piperidine. As a result the bands corresponding to the contacted guanines are absent in the "bound probe" lane and present in the "unbound probe" lane.

Weak protein–DNA interactions are often observed where those DNA molecules not contacted by the protein create residual piperidine cleavage at guanine residues. There is a reduction in the intensity of the bands corresponding to interfered guanines from the bound probe lane relative to the unbound probe lane. This phenomenon is referred to as partial interference. The extent of partial interference can be quantitated by subjecting the autoradiogram to densitometric analysis.

To interpret the methylation interference assay correctly, it is necessary to establish the boundaries of the DNA–protein contacts. This can be achieved using a DNA probe whose guanine residues are not contacted by protein at either its 5′ or 3′ end. These guanine residues will be cleaved during piperidine treatment and will give strong bands on both bound probe and free probe lanes.

RNA Analysis

For investigation of regulated gene expression, measurement of specific mRNAs is an informative starting point for documenting developmental changes or tissue specificity. Analysis by Northern blot, primer extension, or S1 nuclease assay allows quantitative measurements. Nuclear runoffs with intact nuclei can assess mRNA synthesis rates (46). Because RNA is readily digested by RNase activity during preparation, it is necessary to carry out extraction procedures, such as treatment with guanidium isothiocyanate, that denature cellular nucleases and prevent degradation.

Isolation of Total RNA from Muscle Cells and Fetal Muscle Tissue

Total cellular RNA has been isolated from human skeletal myoblasts and myotubes and human fetal muscle tissue by a modification of a procedure described by Strohman *et al.* (47).

1. Collect approximately 3×10^6 cells from one 100-mm dish and lyse in 1 ml of ice-cold, 5.7 *M* guanidine hydrochloride, 0.1 *M* potassium acetate (pH 5.0).
2. Pass the disrupted cells through a 19-gauge syringe needle to shear chromosomal DNA, and precipitate overnight with 0.5 vol of 100% ethanol at $-20°$C.
3. Collect the crude RNA pellet by centrifugation at 10,000 rpm for 10 min and resuspend in 1 ml 5.7 *M* guanidine hydrochloride, 20 m*M* EDTA (pH 8). After vortexing, add potassium acetate (pH 5.0), to 0.1 *M* and precipitate at $-20°$C for 1–12 hr with 0.5 vol of ethanol.
4. Collect the precipitate by centrifugation, dissolve in 200 μl of 20 m*M* EDTA (pH 8.0), and extract with 3 vol of chloroform–butanol (4:1). Reextract with chloroform–butanol, pool the aqueous phases, and reextract. Precipitate the RNA solutions by the addition of 2 vol of 4.5 *M* sodium acetate at $-20°$C overnight.
5. Collect the RNA pellet by centrifugation (10,000 rpm, 10 min), resuspend in 400 μl H_2O, and then precipitate by addition of 2.5 vol of 100% ethanol and 0.1 vol of 3 *M* sodium acetate.

RNA Isolation from Adult Muscle Tissue

1. Pulverized tissue samples on dry ice, resuspend in 3–4 vol of lysis buffer [8 *M* guanidium hydrochloride, 10 m*M* sodium acetate (pH 5.0), 1 m*M* DTT] and homogenize immediately with a Brinkmann Polytron (Westbury, NY) for 30 sec.
2. Sediment debris and draw the clear supernatant through a 21-gauge needle to shear the high molecular weight DNA.
3. Selectively precipitate RNA with 0.5 vol of ethanol at $-20°$C overnight and collect the pellet by centrifugation (10,000 rpm, 10 min at 4°C).
4. Resuspended the RNA pellet in 2–4 vol of lysis buffer and purify by phenol–chloroform extraction followed by ethanol precipitation. (It is important to use molecular biology-grade phenol, such as Boehringer Mannheim Cat. No. 100300, to avoid degrading the RNA.)

Quantitation of Muscle Gene Transcripts by Northern Analysis

To examine the trends in sarcomeric gene transcript accumulation during myogenic differentiation, we have employed Northern analysis of RNA samples from human skeletal muscle cells. For measuring levels of highly abundant transcripts (such as sarcomeric gene transcripts) Northern blot analysis is sufficiently sensitive to give a readily detectable signal. For sarcomeric actin genes we employ human 3′ UTR probes that are isotype specific but conserved across species. Cardiac α-actin was rapidly induced from low confluency primary myoblasts to 1-day-old myotubes and remained relatively constant through a 15-day culture period (48). Skeletal α-actin mRNA levels reached a maximum at day 1 and then decreased to day 5, after which they increased slightly to day 15. On days 0 and 1 sarcomeric *MHC* transcripts increased from 7 to 42% of maximum, after which the transcript levels increased only slightly until day 15.

S1 Nuclease Protection Analyses

S1 nuclease protection analysis is more sensitive than Northern analysis for measurement of levels of gene expression (12). In this assay specific mRNA transcripts are hybridized quantitatively with an excess of radiolabeled DNA probe. The hybridized double-stranded molecules are not susceptible to endonucleolytic cleavage by S1 nuclease, which cleaves single-stranded DNA. The amount of protected labeled DNA is a direct measurement of the concentration of the specific mRNA and is assayed by acrylamide gel electrophoresis and radionuclide detection.

If the radiolabeled DNA fragment is longer than the mRNA that protects it, then the size of the protected fragment can also be used to determine the end points of the DNA.

Materials

S1 nuclease
T4 DNA polymerase
Hybridization buffer: 80% (v/v) formamide, 10 m*M* piperazine, *N,N′*-bis(2-ethanesulfonic acid) (PIPES) (pH 6.4), 1 m*M* EDTA
Incubation buffer: 260 m*M* NaCl, 40 m*M* sodium acetate (pH 4.6), 1 m*M* $ZnSO_4$, 100 μg sonicated calf thymus DNA/ml

Methods

1. Label the linearized DNA with T4 DNA polymerase in a two-step replacement reaction with phage T4 DNA polymerase (49). Digest 1.5-μg samples of DNA with T4 DNA polymerase (2 U) for 5, 15, or 30 min at 37°C. Carry out replacement

synthesis by incubation at 37°C for 90 min with 20 μCi of [α-^{32}P]dCTP (>400 Ci/mmol) and 100 μM concentrations each of dATP, dGTP, and dTTP followed by a 20-min cold chase with 100 μM dCTP to ensure maximal incorporation of the radiolabeled dCTP.

2. Determine the maximally labeled region of probe by restriction digestion of aliquots of these samples and separation by agarose gel electrophoresis. Select the linearized probe that is maximally labeled in the desired region and remove unincorporated nucleotides on a Sephadex G-50 spun column.

3. For hybridization, combine 100 μg of RNA with 100 ng of linearized DNA (containing 1.5 ng of cDNA fragment) in 30 μl of hybridization buffer.

4. Denature at 80°C for 5 min followed by hybridization at 50°C for 18 hr.

5. Add 200 μl of incubation buffer with 50 U of S1 nuclease (Sigma) and digest for 30 min at 37°C.

6. Precipitate samples with 2-propanol in the presence of carrier tRNA and then analyze on an 8% (w/v) acrylamide urea sequencing gel.

7. After exposure of the dried gel at −80°C with intensifying screens, quantitate bands on the autoradiogram by densitometry.

Primer Extension Analysis

For measurements of transcript levels as well as locating the transcriptional start site of human muscle genes, we have used primer extension analysis as an alternative technique to S1 nuclease analysis (50). Primer extension makes use of a high specific activity 5′-labeled probe and can detect mRNAs with less background than by S1 nuclease analysis and with greater sensitivity than by Northern analysis. Primer extension analysis using total RNA employed the method of Bodner and Karin (51). A radioactively labeled primer whose sequence was taken from the appropriate cDNA is first annealed to RNA and then extended using avian myeloblastosis virus (AMV) reverse transcriptase and dNTPs in the presence of actinomycin D to prevent hairpin loop formation. This procedure creates extension products from 100 to 400 bp. For size standards, a sequencing ladder or kinase-labeled *Hpa*II-digested DNA is used.

Materials

RNasin (Promega, Madison, WI)

Primer extension buffer: 9.5 mM Tris-HCl (pH 8.7), 9.5 mM $MgCl_2$, 3 mM dithiothreitol (DTT), 340 ng/ml actinomycin D, 0.25 mM concentrations of each dNTP

Stop solution: 25 mM EDTA, 250 mM NaCl

AMV reverse transcriptase (Seikagaku America, Inc., St. Petersburg, FL)

RNase A
Salmon sperm DNA

Methods

1. Suspend 25 μl of total RNA in a solution containing 5 ng of 5′ ^{32}P-labeled oligonucleotide primer, 10 m*M* Tris-HCl (pH 7.9), 1 m*M* EDTA, 0.25 *M* KCl, and 0.2 μl of RNasin. Allow annealing to continue for 1 hr at 48°C.
2. Add 24 μl of primer extension buffer with 0.02 μg/μl AMV reverse transcriptase and 0.015 μg/μl RNasin and incubate the extension reaction at 42°C for 1 hr.
3. Terminate the reaction and remove RNA by the addition of 80 μl of stop solution with 5 μl RNase A (10 mg/ml) and 1 μl (2 μg) of sonicated salmon sperm DNA; incubate for 45 min at 37°C.
4. Extract samples with phenol–chloroform, precipitate, wash with 70% ethanol, and dry. Resuspend samples in sequencing loading buffer and run extension products on an 8% (w/v) polyacrylamide-urea sequencing gel.

Direct DNA Injection *in Vivo*

It was originally reported by Wolff *et al.* (52) that mouse skeletal muscle can take up pure DNA *in vivo* following direct injection of DNA solutions into the muscle. This observation was extended to rat heart, in which it was observed through β-galactosidase staining that the cardiomyocyte was the cell type that takes up the nucleic acids (53, 54). The hormone responsiveness of the rat α-cardiac myosin heavy chain gene (α-*MHC*) has been demonstrated by coinjection of promoter–luciferase constructs with an RSV–*CAT* plasmid as an internal control (55). Activity of an α-*MHC*–luciferase construct was detectable in heart but not in skeletal muscle and was significantly increased in response to thyroid hormone treatment. The mechanism of DNA uptake through direct injection is not clear but it is known that the phenomenon is unique to muscle. It has been postulated that DNA uptake may be mediated by endocytosis through the T tubular system (56). Denervation of skeletal muscle, which increases the number and complexity of the T tubules, may be responsible for a corresponding increase in pRSV-luciferase expression (56).

The activity of promoter–reporter gene constructs in rat cardiac tissue *in vivo* has been measured by a number of groups. Typically 100 μg or less of plasmid DNA in 50–100 μl of PBS with 5% (w/v) sucrose is injected into the apical portion of the beating left ventricle using a narrow (27- to 30-gauge) needle. Expression of transfected genes has been shown to be a relatively low-frequency event (<1% of myocytes in the area of the injection).

To assess the activity of promoter-*CAT* constructs in rat heart *in vivo,* we coinject 50 μg of test promoter–*CAT* with 50 μg of RSV-luciferase and measure CAT and luciferase activities 7 days later. By taking the ratio of CAT activity to luciferase activity we can normalize for quantity of DNA taken up.

Direct injections of human troponin C slow promoter constructs indicate that heart-specific transcription is regulated by a 225-bp region upstream of the transcriptional start site. This observation parallels reported findings for the mouse *TnC* slow gene (57) and implies that *in vivo* introduction of promoter constructs into cardiocytes results in tissue-specific gene regulation, as it does *in vitro*. Consistent with this proposal we have also found that *TnC* fast enhancer–promoter constructs are not expressed *in vivo,* whereas *TnC* fast enhancer–SV40 promoter constructs are active; a pattern we had already observed *in vitro*.

One drawback of the direct injection technique is interanimal variability in CAT : luciferase ratios of up to two-fold. For this reason the approach is best suited to the testing of a few key promoter constructs rather than a fine deletion series. An advantage of the direct injection technique is that it enables the evaluation of physiological parameters *in vivo* and their influence on tissue-specific gene regulation.

We have also used the direct injection technique to introduce DNA into tongues of adult mice. The muscle-specific *TnC* fast promoter was found to drive high-level expression of *CAT* and *lacZ* reporter genes at either 4 or 7 days after injection. Injection of a *lacZ* reporter gene construct resulted in β-galactosidase staining of muscle fibers radiating a considerable distance from the region of the needle track.

References

1. P. J. Mitchell and R. Tjian, *Science* **245,** 371 (1989).
2. F. M. Ausubel, R. Brent, R. E. Kingston, D. D. Moore, J. G. Seidman, J. A. Smith, and K. Struhl, *in* "Current Protocols in Molecular Biology." Wiley (Interscience), New York, 1990.
3. S. K. Nordeen, P. P. Green, and D. M. Fowlkes, *DNA* **6,** 173 (1987).
4. B. Seed and J. Sheen, *Gene* **67,** 271 (1988).
5. J. Alam and J. L. Cook, *Anal. Biochem.* **188,** 245 (1990).
6. R. F. Selden, K. Howie Burke, M. E. Rowe, H. M. Goodman, and D. D. Moore, *Mol. Cell. Biol.* **6,** 3173 (1986).
7. J. Berger, J. Hauber, R. Hauber, R. Geiger, and B. R. Cullen, *Gene* **66,** 1 (1988).
8. D. W. Moss, *Clin. Biochem.* **20,** 225 (1987).
9. J. Sanes, J. Rubenstein, and J. Nicolas, *EMBO J.* **5,** 3133 (1986).
10. G. P. Nolan, S. Fiering, J.-F. Nicolas, and G. L. A. Herzenberg, *Proc. Natl. Acad. Sci. U.S.A.* **85,** 2603 (1988).
11. G. H. Smith, D. Callahan, J. A. Zwiebel, S. M. Freeman, R. H. Bassin, and R. Callahan, *J. Virol.* **65,** 6365 (1991).
12. R. Gahlmann and L. Kedes, *J. Biol. Chem.* **265,** 12520 (1990).
13. J. H. McCutchan and J. S. Pagano, *J. Natl. Cancer. Inst.* **41,** 351 (1968).
14. F. Graham and A. van der Eb, *Virology* **52,** 456 (1973).

15. H. Potter, L. Weir, and P. Leder, *Proc. Natl. Acad. Sci. U.S.A.* **81,** 7161 (1984).
16. T. M. Ou-Lee, R. Turgeon, and R. Wu, *Proc. Natl. Acad. Sci. U.S.A.* **83,** 6815 (1986).
17. J. Miller, W. J. Dower, and L. S. Tompkins, *Proc. Natl. Acad. Sci. U.S.A.* **85,** 856 (1988).
18. T. Parker, K.-L. Chow, R. Schwarz, and M. Schneider, *Proc. Natl. Acad. Sci. U.S.A.* **87,** 7066 (1990).
19. R. Tsika, J. Bahl, L. Leinwand, and E. Morkin, *Proc. Natl. Acad. Sci. U.S.A.* **87,** 379 (1990).
20. D. Yaffe, *Proc. Natl. Acad. Sci. U.S.A.* **61,** 477 (1968).
21. J. Grichnik, D. Bergsma, and R. Schwartz, *Nucleic Acids Res.* **14,** 1683 (1986).
22. H. M. Blau, *Cell* **53,** 673 (1988).
23. D. Yaffe and O. Saxel, *Nature (London)* **270,** 725 (1977).
24. D. Yaffe and O. Saxel, *Differentiation* **7,** 159 (1977).
25. A. Minty, H. Blau, and L. Kedes, *Mol. Cell. Biol.* **6,** 2137 (1986).
26. R. Hickey, A. Skoultchi, P. Gunning, and L. Kedes, *Mol. Cell. Biol.* **6,** 3287 (1986).
27. P. Simpson and S. Savion, *Circ. Res.* **50,** 101 (1982).
28. N. H. Bishopric, P. C. Simpson, and C. P. Ordahl, *J. Clin. Invest.* **80,** 1194 (1987).
29. J. T. Kadonaga, K. A. Jones, and R. Tjian, *Trends Biochem. Sci.* **11,** 20 (1986).
30. M. P. Scott and A. J. Weiner, *Proc. Natl. Acad. Sci. U.S.A.* **81,** 4115 (1984).
31. R. T. Sturm, T. Baumruker, B. R. Franza, and W. Herr, *Genes Dev.* **1,** 1147 (1987).
32. R. Sturm, G. Das, and W. Herr, *Genes Dev.* **2,** 1582 (1988).
33. H. A. Ingraham, R. Chen, and H. J. Managalam, *Cell* **55,** 519 (1988).
34. K. Struhl, *Trends Biochem. Sci.* **10,** 461 (1989).
35. M. Ptashne, *Nature (London)* **135,** 683 (1988).
36. M. Ptashne and A. A. F. Gann, *Nature (London)* **346,** 329 (1990).
37. S. J. Tapscott and H. Weintraub, *J. Clin. Invest.* **87,** 1133 (1991).
38. V. Sartorelli, K. A. Webster, and L. Kedes, *Genes Dev.* **4,** 1811 (1990).
39. A. J. Courey and R. Tjian, *Cell* **55,** 887 (1988).
40. T. Miwa and L. Kedes, *Mol. Cell. Biol.* **7,** 2803 (1987).
41. T. A. Gustafson, B. E. Markham, J. J. Bahl, and E. Morkin, *Proc. Natl. Acad. Sci. U.S.A.* **84,** 3122 (1987).
42. K.-L. Chow and R. J. Schwartz, *Mol. Cell. Biol.* **10,** 528 (1990).
43. M. Fried and D. M. Crothers, *Nucleic Acids Res.* **9,** 6505 (1981).
44. E. Schreiber, P. Matthias, M. M. Muller, and W. Schaffner, *Nucleic Acids Res.* **15,** 6419 (1989).
45. V. Sartorelli, N. A. Hong, N. A. Bishopric, and L. Kedes, *Proc. Natl. Acad. Sci. U.S.A.* **89,** 4047 (1992).
46. M. E. Greenberg, L. A. Greene, and E. B. Ziff, *J. Biol. Chem.* **260,** 14101 (1985).
47. R. C. Strohman, P. S. Moss, J. Micon-Eastwood, D. Specter, A. Przybyla, and B. Patterson, *Cell* **10,** 265 (1977).
48. P. Gunning, E. Hardeman, R. Wade, P. Ponte, W. Bains, H. Blau, and L. Kedes, *Mol. Cell. Biol.* **7,** 4100 (1987).
49. J. Sambrook, E. F. Fritsch, and T. Maniatis, "Molecular Cloning: A Laboratory Manual." Cold Spring Harbor Lab. Press, Cold Spring Harbor, New York, 1989.
50. T. Schreier, L. Kedes, and R. Gahlmann, *J. Biol. Chem.* **265,** 12520 (1990).
51. M. Bodner and M. Karin, *Cell* **50,** 267 (1987).

52. J. A. Wolff, R. W. Malone, P. Williams, W. Chong, G. Acsadi, A. Jani, and P. L. Felgner, *Science* **247,** 1465 (1990).
53. H. Lin, M. S. Parmacek, G. Morle, S. Bolling, and J. Leiden, *Circulation* **82,** 2217 (1990).
54. G. Acsadi, S. Jiao, A. Jani, D. Duke, P. Williams, W. Chong, and J. A. Wolff, *New Biol.* **3,** 71 (1991).
55. R. N. Kitsis, P. M. Buttrick, E. M. McNally, M. L. Kaplan, and L. A. Leinwand, *Proc. Natl. Acad. Sci. U.S.A.* **88,** 4138 (1991).
56. J. A. Wolff, P. Williams, G. Acsadi, S. Jiao, A. Jani, and W. Chong, *BioTechniques* **11,** 474 (1991).
57. M. S. Parmacek, A. J. Vora, T. Shen, E. Barr, F. Jung, and J. M. Leiden, *Mol. Cell. Biol.* **12,** 1967 (1992).

[18] Using *Drosophila* P-Element-Mediated Germ Line Transformation to Examine Chromatin Structure and Expression of *in Vitro*-Modified Genes

Qin Lu, Lori L. Wallrath, and Sarah C. R. Elgin

Introduction

Gene activation *in vivo* requires interactions between template, specific activation factors, and the general transcription apparatus. To initiate transcription, RNA polymerase and transcription factors must gain access to their binding sites on the DNA of the promoter region of a gene. The accessibility of these binding sites is mediated by the local chromatin structure, that is, the local pattern of protein–DNA interactions. When the DNA of a promoter/enhancer region is packaged into a nucleosome array (as is the bulk of eukaryotic DNA), critical binding sites can be blocked, resulting in gene repression. This implies that *in vivo* the binding sites for regulatory factors of active genes must be in an accessible, nucleosome-free configuration to allow gene expression (reviewed in Ref. 1). These nucleosome-free regions can be detected as hypersensitive sites on digestion of intact nuclei with DNA cleavage reagents such as DNase I, methidiumpropyl-EDTA·Fe(II), and micrococcal nuclease. Hypersensitive sites are generally located at the 5′ or 3′ boundaries of genes, or at other locations where function requires binding of nonhistone chromosomal proteins (reviewed in Refs. 2 and 3).

Although experiments utilizing such techniques as the gel shift assay or DNase I footprinting (4) have been successfully applied to identify the sequence elements that are bound by specific regulatory proteins *in vitro,* functional testing is required to establish the significance of these protein–DNA interactions *in vivo*. Functional analyses include examining the chromatin structure and gene expression of a particular gene (1) with mutations within the putative regulatory element (see, e.g., Ref. 5) and (2) in a host in which the gene encoding the proposed trans-acting regulatory protein is altered or missing (see, e.g., Refs. 6–8). The extensive genetic information available (9) and the ability to return an altered gene to the genome (10) make *Drosophila melanogaster* an excellent system for these types of functional studies.

In this article we describe the use of *Drosophila* to examine the chromatin structure and gene expression of transgenes containing mutations created *in vitro* within the promoter region of a test gene. We use the *Drosophila melanogaster hsp26* gene to demonstrate that mutations that result in alterations in chromatin structure can result in altered transcriptional activity. Particular constructs containing altered

Methods in Molecular Genetics, Volume 1

hsp26 sequences, fused in frame to the *Escherichia coli lacZ* gene as a reporter, have been introduced into the *D. melanogaster* genome by P-element-mediated germ line transformation (10). We describe protocols to isolate nuclei from embryos, larvae, and adults of transformed *Drosophila* lines. Protocols for analyzing the chromatin structure of a gene in isolated nuclei using DNase I, methidium-propyl-EDTA·Fe(II), micrococcal nuclease, and restriction enzymes are also described. Restriction enzyme analysis is particularly useful in providing a quantitative assessment of the accessibility of a DNA sequence in isolated nuclei.

Methods

Generation of Transformants Carrying Mutant Genes

Specific mutations within a cloned test gene can be made *in vitro* by using standard protocols (11). An *in vitro*-altered gene is then inserted into an appropriate site of a P-element transformation vector. These vectors have a readily detectable reporter gene and a cloning site, together flanked by the P-element repeats that permit integration into the host genome in the presence of transposase (12, 13). Two commonly used reporter genes are the *rosy* gene (encoding xanthine dehydrogenase) and the *white* gene (10, 12, 14), both of which are required for wild-type eye pigmentation in the adult fly. The P-element constructs are integrated into the germ line of *Drosophila* by coinjection with a helper plasmid (pπ25.7 "wings clipped") of appropriately staged embryos that are deficient for the reporter gene activity. The helper plasmid encodes the transposase required for P-element integration, but cannot itself integrate into the genome (12, 15). Transformants, identified by wild-type eye pigmentation, are recovered from the progeny of the injected embryos. Alternatively, the P-element construct can be injected into a host strain that contains a stably integrated transposase gene, designated P[ry$^+$ Δ2-3](99B) (16); the transposase gene can subsequently be removed from the genome by appropriate genetic crosses to obtain stable transformants. The presence of a single copy of the transgene with the original sequence organization can be verified by Southern blot analysis of genomic DNA from transformed lines. Transformants carrying single insertions of the P-element construct are maintained as homozygous stocks or, if necessary, as a heterozygous stock carrying the insertion over a balancer chromosome (12). The level of gene expression, and in some cases its tissue specificity, can be affected by the DNA flanking the site of insertion. To overcome this "euchromatic position effect" (17), it is important to establish a number of independent lines (from 5 to 10 different insertion sites) for each construct. Alternatively, P[ry$^+$ Δ2-3](99B) can be used to mobilize a P-element insert to other integration sites (16). The heat shock genes (such as *hsp26,* used here) appear to be relatively insensitive to euchromatic position effect. Of 5–10 independent transformants, levels of expression rarely differ by more than ±15%. At least two lines homozygous for a particular transgene that show average transgene expression are used for chromatin structure analyses.

Collection of Larvae or Adults for Preparation of Nuclei

Transformed fly lines are maintained at 23°C in bottles (6-oz disPo bottles; Baxter Healthcare, McGaw Park, Ill) with standard cornmeal medium supplemented with active yeast (18). The life cycle of flies maintained at 23°C is approximately 11–12 days. Embryos hatch to become first instar larvae after 24 hr. After two more molts, the third instar larvae crawl up the walls of the bottle and pupate on the sixth day; adult flies will emerge on the eleventh day.

Because *Drosophila* embryos have relatively low levels of degradative enzymes, they are an ideal source for preparation of nuclei for chromatin structure analysis. Embryos can be collected in gram quantities from a population cage (containing approximately 50,000 flies) as described (18). However, when experiments require the analysis of a number of P-element-transformed lines, maintaining population cages becomes impractical. Unless embryos are specifically required by the study (e.g., in examining maternal effects), larvae or adults are the preferred source for nuclei. In this study we have performed chromatin structure analyses using nuclei isolated from transformants containing variant *hsp26/lacZ* transgenes. Because the heat shock genes (and heat shock-inducible transgenes) respond to heat shock induction equally well throughout development, larvae or adults are a convenient source of nuclei for studying the chromatin structure and expression of a heat shock gene.

The following procedure can be used to obtain maximum numbers of larvae from a limited number of adult flies. Approximately 2000 flies (3 to 5 days after emerging) are maintained in 4 bottles; the flies are transferred twice to fresh bottles after 4 days, and are discarded on day 12. Approximately 4 g of third instar larvae can be collected from the walls of these bottles by using a spatula. Non-heat-shocked larvae are collected, frozen in liquid nitrogen, and stored at −70°C. For heat shock induction, the collected larvae are transferred to a 15-ml beaker kept humid with a disk of wet filter paper. The beaker is covered with aluminum foil and placed in an incubator at 37°C for 45 min. The heat-shocked larvae are frozen in liquid nitrogen and stored at −70°C.

Live adult flies are collected after anesthetizing with CO_2, and transferred to a 50-ml screw-cap tube. Non-heat-shocked flies are frozen by immersing the tube in liquid nitrogen. For heat shock induction, flies are anesthetized with CO_2 and collected as above. The cap of the tube is punctured several times with a needle to allow air to circulate, and the tube is placed in a 37°C incubator for 45 min. After heat shock, the tube is placed into liquid nitrogen. Flies can be stored at −70°C in sealed tubes.

Preparation of Drosophila Nuclei

All buffers are prepared fresh and kept on ice. We recommend wearing gloves during the entire procedure, particularly during steps 4 and 5, to avoid contamination from degradative enzymes present on the hands. The procedure for isolating nuclei from larvae has been described previously (5).

Isolation of Nuclei from Larvae

1. One gram of larvae is pulverized in liquid nitrogen, using a mortar and pestle.
2. After the liquid nitrogen has evaporated, the homogenate is transferred to a beaker containing 3 ml of buffer A^+ [60 m*M* KCl, 15 m*M* NaCl, 13 m*M* ethylenediaminetetraacetic acid (EDTA), 0.1 m*M* ethylene glycol-bis(β-aminoethyl ether)-*N,N,N′,N′*-tetra acetic acid (EGTA), 15 m*M*, Tris-HCl (pH 7.4), 0.15 m*M* spermine, 0.5 m*M* spermidine, 0.5 m*M* dithiothreitol (DTT), 0.5% Nonidet P-40 (NP-40)] and mixed.
3. The suspension is transferred to a 7-ml ground glass homogenizer (Pyrex) and disrupted with five strokes of the pestle.
4. The homogenate is transferred to a 15-ml Dounce homogenizer (Wheaton, Millville, NJ), disrupted with 10 strokes of the pestle (type B), and filtered through two layers of 89-nm nylon mesh cloth (Nitex).
5. Buffer A^+ (2 ml) is used to rinse the ground glass homogenizer and the Dounce homogenizer, and the suspension is filtered through the same two layers of nylon mesh cloth; it is necessary to squeeze the cloth to complete filtration.
6. The filtrate is loaded over 1 ml of buffer AS [60 m*M* KCl, 15 m*M* NaCl, 1 m*M* EDTA, 0.1 m*M* EGTA, 15 m*M* Tris-HCl (pH 7.4), 0.15 m*M* spermine, 0.5 m*M* spermidine, 0.5 m*M* dithiothreitol, 0.3 *M* sucrose] and centrifuged in a Sorvall (Norwalk, CT) HB-4 swinging bucket rotor at 4°C, 3000 rpm (1500 *g*) for 5 min.
7. The recovered pellet is resuspended in 3 ml of buffer A^+, transferred to a 7-ml Dounce homogenizer (Wheaton), and dispersed with five strokes of the pestle (type A). The mixture is loaded over 1 ml of buffer AS, and centrifuged as before.
8. The recovered pellet is resuspended in 3 ml of buffer A [60 m*M* KCl, 15 m*M* NaCl, 1 m*M* EDTA, 0.1 m*M* EGTA, 15 m*M* Tris-HCl (pH 7.4), 0.15 m*M* spermine, 0.5 m*M* spermidine, 0.5 m*M* dithiothreitol], transferred to the 7-ml Dounce homogenizer, and dispersed with five strokes of the pestle (type B).
9. Nuclei are collected by centrifugation in the same rotor at 2000 rpm (700 *g*)

for 5 min and are resuspended in the appropriate buffer for a given digestion as described below (see *Treatment of Nuclei with Cleavage Reagents).*

Isolation of Nuclei from Adult Flies

The following protocol for the isolation of nuclei from adult flies is a modification of the isolation procedure for larval nuclei described above, using the same buffers.

1. One gram of adult flies is pulverized in liquid nitrogen, using a mortar and pestle.
2. After the liquid nitrogen has evaporated, the homogenate is transferred to a beaker containing 5 ml of buffer A^+ and mixed.
3. The suspension is transferred to a 7-ml ground glass homogenizer (Pyrex) and disrupted with 10 strokes of the pestle.
4. The homogenate is then transferred to a 15-ml Dounce homogenizer (Wheaton) and disrupted with 20 strokes of the pestle (type B). The homogenate is filtered through one layer of Miracloth (Habco Products, NJ). It is necessary to squeeze the Miracloth to complete filtration.
5. Buffer A^+ (3 ml) is used to rinse the ground glass homogenizer and the Dounce homogenizer, and the suspension is filtered through the same Miracloth.
6. The filtrate is loaded over 2 ml of buffer AS and centrifuged in a Sorvall HB-4 swinging bucket rotor at 4°C, 3000 rpm (1500 *g*) for 5 min.
7. The recovered pellet is resuspended in 5 ml of buffer A^+, transferred to a 7-ml Dounce homogenizer (Wheaton), and dispersed with 10 strokes of the pestle (type A). The mixture is loaded over 2 ml of buffer AS and centrifuged as before.
8. The recovered pellet is resuspended in 5 ml of buffer A, transferred to the 7-ml Dounce homogenizer, and dispersed with 10 strokes of the pestle (type B).
9. The nuclei are collected by centrifugation in the same rotor at 3000 rpm (1484 *g*) for 5 min, as above. The pellet is washed with another 5 ml of buffer A and centrifuged at 4000 rpm (2600 *g*) for 5 min. The pellet is resuspended in the appropriate buffer for a given digestion as described below.

Nuclei can also be isolated from 0.5 g of embryos, using the procedure described for isolation of larval nuclei. Embryos, larvae, and adult flies can be used fresh or kept frozen at −70°C for several years. No difference was detected using nuclei isolated from fresh or frozen material in the low-resolution chromatin structure analyses described below.

Because larvae and adults contain relatively high levels of endogenous nuclease activity, minimal levels of divalent cations are used in the nuclear isolation buffers. To prevent the swelling and consequent rupture of nuclei due to low levels of divalent cations, it is recommended that spermine and spermidine be included in the nuclear isolation buffers (19). However, it has been reported that the presence of polyamines in the nuclear isolation buffers will affect chromatin structure (20, 21).

Use of polyamines in the nuclear isolation buffer can lead to selective extraction of some chromosomal proteins, such as the high mobility group (HMG) proteins (see, e.g., Ref. 22). Moreover, it has been reported that addition of polyamines to nuclear isolation buffers causes a loss of transcription factors at the human X-linked phosphoglycerate kinase locus, leaving the promoter with a reactivity similar to that of naked DNA (23). The TATA box footprint of *Drosophila hsp26*, observed in a high-resolution analysis, is also lost if nuclear isolation buffers containing polyamines are used (G. H. Thomas and S. C. R. Elgin, unpublished observations, 1988).

To investigate the possible impact of the polyamines on the analysis of nucleosome arrays, we prepared nuclei from both larvae and adults using buffers with or without spermine and spermidine. We observed no difference in the pattern of DNase I-hypersensitive (DH) sites or in the level of accessibility of a restriction enzyme site in the presence or absence of polyamines in the nuclear isolation buffers used here. This indicates that the basic nucleosome array has not been perturbed and that the nucleosomes have maintained the same positions as observed when nuclear isolation buffers containing no polyamines are used. This observation does not, however, contradict the above findings. Caution should be taken in using nuclear isolation buffers containing polyamines, especially when performing high-resolution mapping of the binding sites of transcription factors or of the polymerase complex.

Treatment of Nuclei with Cleavage Reagents

The chromatin structure of a particular chromosomal region in nuclei can be detected by examining the pattern of cleavage by various DNA cleavage reagents, each of which has specific properties. Because they provide different information, more than one cleavage reagent is normally used.

The extent of digestion of a given site by general DNA cleavage reagents depends on the frequency of cleavage and on the detection strategy. Hypersensitive sites within a region of several kilobases can be detected by an indirect end-labeling experiment at low resolution using probes or primers that are relatively distant from the region of interest. Under these conditions, a limited digestion is required to obtain large fragments that hybridize to the probe. On the other hand, high-resolution analyses often involve using probes or primers that are close to the region of interest. Relatively extensive digestion is required to provide sufficient cleavage within the region of interest. The actual extent of digestion that is most appropriate for each cleavage reagent and/or each detection strategy used is usually determined empirically.

Because all of the general cleavage reagents will show some sequence bias in cleaving DNA, a "naked DNA" control should always be performed to avoid misinterpretation of the results. The control naked DNA is treated with the particular cleavage reagent to the same extent as the DNA in nuclei. Differences in the relative intensities between bands of the naked DNA control and the chromatin digests indicate aspects of the chromatin structure. Compared to DNA in nuclei, naked genomic DNA is extremely susceptible to general DNA cleavage reagents. The actual ratio of cleavage reagent to DNA needed to generate the appropriate size fragments from naked DNA needs to be determined empirically.

DNase I Digestion

DNase I can be used to detect nucleosome-free regions of chromatin. DNase I-hypersensitive sites (usually 50 to 200 bp) can be mapped within a region of several kilobases of DNA by limited digestion of nuclei with DNase I and low-resolution mapping techniques (24, 25). Because DNase I attacks the minor groove of DNA (26), DNase I can also be used to determine whether a particular stretch of DNA is wrapped around the surface of a protein complex. When DNA is associated with a histone octamer, a 10- to 11-bp interval cleavage pattern spanning approximately 140 bp will be detected by high-resolution analysis (25, 27). The following is a protocol for DNase I treatment of nuclei.

1. The pellet of isolated larval or adult *Drosophila* nuclei (see above) is resuspended in 1 ml of DNase I digestion buffer [60 m*M* KCl, 15 m*M* NaCl, 15 m*M* Tris-HCl (pH 7.4), 0.25 *M* sucrose, 3 m*M* $MgCl_2$, 0.05 m*M* dithiothreitol].
2. Aliquots (250 μl) of this nuclear suspension (containing approximately 50 μg of genomic DNA) are incubated with 0, 2, 4, or 6 μl DNase I (20,000 units/ml; Boehringer Mannheim Biochemicals, Indianapolis, IN) on ice for 3 min with occasional agitation. For nuclei isolated from 0.5 g of embryos, 250-μl aliquots (approximately 50 μg of genomic DNA) of the nuclear suspension are incubated with 0, 0.25, 0.5, or 1 μl of DNase I (20,000 units/ml) on ice for 3 min with occasional agitation. These amounts of DNase I have been established empirically; the reasons for the difference in DNase I sensitivity between nuclei isolated from larvae, adults, and embryos are not known.
3. The digestion is terminated by the addition of 5 μl of 0.4 *M* EDTA, and the DNA products are purified and analyzed as described below.

Micrococcal Nuclease Digestion

Micrococcal nuclease (MNase) is the enzyme most commonly used to detect a nucleosome array, because cleavage of DNA by MNase is inhibited when DNA is associated with a histone octamer; the enzyme therefore cuts linker DNA preferentially (28). However, MNase shows a significant sequence bias, preferentially cleaving DNA at the sequence d(A-T)*n*, where $n \geq 1$, flanked by a 5′ dC or dG (29–32). Consequently, an MNase digestion pattern on purified genomic DNA is an essential control for any mapping experiments. Limited MNase digestion allows detection of nucleosomal arrays in chromatin, whereas extensive MNase digestion results in production of mononucleosomes and dinucleosomes as the majority of the product. The following is a protocol for MNase treatment of nuclei.

1. The pellet of isolated larval or adult nuclei (see above) is resuspended in 1 ml of MNase digestion buffer [60 m*M* KCl, 15 m*M* NaCl, 15 m*M* Tris-HCl (pH 7.4), 0.5 m*M* dithiothreitol, 0.25 *M* sucrose, 1 m*M* $CaCl_2$].

2. For limited MNase digestion, 250-μl aliquots of the nuclear suspension (containing approximately 50 μg of genomic DNA) are incubated with 0, 0.01, 0.05, and 0.1 units of MNase (Sigma, St. Louis, MO) at 25°C for 3 min with occasional agitation. For extensive MNase digestion, 250-μl aliquots of nuclear suspension are incubated with 0, 0.5, 1.5, or 4.0 units of MNase at 25°C for 3 min with occasional agitation.

3. The digestion is terminated by the addition of 5 μl of 0.4 *M* EDTA, and the DNA products are purified and analyzed as described below.

Methidiumpropyl-EDTA·Fe(II) Cleavage

Methidiumpropyl-EDTA (MPE) is a synthetic compound that, when complexed with the ferrous ion in equal molar amounts, forming MPE·Fe(II), introduces both nicks and double-stranded cuts in DNA (33). Compared to MNase and DNase I, MPE·Fe(II) cleaves DNA with little sequence bias (34–36). Its relatively small size makes it extremely powerful in detecting protein–DNA interactions in chromatin both *in vivo* and *in vitro* (37).

The synthesis and purification of MPE has been described (33, 36). Currently, MPE is not commercially available; however, samples for experimental purposes can be obtained from Dr. P. B. Dervan (Division of Chemistry and Chemical Engineering, California Institute of Technology, Pasadena, CA). Methidiumpropyl-EDTA can be stored at −20°C in solid or in aqueous solution. The following protocol for digestion of nuclei with MPE·Fe(II) is modified from that of Cartwright and Elgin (37).

1. The pellet of isolated larval or adult nuclei (see above) is resuspended in 900 μl of MPE digestion buffer [15 m*M* Tris-HCl (pH 7.4), 15 m*M* NaCl, 60 m*M* KCl, 1 m*M* EDTA, 0.25 *M* sucrose]. The nuclei are warmed to 25°C.

2. The MPE·Fe(II) cleavage reagent is prepared by mixing 12 μl of 2.6 m*M* MPE with 12 μl of 2.6 m*M*, freshly prepared aqueous ferrous ammonium sulfate; 76 μl of MPE solution [15 m*M* Tris-HCl (pH 7.4), 15 m*M* NaCl, 60 m*M* KCl, 0.25 *M* sucrose] is then added. Prior to the cleavage reaction, the MPE·Fe(II) complex is activated by the addition of dithiothreitol to 10 m*M*, and the solution is warmed to 25°C.

3. The activated MPE·Fe(II) complex (100 μl) is added to the nuclear suspension (900 μl), and the reaction is carried out at 25°C with occasional agitation.

4. Aliquots (250 μl) of the nuclear solution (approximately 50 μg of genomic

DNA) are withdrawn at 0, 1, 2, and 3 min. The reaction is terminated by the addition of 25 μl of 50 m*M* bathophenanthroline disulfonate (Sigma), and the DNA products are purified and analyzed as described below.

Restriction Enzyme Digestion

The cleavage reagents used above cleave DNA nonspecifically, and under conditions of limited digestion will provide information on protein–DNA interactions in nuclei. However, the pattern of products reflects the differential sensitivity; as the reaction progresses, much of the DNA will be cleaved, and the initial pattern lost. Therefore, these nonspecific cleavage reagents can only be used to generate a *qualitative* assessment of the sensitivity of a region. In contrast, restriction enzymes cleave DNA in a sequence-specific fashion, and in combination with appropriate detection techniques (see below) can be used to *quantitate* the accessibility of a particular site within isolated nuclei (Fig. 1) (5, 38). The degree of accessibility of a given site reflects features of the chromatin structure at this site. Hypersensitive sites (nucleosome-free regions) show considerable accessibility (50–80%) whereas DNA packaged into nucleosomes shows greatly reduced cleavage (6–8%), allowing one to map the borders of hypersensitive regions and the positions of nucleosomes (38–40).

Treatment of isolated nuclei with a particular restriction enzyme requires that the enzyme be present in excess amounts to ensure that all of the available sites are cleaved; this is essential in obtaining a quantitative assessment of accessibility. To ensure that a particular enzyme is in excess concentration, a titration experiment should be carried out in each case. A series of samples of nuclei are incubated with increasing amounts of restriction enzyme; the minimal amount of restriction enzyme that shows maximal cleavage at the site proximal to the probe used for indirect end labeling analysis (see below) is the ideal concentration for that case. For most restriction enzymes, using the protocol below, 200 to 400 units is sufficient for maximal cleavage for a nuclear suspension containing approximately 20 μg of genomic DNA.

1. The nuclear pellet is resuspended in 1.47 ml of a restriction enzyme digestion buffer (manufacturer specifications).
2. Phenylmethylsulfonyl fluoride, leupeptin, and aprotinin are added to final concentrations of 0.5 m*M*, 5 μg/ml, and 0.1 unit/ml, respectively, to minimize proteolysis during the prolonged incubation.
3. Restriction enzyme is added to 250-μl aliquots of suspended nuclei (containing approximately 20 μg of genomic DNA) and the mixture incubated at 37°C for 45 min.
4. The reaction is terminated by the addition of 5 μl of 0.4 *M* EDTA, and the DNA is purified as described in the next section.

Quantitative measurements of the susceptibility of different sites to restriction enzymes provide a powerful means to measure changes in chromatin configurations in nuclei. It is noted that restriction enzyme digestion of a particular site in isolated nuclei has never been observed to reach 100% cleavage; a maximal cleavage of 75 to 85% has been obtained for sites that are nucleosome free by other criteria (38, 41, 42; L. L. Wallrath, unpublished observations, 1992). It has been suggested that

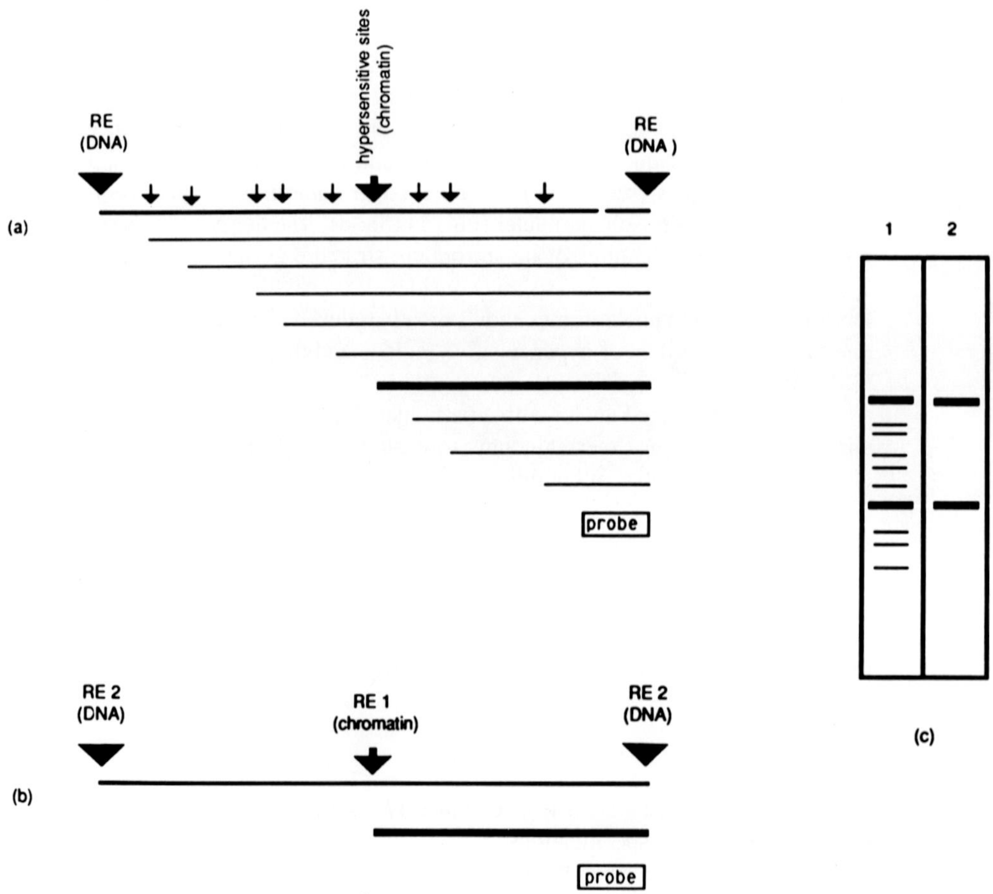

FIG 1 Schematic diagram of products resulting from treatment of nuclei with general cleavage reagents or with restriction enzymes. (a) Potential products resulting from treatment of isolated nuclei with a limited amount of a general cleavage reagent, as detected by the probe given. The top line represents genomic DNA; a hypersensitive site is marked with a bold arrow, whereas general background cleavage sites are marked by thin arrows. For indirect end-labeling analysis, DNA is purified from nuclei that have been treated with a general cleavage reagent, and is completely digested with a restriction enzyme (RE), with sites as

the incomplete digestion of a restriction site within a hypersensitive site may be due to mitotic condensation in a subpopulation of nuclei (38, 41).Cleavage by various restriction enzymes along the *Drosophila hsp26* gene detects a minimal cleavage of 6–8%. This incomplete protection may be due to disruption of nucleosome arrays during nuclear preparation or may simply reflect the dynamic nature of the protein–DNA complexes in chromatin.

Purification of DNA

After termination of the DNA cleavage reaction, nuclei are centrifuged for 30 sec in an Eppendorf centrifuge, and the pellet is resuspended in 250 μl of sarcosyl lysis buffer [50 m*M* Tris-HCl (pH 8.0), 100 m*M* EDTA, 0.5% (w/v) sodium laurylsarcosine (Sigma)]. Three microliters of proteinase K (10 mg/ml) is added, the nuclei are incubated at 37°C for at least 1.5 h with occasional mixing, and 27 μl of 3 *M* NaCl is added. The DNA is extracted with 400 μl of phenol–chloroform–isoamyl alcohol (24:24:1). Ten microliters of the aqueous phase is loaded and fractionated on a 1% (w/v) agarose gel to check the extent of digestion (e.g., see Fig. 2). 2-Propanol (400 μl) is added to precipitate the DNA. The DNA is collected by centrifugation for 10 min in an Eppendorf centrifuge and the pellet is washed with 70% ethanol. The pellet is vacuum dried, resuspended into 120 μl of RNase A digestion buffer [50 m*M* Tris-HCl (pH 8.0), 10 m*M* EDTA, 100 m*M* NaCl] with 3 μl of RNase A (10 mg/ml), and incubated at 37°C for 3 hr with occasional mixing. Twenty-seven microliters of 3 *M* NaCl is then added to the DNA. The DNA is extracted with 200 μl of phenol–chloroform–isoamyl alcohol (24:24:1) and

---

marked by the large arrowheads. The lines below the top line represent the final products detected, with the major product (indicating a hypersensitive site) shown as a bold line. (b) Products resulting from treatment of isolated nuclei with excess amounts of a restriction enzyme having a site within an accessible region (i.e., hypersensitive site). The top line represents genomic DNA showing a site, marked as a bold arrow, cleaved by the restriction enzyme used to treat nuclei (RE 1, chromatin). For indirect end-labeling analysis, the DNA is then purified and completely digested with another restriction enzyme, having sites as marked by large arrowheads (RE 2, DNA). The line below the top line represents the final product. (c) Schematic representation of an autoradiograph of the products shown in (a) and (b) detected by hybridization with the probe indicated. Lane 1 shows products from (a), with the top bold band representing the parental restriction fragment resulting from no cleavage between the RE sites in chromatin. The bottom bold band is the fragment generated from the hypersensitive site. The thin bands indicate the minor products of general cleavage. Lane 2 shows the products of (b), with the top bold band representing the parental restriction fragment resulting from absence of cleavage by RE 1 in nuclei. The bottom bold band indicates the product generated by cleavage by RE 1 in nuclei.

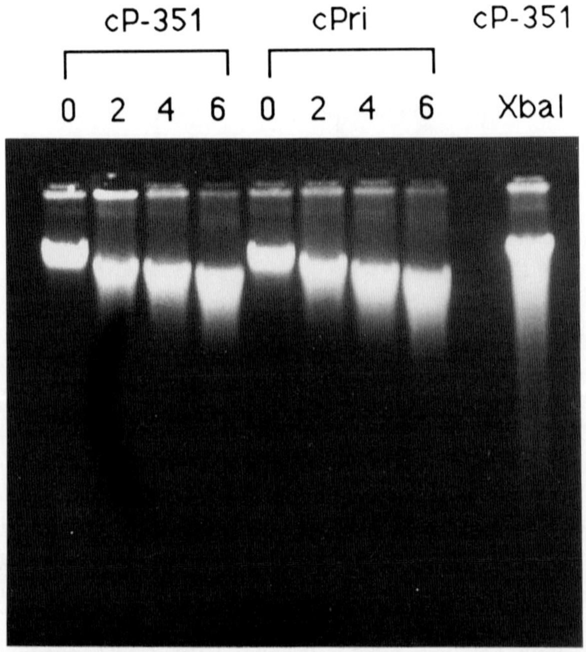

FIG 2. Digestion of isolated nuclei with DNase I or *Xba*I. Nuclei were isolated from non-heat-shocked larvae containing transgene cP-351 or transgene cPri, and treated with increasing amounts of DNase I or excess amounts of restriction enzyme *Xba*I. Samples of the purified DNA were fractionated on an agarose gel and stained with ethidium bromide. Numbers at the top of the lanes indicate the relative amount of DNase I used (microliters of a 20,000-U/μl stock) (see Methods for details). The lane labeled *Xba*I shows the cleavage of the genomic DNA in chromatin by restriction enzyme *Xba*I.

then with 1 ml of ether. The DNA is precipitated with 500 μl of ethanol, rinsed with 70% ethanol, and resuspended in 50 μl of TE [10 m*M* Tris-HCl (pH 8.0), 1 m*M* EDTA]. The concentration of purified DNA in each tube should be approximately 1 mg/ml. It should be noted that third instar larvae contain high percentages of DNA from polytene tissues; adult flies contain DNA mostly from diploid nuclei (43).

Genomic DNA (for the naked DNA control) can be isolated from adult flies or larvae as described above. However, a quicker and simpler protocol is frequently used. The following procedure is modified from that of Lis *et al.* (44). Adult flies (200 to 500) are collected in a 50-ml tube and then pulverized in liquid nitrogen with a mortar and pestle. After the liquid nitrogen has evaporated, the homogenate is transferred to a 15-ml Dounce homogenizer (Wheaton) containing 10 ml homogenization buffer [1 *M* Tris-HCl (pH 9.2), 50 m*M* EDTA, 0.5% (w/v) sodium dodecyl sulfate (SDS), 0.2 *M* sucrose], and disrupted with 20 strokes of the pestle (type B). The homogenate is transferred to a 50-ml tube and incubated at 65°C for 30 min with occasional mixing. 2.25-ml of 8 *M* NaOAc is added and the contents mixed and incubated on ice for 15 min; the tube is then centrifuged in a Sorvall SS-34 rotor at 10,000 rpm (12,000 *g*) for 10 min. The supernatant is carefully transferred to a 50-ml tube, 35 ml of ethanol is added, and the DNA precipitated on ice for 5 min. The DNA is then pelleted by centrifugation at 5000 rpm for 5 min, rinsed with 10 ml 70% ethanol, desiccated, and resuspended in 2 ml of RNase A digestion buffer [50 m*M* Tris-HCl (pH 8.0), 10 m*M* EDTA, 100 m*M* NaCl]. Twenty microliters of RNase A (10 mg/ml) is then added, and the tube is incubated at 37°C for 2 hr. Twenty microliters of proteinase K is added, and the tube is incubated at 37°C for another 2 hr. The DNA is extracted with 2 ml of phenol–chloroform–isoamyl alcohol (24:24:1), 2 ml of chloroform–isoamyl alcohol (24:1), and precipitated with 6 ml of ethanol on ice for 5 min. The DNA is pelleted by centrifugation in a Sorvall SS-34 rotor at 10,000 rpm (12,000 *g*) for 10 min, washed with 5 ml of 70% ethanol, and resuspended in 1 ml of low-salt buffer [0.2 *M* NaCl, 20 m*M* Tris-HCl (pH 7.4), 1 m*M* EDTA]. The DNA is further purified by passing it through an Elutip column, according to the instructions of the manufacturer (Schleicher & Schuell, Keene, NH).

Chromatin Structure Analysis

The cleavage pattern of the DNA within nuclei treated as above can be determined by using several techniques (reviewed in Ref. 45). Both of the mapping techniques described here depend on measuring the distance along the DNA between a fixed reference point and the site of experimental cleavage. These techniques can be used for either low- or high-resolution analysis.

Indirect End-Labeling Analysis

Indirect end-labeling analysis was developed in 1980 by two independent groups (46, 47). This technique can be used to map cleavage patterns for any single-copy gene provided that the gene has been cloned and the restriction map is known. After the nuclei have been treated with an appropriate cleavage reagent, the DNA is purified. The purified DNA (10–15 μg) is then digested to completion with a specific restriction enzyme, chosen such that two restriction sites bracket the region of interest, with no sites for this enzyme within the region of interest (Fig. 1). The bracketing restriction sites provide the reference points for mapping specific cleavage events. The DNA is fractionated (on an agarose or acrylamide gel, depending on the resolution required), and blotted to a nylon membrane. Hybridization of the membrane with a cloned DNA fragment (500 bp to 2 kb) that abuts the reference point (restriction site) will allow detection of specific cleavage fragments from the region of interest that have this end point (Fig. 1). The size of the fragments (evaluated using size markers) gives the position of the cleavage site relative to the end point (Fig. 1). Probes used can be from either end of the region of interest; probes can be double-stranded DNA (e.g., see Fig. 6), single-stranded DNA, or RNA (see, e.g., Refs. 25 and 48). A detailed protocol for high-resolution analysis using this technique has been described (49, 50).

Primer Extension

Specific cleavage patterns can also be detected by primer extension techniques (51). Primer extension techniques involve using an oligonucleotide close to the region of interest that is complementary to one of the DNA strands. The binding of the oligonucleotide to the desired site provides the reference point for subsequent extension along the genomic template by a DNA polymerase. This primer-mediated DNA synthesis is terminated when a nick on the template strand of the DNA (generated by a cleavage reagent) is encountered. The extension products can be labeled by using a radioactive primer and the products detected after fractionation on an agarose or sequencing gel by subsequent autoradiography (51). The amount of products from a primer extension reaction can be greatly increased by using a thermally stable polymerase that will perform multiple rounds of primer extension reactions in a polymerase chain reaction (PCR) machine (52). Alternatively, nonradioactive products can be fractionated on a gel, electroblotted onto nylon membrane, and hybridized with labeled probes (53). Detailed protocols for these techniques have been described (50, 52, 53). A modification of this procedure involves ligation of a linker to the end of the extended products. Amplification of the sequence between two primers, one that hybridizes to the linker and the one internal to that used for the original extension, yields significant amounts of product. This technique is called ligation-mediated polymerase chain reaction (LMPCR) (54, 55). However, unbiased results require uniform efficiency of ligation of the linker and subsequent amplification of all such DNA fragments.

Detection of Gene Expression

P-element constructs are usually designed to facilitate subsequent chromatin structure analyses and to allow easy monitoring of expression of the transgenes. The gene of interest is frequently fused to a unique sequence that can serve as a probe for Northern analyses. A modified version of the gene of interest (having a different-sized transcript) can also be used to discriminate its product from that of the endogenous gene on Northern blot analysis. Alternatively, a fusion gene is constructed with the regulatory region and transcription start sites of the gene of interest spliced to the *E. coli lacZ* gene or Tn*9 CAT* gene. The expression of these transgenes can be monitored by enzyme activity assays, for example, Chlorophenol red-β-D-galactopyranoside (CPRG) assays for β-galactosidase activity (56) or CAT assays for chloramphenicol acetyltransferase activity (57), as well as by Northern analysis.

Examples

We have used the protocols described above to study the chromatin structure and gene expression of the *D. melanogaster hsp26* gene. *hsp26* sequences with and without mutations within the promoter were fused in frame with the *E. coli lacZ* gene (Fig. 3). These fusion genes were inserted into a P-element vector containing the *rosy* (xanthine dehydrogenase) gene as a reporter (58). The P-element con-

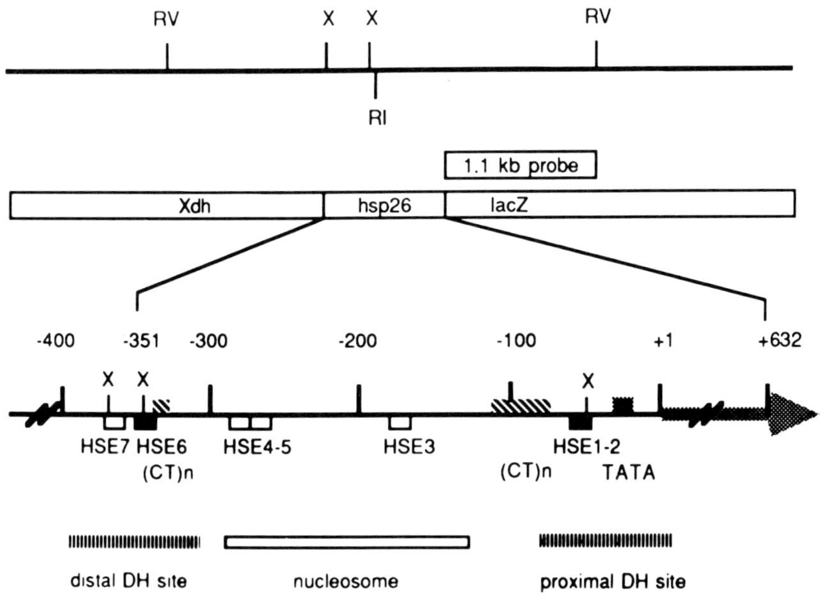

FIG 3 Mapping strategy for chromatin structure analysis of *hsp26*. The structure of cP-351 is shown. The *hsp26* sequence from −351 to +632 is fused in frame to the *E. coli lacZ* gene as previously described (58). The 1.1-kb DNA fragment used as probe in the indirect end-labeling analysis is the *Eco*RI/*Eco*RV fragment isolated from plasmid pMC1403, indicated as a box (72). A partial restriction map of the *hsp26* sequence (−351 to +632) is enlarged below with the $(CT)_n{\cdot}(GA)_n$ regions, the TATA box, and the two required heat-shock elements (HSE 1–2 and HSE 6; filled boxes) diagrammed. HSEs 3–5 (shown as open boxes) have similarity to the HSE consensus sequence but are not involved in heat-shock induction; HSE 7 is accessible in isolated nuclei, but is not required for heat-shock induction (25, 56, 59–61). The proximal and distal DNase I-hypersensitive (DH) sites and a nucleosome-associated region as determined by previous analyses are marked below. RI, *Eco*RI, RV, *Eco*RV; X, *Xba*I.

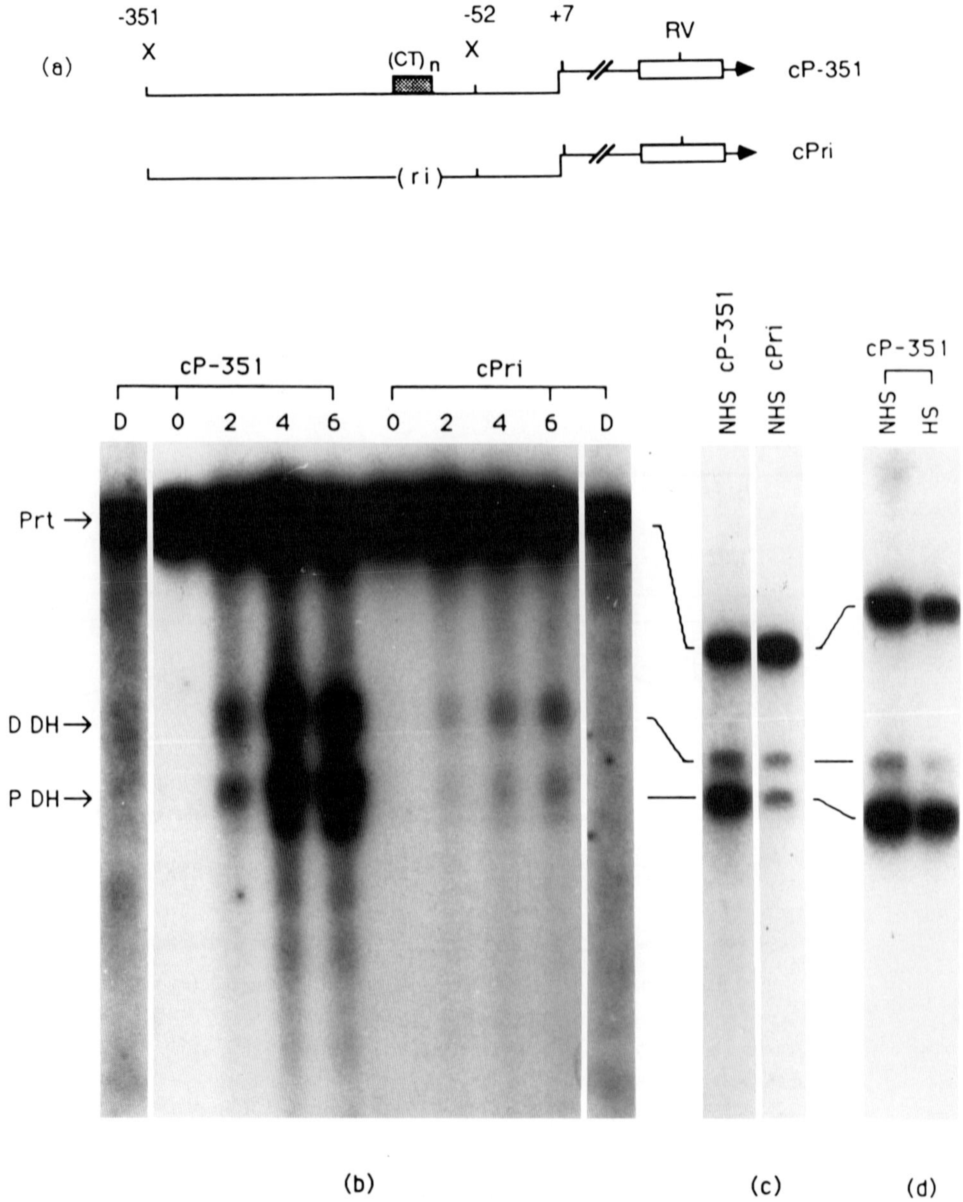

FIG 4 Patterns of DNase I-hypersensitive sites and quantitation of the accessibility of the *Xba*I site within the proximal DH site in nuclei isolated from larvae of *hsp26* transformants. (a) Schematic drawing of transgenes cP-351 and cPri. In transgene cP-351, the $(CT)_n$ represents the $(CT)_n \cdot (GA)_n$ repeats located between -135 and -85 with respect to the transcription start site, marked where the line bends up. In transgene cPri, "ri" represents a random

structs were introduced into the *Drosophila* germ line by coinjection of the construct DNA with plasmid pπ25.7 "wings clipped" into embryos from a fly stock deficient for the *xanthine dehydrogenase* gene (12). Transformants from progeny of the injected embryos were identified by wild-type eye color. The heat shock-induced expression of the transgenes was determined by measuring the β-galactosidase activity using CPRG assays (56, 59), and the chromatin structure of the transgenes was investigated using the protocols described above.

Transgene cP-351, which serves as a wild-type control for a series of transgenes with modified regions within the promoter, contains *hsp26* sequences from −351 to +632 fused in frame to the *E. coli lacZ* gene (Fig. 3). Earlier studies have shown this fragment to be sufficient for heat-inducible transcription (56, 59–61). We have analyzed the chromatin structure of this transgene, using nuclei of larvae or adult flies, by examination of its DH site patterns and restriction enzyme cleavage susceptibility. Nuclei were isolated from non-heat-shocked larvae or adult flies carrying one copy of cP-351, and were treated with DNase I (Fig. 2). Fifteen-microgram samples of the purified DNA were completely digested with *Eco*RV (Fig. 3), fractionated on a 1.2% (w/v) agarose gel and transferred to Nytran membrane (Schleicher & Schuell) by Southern blotting (11). The membrane was hybridized with a 1.1-kb *lac*ZDNA fragment (Fig. 3) labeled with [α-^{32}P]dCTP, using random hexamer priming (62). The conditions for hybridization and washing were as previously described (24). As shown in Fig. 4b and 5a, two strong DNase I-hypersensitive

---

insert of salmon sperm DNA that was substituted for the $(CT)_n \cdot (GA)_n$ repeats of cP-351 (59). The open box indicates the *E. coli lacZ* gene. X and RV represent the *Xba*I and *Eco*RV sites, respectively. (b) The DNase I-hypersensitive site patterns of transgenes cP-351 and cPri. Nuclei were isolated from non-heat-shocked larvae and treated with DNase I (samples of the purified DNA are shown in Fig. 2). Purified DNA was restricted to completion with *Eco*RV, fractionated on a 1.2% (w/v) agarose gel, transferred to a nylon membrane, and the membrane was hybridized with the 1.1-kb *lacZ* probe (Fig. 3) and autoradiographed. Numbers at the top of the lanes indicate the relative amount of DNase I used (microliters of a 20,000-U/μl stock) (see Methods for details). The lanes marked D show naked DNA control samples digested with DNase I. Pri represents the parental band generated by *Eco*RV digestion. P DH and D DH are the proximal and distal DH sites, respectively. Transgenes are identified above the lanes. (c) *Xba*I accessibility in transgenes cP-351 and cPri. Nuclei were isolated from non-heat-shocked (NHS) larvae and treated with excess amounts of *Xba*I. The purified DNA was subjected to indirect end-labeling analysis as described above for the samples treated with DNase I. The accessibility detected was 52% for cP-351, and 8% for cPri. (d) *Xba*I accessibility in transgene cP-351 before and after heat shock. Nuclei were isolated from non-heat-shocked (NHS) and heat-shocked (HS) larvae and treated with excess amounts of *Xba*I. The purified DNA was digested to completion with *Eco*RV and subjected to indirect end-labeling analysis as described above for samples treated with DNase I. The accessibility detected before and after heat shock was 54 and 62%, respectively.

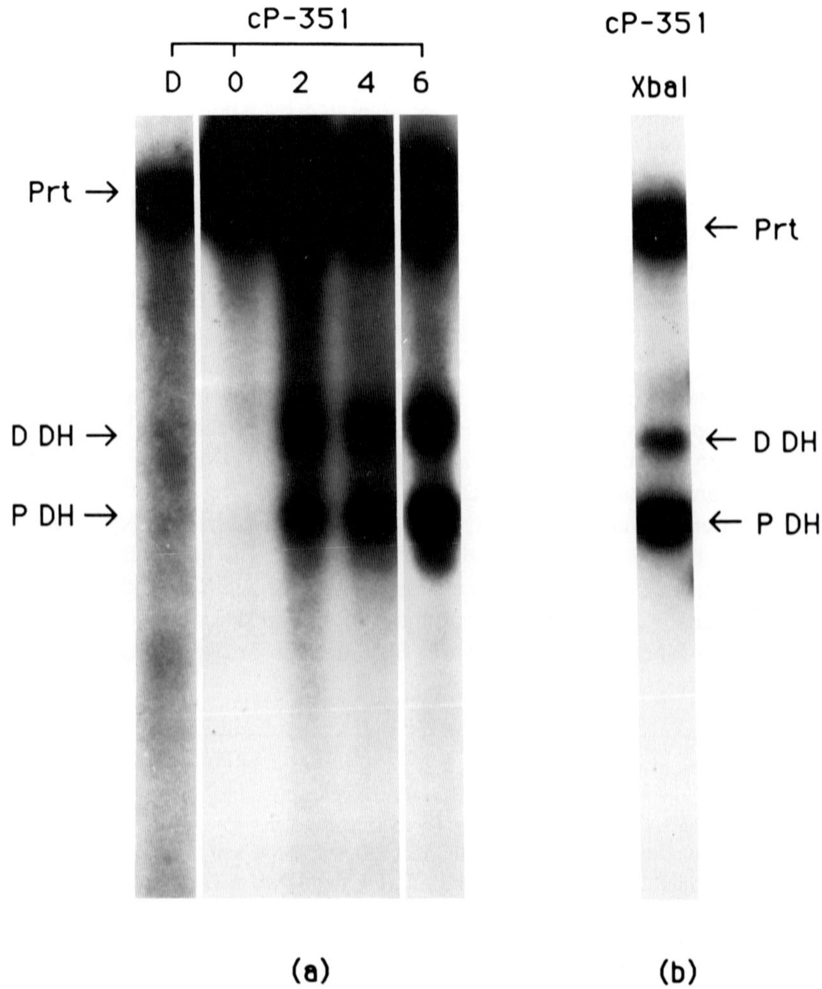

FIG 5 Patterns of DNase I-hypersensitive sites and quantitation of the accessibility of the *Xba*I site within the proximal DH site in nuclei isolated from adults. (a) The DNase I-hypersensitive site pattern of transgene cP-351. Nuclei were isolated from non-heat-shocked adults and treated with DNase I. The purified DNA was subjected to indirect end-labeling analysis as described in Fig. 4. Other designations are as described for Fig. 4. (b) *Xba*I accessibility in transgene cP-351. Nuclei were isolated from non-heat-shocked adults and treated with an excess amount of *Xba*I. The purified DNA was digested to completion with *Eco*RV and subjected to indirect end-labeling analysis as described above for samples treated with DNase I. The accessibility detected was 31%.

sites (DH sites), designated the proximal (P DH) and distal (D DH) DH sites, were located in the promoter region of *hsp26* in transgene cP-351. We also prepared nuclei from 0- to 18-hr embryos, using the same procedure as for the isolation of larval nuclei, to determine the chromatin structure of the endogenous *hsp26* gene. The DH site pattern detected (Fig. 6b) and *Xba*I accessibility (see below) at the proximal DH site (Fig. 6c) were the same as those detected for the endogenous *hsp26* gene, using nuclei isolated from embryos by this and other nuclear preparation protocols (24, 25, 63; L. L. Wallrath, unpublished observations, 1992). The results indicate that sequences from −351 to +632 of *Drosophila hsp26* gene contain sufficient DNA information for the formation of DNase I-hypersensitive sites and that local protein–DNA interactions are the major determinants of this chromatin structure.

The proximal and distal DH sites of the *hsp26* promoter map to the locations of the consensus heat shock elements (HSEs) that contain *Xba*I restriction enzyme recognition sequences (Fig. 3). The accessibility of *Xba*I to these sites can be used to quantitate accessibility of the DNA in the DH sites. Nuclei were isolated from non-heat-shocked third instar larvae or adult cP-351 transformants and incubated with *Xba*I (Fig. 2). The DNA was purified, digested to completion with *Eco*RV, fractionated on a 1% (w/v) agarose gel, hybridized with the 1.1-kb *lacZ* probe (Fig. 3), and exposed to X-ray film. The intensity of each band was determined using a scanning densitometer (Molecular Dynamics, Sunnyvale, CA). The percentage cleavage at the proximal *Xba*I site (which is most proximal to the probe) was determined as the fraction of the intensity of the band generated by cleavage at the proximal *Xba*I site to the total intensity of all the bands within a given lane. As shown in Fig. 4c, the accessibility of the proximal DH site in transgene cP-351 is 52% as measured by using nuclei isolated from non-heat-shocked larvae. Similarly, the accessibility of a site can be determined using embryos or adults as starting material (Figs. 5b and 6c). These results clearly demonstrated that nuclei that maintain a specific chromatin structure can be obtained from either embryos, larvae, or adult flies by using the nuclear preparation protocols described above. *Hsp26* shows relatively little tissue specific expression, and appears to be organized and regulated in a similar manner at all stages of development. However, it is best to use nuclei isolated from flies of the same developmental stage for chromatin structure analyses when comparing transgenes. The chromatin structure of a developmentally regulated gene in different developmental stages is likely to vary; the choice of developmental stage for chromatin structure analyses will depend on the question being addressed.

The proximal *Xba*I site lies within the consensus HSE, which is bound by heat shock factor (HSF) on heat shock induction (25, 64, 65). There is 52% cleavage of the *Xba*I site within the proximal DH site of the cP-351 transgene prior to heat shock; we wanted to determine the accessibility of this site after heat shock induc-

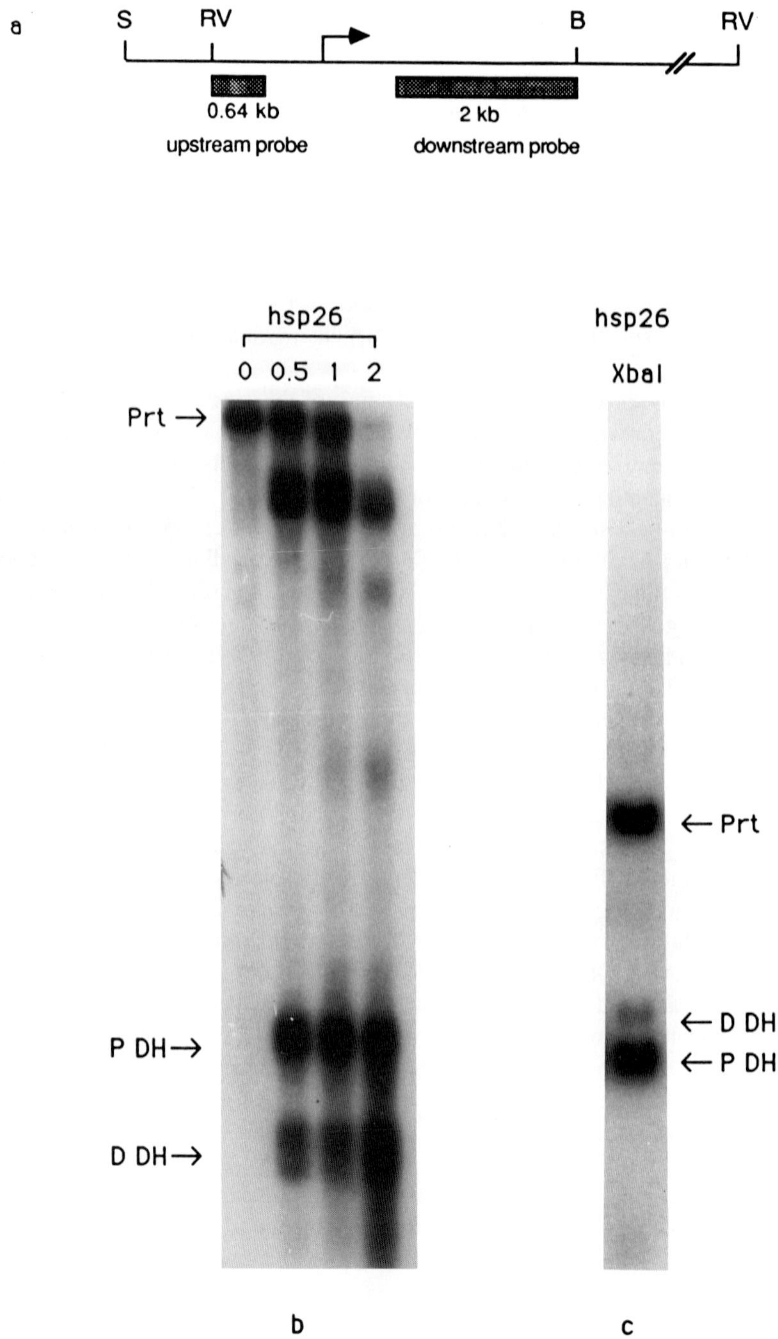
a
S
RV
B
RV
0.64 kb
upstream probe
2 kb
downstream probe
hsp26
0 0.5 1 2
Prt →
P DH→
D DH→
b
hsp26
XbaI
← Prt
← D DH
← P DH
c

tion. As shown in Fig. 4d, the proximal *Xba*I site showed similar accessibility to *Xba*I before and after heat shock induction, indicating that the presence of HSF, seen by DNase I footprinting, does not block access of the restriction enzyme to DNA. This is consistent with observations by Jack *et al.* (38) on the accessibility of a restriction site overlapping the HSE of the *Drosophila hsp70* gene. There are several reasons why a site that is DNase I footprinted after heat shock (25) may be accessible to restriction enzyme digestion. The binding of HSF to HSE is most likely a dynamic process; during the course of the 45-min incubation of the isolated nuclei with the restriction enzyme there may exist moments of accessibility. Alternatively, HSF may be bound to DNA in such a way that the restriction site is still accessible. However, positioned nucleosomes appear to block restriction enzyme cleavage. Changes in the nucleosome array on gene activation, such as those that occur in the yeast *PHO5* or mouse mammary tumor virus (MMTV) long terminal repeat promoters, have been detected by comparing the accessibility of restriction enzyme sites (39, 40, 66, 67).

The accessibility of a restriction site can now be used to compare the chromatin structure of various altered transgenes. In the *Drosophila hsp26* gene there is a sequence containing $(CT)_n \cdot (GA)_n$ repeats in the promoter region that is bound by the GAGA transcription factor (68, 69). To examine the functional importance of this protein–DNA interaction *in vivo,* transgene cPri was generated. cPri is identical to cP-351, except that a 48-bp sequence containing $(CT)_n \cdot (GA)_n$ repeats located between −135 and −85 of the *hsp26* promoter region has been replaced by a random sequence from salmon sperm DNA (59) (Fig. 4a). The heat shock-induced expression of the cPri transgene was reduced to 27% that of cP-351 (59). Chromatin structure analysis using non-heat-shocked larvae indicates that the DH sites are dramatically reduced in cPri (Fig. 4b). The hypersensitivity of the proximal

---

FIG 6 Patterns of DNase I-hypersensitive sites and quantitation of the accessibility of the *Xba*I site within the proximal DH site of the endogenous *hsp26* gene in nuclei isolated from embryos. (a) The drawing shows a partial restriction map of the endogenous *hsp26* gene, with the bent arrow representing the transcription start site. The shaded boxes represent probes used independently for mapping in (b) and (c). B, *Bam*HI; RV, *Eco*RV; S, *Sal*I. (b) The DNase I-hypersensitive site pattern of the endogenous *hsp26* gene. Nuclei were isolated from non-heat-shocked embryos and treated with DNase I. Purified DNA was digested to completion with *Eco*RV and subjected to indirect end-labeling analysis as described for Fig. 4 with the 0.64-kb upstream fragment as a probe. (c) *Xba*I accessibility of sites in the endogenous *hsp26* gene. Nuclei were isolated from non-heat-shocked embryos and treated with excess amounts of *Xba*I. The purified DNA was digested to completion with *Bam*HI and *Sal*I and subjected to indirect end-labeling analysis as described for Fig. 4, using the 2-kb downstream fragment as a probe. The *Xba*I accessibility detected was 50%. Other designations are as described for Fig. 4.

DH site of cPri was quantitated by measuring accessibility to the *Xba*I site. As shown in Fig. 4c, there was 8% cleavage for cPri compared to 52% observed for cP-351, indicating that the sequence containing $(CT)_n \cdot (GA)_n$ repeats plays an integral role in generating the DH sites of this chromatin structure prior to heat shock induction.

Discussion

In this article we have described general procedures for studying the *in vivo* chromatin structure and regulated expression of a *Drosophila* gene by using transgenes containing specific mutations. Previous chromatin structure analyses of transgenes in *Drosophila* involved establishing population cages (approximately 30,000 adults) to collect sufficient amounts of embryos for bulk isolation of nuclei (38, 41, 42). However, this is time consuming and laborious, especially when the chromatin structure of multiple transgenes is to be investigated. We have demonstrated that nuclei for chromatin structure analysis can also be obtained from relatively small numbers of larvae or adult flies.

We have shown that local protein–DNA interactions are the major determinants of chromatin structure of the *hsp26* gene, and that the control transgene, cP-351, contains sufficient sequence information for the formation of the wild-type DH site pattern. Consequently, we can use this approach to study the effect of specific DNA sequence changes on chromatin structure.

Although the chromatin structure of a transgene appears to be primarily determined by local protein–DNA interactions, the genomic environment in which a transgene resides may also have a role in determining the chromatin structure and level of expression of a transgene. Euchromatic position effects may be minimized by including sufficient amounts of flanking DNA to act as a buffer for a particular transgene. In addition, sequences that are known to function as domain boundaries, such as the special chromatin structures (scs) identified by Kellum and Schedl (70), may also be placed to flank the transgene in the P-element construct. These sequences have been shown to block the function of neighboring enhancers (71).

By using the procedures described in this article we have obtained critical information concerning the chromatin structure and expression of the *D. melanogaster* *hsp26* gene. The stretch of alternating C and T residing in the *hsp26* promoter, which binds the GAGA transcription factor *in vitro* (68), is essential for the formation of the appropriate chromatin structure and induced expression *in vivo* (5). Analyses of additional *hsp26/lacZ* transgenes with mutations within other regulatory regions of the *hsp26* promoter have provided further information on the mechanism of establishing the appropriate chromatin structure for gene regulation (unpublished results, 1992). A similar strategy can be used for the dissection and assignment of functional elements for any cloned *Drosophila* gene.

Acknowledgments

We thank members of the Elgin laboratory for critical reading of the manuscript. Work carried out in this laboratory is supported by Grants GM31532 and HD23844 from the National Institutes of Health (NIH) to S. C. R. Elgin. L. L. Wallrath is supported by NIH Fellowship GM14516.

References

1. S. C. R. Elgin, *J. Biol. Chem.* **263,** 19259 (1988).
2. J. C. Eissenberg, I. L. Cartwright, G. H. Thomas, and S. C. R. Elgin, *Annu. Rev. Genet.* **19,** 485 (1985).
3. D. S. Gross and W. T. Garrard, *Annu. Rev. Biochem.* **57,** 159 (1988).
4. F. M. Ausubel, R. Brent, R. E. Kingston, D. D. Moore, J. G. Seidman, J. A. Smith, and K. Struhl, "Current Protocols in Molecular Biology," Vol. 2, Chap. 12. Wiley, New York, 1991.
5. Q. Lu, L. L. Wallrath, B. D. Allan, R. L. Glaser, J. T. Lis, and S. C. R. Elgin, *J. Mol. Biol.* **225,** 985 (1992).
6. L. K. Durrin, R. K. Mann, and M. Grunstein, *Mol. Cell. Biol.* **12,** 1621 (1992).
7. E. Matallana, L. Franco, and J. E. Pérez-Ortin, *Mol. Gen. Genet.* **231,** 394 (1992).
8. S. Y. Roth, M. Shimizu, L. Johnson, M. Grunstein, and R. T. Simpson, *Genes Dev.* **6,** 411 (1992).
9. D. L. Lindsley and G. Zimm, "The Genome of *Drosophila melanogaster.*" Academic Press, San Diego, 1992.
10. G. Rubin and A. C. Spradling, *Science* **218,** 348 (1982).
11. J. Sambrook, E. F. Fritsch, and T. Maniatis, "Molecular Cloning: A Laboratory Manual," 2nd Ed. Cold Spring Harbor Lab. Press, Cold Spring Harbor, New York, 1989.
12. A. C. Spradling, *in* "*Drosophila:* A Practical Approach" (R. D. Roberts, ed.), 175. IRL Press, Oxford, 1986.
13. V. Pirrotta, *in* "Vectors: A Survey of Molecular Cloning Vectors and Their Uses" (R. L. Rodriguez and D. T. Denhardt, eds.), p. 437. Butterworth, Stoneham, Massachusetts, 1988.
14. R. Klemenz, U. Weber, and W. J. Gehring, *Nucleic Acids Res.* **15,** 3947 (1987).
15. R. E. Karess and G. M. Rubin, *Cell* **38,** 135 (1984).
16. H. M. Robertson, C. R. Preston, R. W. Phillis, D. Johnson-Schlitz, W. K. Benz, and W. R. Engels, *Genetics* **118,** 461 (1988).
17. A. C. Spradling and G. M. Rubin, *Cell* **34,** 47 (1983).
18. S. C. R. Elgin and D. W. Miller, *in* "The Genetics and Biology of *Drosophila*" (M. Ashburner and T. R. F. Wright, eds.), p. 96. Academic Press, New York, 1978.
19. R. D. Kornberg, J. W. LaPointe, and Y. Lorch, *in* "Nucleosomes" (P. Wasserman and R. Kornberg, eds.), *Methods Enzymology,* Vol. 170, p. 3. Academic Press, San Diego, 1989.

20. J. E. Morgan, J. W. Blankenship, and H. R. Mathews, *Biochemistry* **26,** 3643 (1987).
21. P. Colson and C. Houssier, *FEBS Lett.* **257,** 141 (1989).
22. C. W. Wagner, K. Hamana, and S. C. R. Elgin, *Mol. Cell. Biol.* **12,** 1915 (1992).
23. G. P. Pfeifer and A. D. Riggs, *Genes Dev.* **5,** 755 (1991).
24. I. L. Cartwright and S. C. R. Elgin, *Mol. Cell. Biol.* **6,** 779 (1986).
25. G. H. Thomas and S. C. R. Elgin, *EMBO J.* **7,** 7291 (1988).
26. D. Suck and C. Oefner, *Nature (London)* **321,** 620 (1986).
27. M. Noll, *J. Mol. Biol.* **116,** 49 (1977).
28. R. D. Kornberg, *Annu. Rev. Biochem.* **46,** 931 (1977).
29. M. A. Keene, V. Corces, K. Lowenhaupt, and S. C. R. Elgin, *Proc. Natl. Acad. Sci. U.S.A.* **78,** 143 (1981).
30. D. Dingwall, G. P. Lomonossoff, and R. A. Laskey, *Nucleic Acids Res.* **9,** 2659 (1981).
31. W. Hörz and W. Altenburger, *Nucleic Acids Res.* **9,** 2643 (1981).
32. J. T. Flick, J. C. Eissenberg, and S. C. R. Elgin, *J. Mol. Biol.* **190,** 619 (1986).
33. R. P. Hertzberg and P. B. Dervan, *J. Am. Chem. Soc.* **104,** 313 (1982).
34. M. W. Van Dyke and P. B. Dervan, *Biochemistry* **22,** 2373 (1983).
35. M. W. Van Dyke, R. P. Hertzberg, and P. B. Dervan, *Proc. Natl. Acad. Sci. U.S.A.* **79,** 5470 (1982).
36. R. P. Hertzberg and P. B. Dervan, *Biochemistry* **23,** 3934 (1984).
37. I. L. Cartwright and S. C. R. Elgin, *in* "Nucleosomes" (P. Wasserman and R. Kornberg, eds.), Methods in Enzymology, Vol. 170, p. 359. Academic Press, San Diego, 1989.
38. R. S. Jack, P. Moritz, and S. Cremer, *Eur. J. Biochem.* **202,** 441 (1991).
39. C. Straka and W. Hörz, *EMBO J.* **10,** 361 (1991).
40. A. Reik, G. Schütz, and A. F. Stewart, *EMBO J.* **10,** 2569 (1991).
41. R. S. Jack and H. Eggert, *EMBO J.* **9,** 2603 (1990).
42. H. Eggert and R. S. Jack, *EMBO J.* **10,** 1237 (1991).
43. M. Ashburner, "*Drosophila:* A Laboratory Handbook," p. 21. Cold Spring Habor Lab. Press, Cold Spring Harbor, New York, 1989.
44. J. T. Lis, J. A. Simon, and C. A. Sutton, *Cell* **35,** 403 (1983).
45. I. L. Cartwright and S. E. Kelly, *BioTechniques* **11,** 188 (1991).
46. C. Wu, *Nature (London)* **286,** 854 (1980).
47. S. A. Nedospasov and G. P. Georgiev, *Biochem. Biophys. Res. Commun.* **92,** 532 (1980).
48. G. M. Church and W. Gilbert, *Proc. Natl. Acad. Sci. U.S.A.* **81,** 1991 (1984).
49. H. P. Saluz and J. P. Jost, "A Laboratory Guide to Genomic Sequencing." Birkhauser, Basel, 1987.
50. M. Hull, G. Thomas, J. M. Huibregtse, and D. R. Engelke, *Methods Cell Biol.* **35,** 383 (1991).
51. S. B. Selleck and J. Majors, *Nature (London)* **325,** 173 (1987).
52. J. D. Axelrod and J. Majors, *Nucleic Acids Res.* **17,** 171 (1989).
53. S. E. Kelly and I. L. Cartwright, *Mol. Cell. Biol.* **9,** 332 (1989).
54. P. R. Mueller and B. J. Wold, *Science* **246,** 780 (1989).
55. P. A. Garrity and B. J. Wold, *Proc. Natl. Acad. Sci. U.S.A.* **89,** 1021 (1992).
56. J. A. Simon and J. T. Lis, *Nucleic Acids Res.* **15,** 2971 (1987).

57. C. S. Thummel, A. M. Boulet, and H. D. Lipshitz, *Gene* **74,** 445 (1988).
58. G. Rubin and A. C. Spradling, *Nucleic Acids Res.* **11,** 6341 (1983).
59. R. L. Glaser, G. H. Thomas, E. S. Siegfried, S. C. R. Elgin, and J. T. Lis, *J. Mol. Biol.* **211,** 751 (1990).
60. R. S. Cohen and M. Meselson, *Cell* **43,** 737 (1985).
61. D. Pauli, A. Spierer, and A. Tissieres, *EMBO J.* **5,** 755 (1986).
62. A. P. Feinberg and B. Vogelstein, *Anal. Biochem.* **132,** 6 (1983).
63. C. Wu, P. M. Bingham, K. J. Livak, R. Holmgren, and S. C. R. Elgin, *Cell* **16,** 797 (1979).
64. J. Topol, D. M. Ruden, and C. S. Parker, *Cell* **42,** 527 (1985).
65. C. Wu, S. Wilson, B. Walker, I. Dawid, T. Paisley, V. Zimarino, and H. Ueda, *Science* **238,** 1247 (1987).
66. K.-D. Fascher, J. Schmitz, and W. Hörz, *EMBO J.* **9,** 2523 (1990).
67. T. Archer, M. G. Cordingley, R. G. Wolford, and G. L. Hager, *Mol. Cell. Biol.* **11,** 688 (1991).
68. D. S. Gilmour, G. H. Thomas, and S. C. R. Elgin, *Science* **245,** 1487 (1989).
69. M. D. Biggin and R. Tjian, *Cell* **53,** 699 (1988).
70. R. Kellum and P. Schedl, *Cell* **64,** 941 (1991).
71. R. Kellum and P. Schedl, *Mol. Cell. Biol.* **12,** 2424 (1992).
72. M. Casadaban, A. Martinez-Arias, S. K. Shapira, and J. Chou, *in* "Recombinant DNA," Part B (R. Wu *et al.*, eds.), Methods in Enzymology, Vol. 100, p. 293. Academic Press, Orlando, Florida, 1983.

[19] Polymerase Chain Reaction from Preimplantation Mouse Embryos: Expression Studies and Genotype Analysis

Eric Lader and Karen Artzt

Introduction

Seldom in the history of experimental biology has a tool as powerful as the polymerase chain reaction (PCR), and a phenomenon as recalcitrant to study as the regulation of gene expression in early mammalian embryos, collided at such an ideal time for developmental biologists. The PCR, coupled with reverse transcription, has revolutionized the study of embryonic gene expression. The inherent paucity of material in embryos is no longer an insurmountable obstacle. It is now possible to study the expression of any gene in individual preimplantation embryos irrespective of message abundance and, when necessary, to genotype the embryo at a chosen genomic locus. Many intriguing but traditionally unapproachable questions can now be directly addressed. Some are as basic as determining precisely when the transcription of a particular embryonic gene is initiated, others are more esoteric. For example, if there are alternate alleles at a locus, differing by only as much as a single base substitution in an exon, primers can be designed that will amplify transcripts originating from one but not the other copy of the gene. Imprinting, the differential transcription of maternal and paternal loci, can thus be studied at the transcriptional level in individual preimplantation embryos. Expression studies of a spectrum of individual cell surface receptors, signaling molecules, cadherens, and growth and transcription factors expressed in early embryos will enhance and expand our current models of the mechanisms of compaction, cell–cell interaction, implantation, and differentiation.

In this article we describe in detail the technique for isolation of mouse preimplantation embryos and provide a simplified protocol for the analysis of gene expression in individual mouse embryos using reverse transcription coupled to the polymerase chain reaction (rtPCR). To illustrate the utility of this system, we will describe how it was applied to explore the expression of the transcription factor *Oct-4* (also known as *Oct-3* or *Otf-3g*) in embryos that were simultaneously typed for the allelic state of a closely linked preimplantation-lethal mutation.

Methods in Molecular Genetics, Volume 1

Methods

Experimental Animals

The mouse room should have a controlled light cycle with a dark period from 7 pm to 5 am (1). The frequency with which female mice enter into estrus is variable. It has been reported to be on average once every 3–5 days (1, 2), but has been observed to be as frequent as every 2 days (3). Females in estrus mate 4–6 hr after the onset of darkness and ovulate several hours later (2). To ensure a steady supply of embryos, males should be kept caged with as many as five females. This increases the chance that at least one female per cage will be in estrus and will mate during any given night. In lieu of direct observation of mating, the timing of embryo harvesting is based on the detection of a vaginal plug in the morning; this is considered day 0. Embryos in litters from different natural matings may vary in age, probably due to differences in time of mating. Additionally, stage differences are sometimes found within litters, probably due to intralitter variation in time of fertilization. For example, within a litter there may be two-cell embryos about to divide and four-cell embryos that have just divided. However, because rtPCR does not require embryos to be pooled into groups for analysis, it is sufficient for most studies to note and record these stage distinctions for each embryo, that is, as follows: early two-cell embryo (isolated from a mixed cohort of one- and two-cell embryos), mid-two-cell embryo (all at the two-cell stage), or late two-cell embryo (from a litter of two- and four-cell embryos). A record of plugs, dissection times and dates, positive and negative pregnancies, and number and stage of embryos isolated should be kept. If necessary, the variations associated with natural matings can be minimized by using *in vitro* fertilization and embryonic development *in vitro*. However, this requires a greater commitment in both labor and expertise. A protocol for *in vitro* fertilization can be found in Hogan *et al.* (4).

We use the schedule in Table I as a guideline for dissection of our C3H/DiSn × CF1 F_1 hybrid mice. Timing may vary for other strains.

TABLE I Dissection Times for Preimplantation Embryos

Stage	Time	Day
One cell	3 pm	0
Two cell	9 am	1
Four cell	12 am–5 am	2
Precompacted eight cell	12 pm	2
Compacted eight cell	6 pm	2

Materials and Reagents

Medium 199 (M199) (Cat. No. 380-2340AG; GIBCO, Grand Island, NY) Bovine serum albumin (BSA): 35% (Cat. No. A-7409; Sigma, St. Louis, MO)

Pipettes

Squibb mineral oil (equilibrated with M199)

Dideoxynucleotide triphosphates (2.5 m*M*): 2.5 m*M* mix of dATP, dTTP, dGTP, and dCTP

Embryo extraction buffer: 10 m*M* Tris-HCl (pH 8.3), 10 m*M* NaCl, 3 m*M* $MgCl_2$, 0.5% Nonidet P-40 (NP-40)

rtPCR buffer (10×): 0.1 *M* Tris-HCl (pH 8.3), 0.5 *M* KCl, 10–40 m*M* $MgCl_2$

Reverse transcription (RT) components: Oligo[$d(T)_{18}$] (150 ng), 1 μl 10× rtPCR buffer, 4 μl 2.5 m*M* dNTPs, 1 μl 10 m*M* dithiothreitol (DTT), 1 μl (20 units) RNasin (Promega, Madison, WI), 1 μl avian myeloblastosis virus (AMV) reverse transcriptase (USB) (15 units), cDNA, and H_2O to 10 μl

PCR components: cDNA template, 4 μl 2.5 m*M* dNTPs, 5 μl 10× PCR buffer, 15–30 p*M* concentrations of each primer, 1.25 units *Taq* DNA polymerase (Perkin Elmer Cetus, Norwalk, CT), H_2O to 50 μl. Overlay with 50 μl mineral oil

Embryo Isolation

During ovulation a mouse ovary releases on average six eggs surrounded by numerous cumulus cells. Known collectively as the cumulus mass, the eggs and surrounding cells are swept into the oviduct by the action of the ciliated opening of the infundibulum and transported to the ampulla, where fertilization occurs. The embryos travel down the oviduct and reach the uterus in about 3 days, at the blastocyst stage. Therefore, if preblastocyst embryos are required for analysis, they must be isolated from oviducts. For each mouse dissection, prepare one 62-mm dish with M199 for oviduct isolation and flushing, and one 62-mm dish with five 50-μl drops of M199 under mineral oil for washing embryos. Because the embryos are handled for only a brief period of time before nucleic acid extraction, the choice of medium is not critical. Both M199 and Brinster's medium [both supplemented with 0.4% (w/v) BSA and 0.1 m*M* sodium pyruvate] have been used with success.

The oviducts can be isolated without a microscope, but dissection is more easily performed using a microscope at low magnification (×7) and under incident illumination (Fig. 1). Grasp the uterine horn with a No. 5 forceps and pull it taut. With another forceps, gently tear the layer of connective tissue, the mesometrium, away from the oviduct and surrounding area. With the uterus still held taut, cut between the oviduct and the ovary with a fine scissors. Finally, gently hold the oviduct and cut the uterus close to the oviduct. Place the oviduct in a dish of medium.

On the morning when the vaginal plug appears, one-cell embryos can be found in the cumulus mass, located in the obviously swollen ampulla of the oviduct (Fig. 2A). To recover the cumulus mass, tear the wall of the oviduct at the ampulla with a 30-gauge needle. The cumulus mass will squeeze out of the hole without further manipulation (Fig. 2B). To isolate the embryos, incubate the cumulus mass in M199 (without BSA) containing hyaluronidase [300 μg/ml; Cat. No. H3884 (Sigma, St. Louis, MO)] until the cells are dispersed, then wash the embryos in several changes of medium.

To isolate older embryos, the oviducts need to be flushed with medium. This is accomplished at higher power ($\times 35$), and embryos are washed and collected at the highest available magnification ($\times 75$) (both with transillumination). Experimentation with the angle of incidence of illumination is crucial for success in both flushing the oviducts and embryo collection. The aim is to get high contrast in the image, because low contrast will make the infundibulum of the oviduct and the embryos difficult to locate.

A siliconized, pulled Pasteur pipette is used to flush the oviducts. These can be stretched after heating over the flame of an alcohol burner, and then broken at the stretched segment by bending. The diameter of the pipette tip should be large enough to fit into the infundibulum of the oviduct snugly, but as small as possible to allow precise control in embryo handling (i.e., about 150 μm). Alternatively, a pipette with a large-diameter tip can be used for flushing, and one with a small tip for embryo transfers.

To flush the oviduct for later stage embryos, manipulate it with two No. 5 Dumont forceps to locate the infundibulum. The infundibulum resembles the stigma of a flower, somewhat thickened, with the inner surface rolled outward, and an opening in the middle (Fig. 3A). If it is difficult to locate the infundibulum, experiment with the intensity and angle of the lighting. Hold the infundibulum gently with one forceps, fill the pipette with medium, insert the tip of the pipette into the opening of the infundibulum, and gently blow 5–10 μl of medium into the oviduct (Fig. 3B). The oviduct should swell, and a small cloud of debris should come out the cut uterine end. Blowing too forcefully will cause the embryos to scatter. If the oviduct does not swell, the tip of the pipette may not be in the infundibulum, or may be jammed against the wall of the oviduct. Pull the pipette out slightly and try again. If an oviduct will not swell it might have been crushed or torn during dissection. The speed and ease of flushing increases with practice. If flushing proves to be too difficult, an alternative method is to hold the oviduct with one forceps and

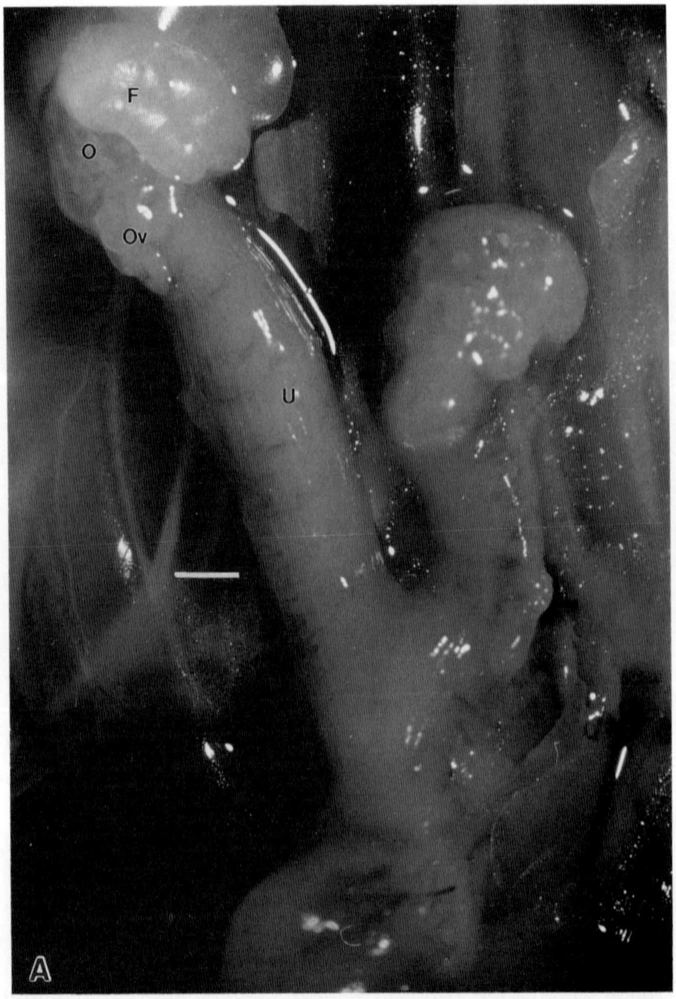

FIG 1 The reproductive tract from an adult, day-0 pregnant female. F, Fat pad; U, uterus; O, oviduct; Ov, ovary. Magnification: (A) ×7; (B) ×20. Bar: 1 mm.

stroke toward the uterine end with another forceps to force out the contents of the oviduct. In our hands, this method produces more debris and lowers the yield. Once both oviducts are flushed, switch to high power and scan the debris for embryos.

Embryos blown from the oviduct quickly settle to the bottom of the dish in a group. They are encased within their refractile zona pellucidae and are easy to identify. Evaluate them for appearance and cell number; they should be at the stage

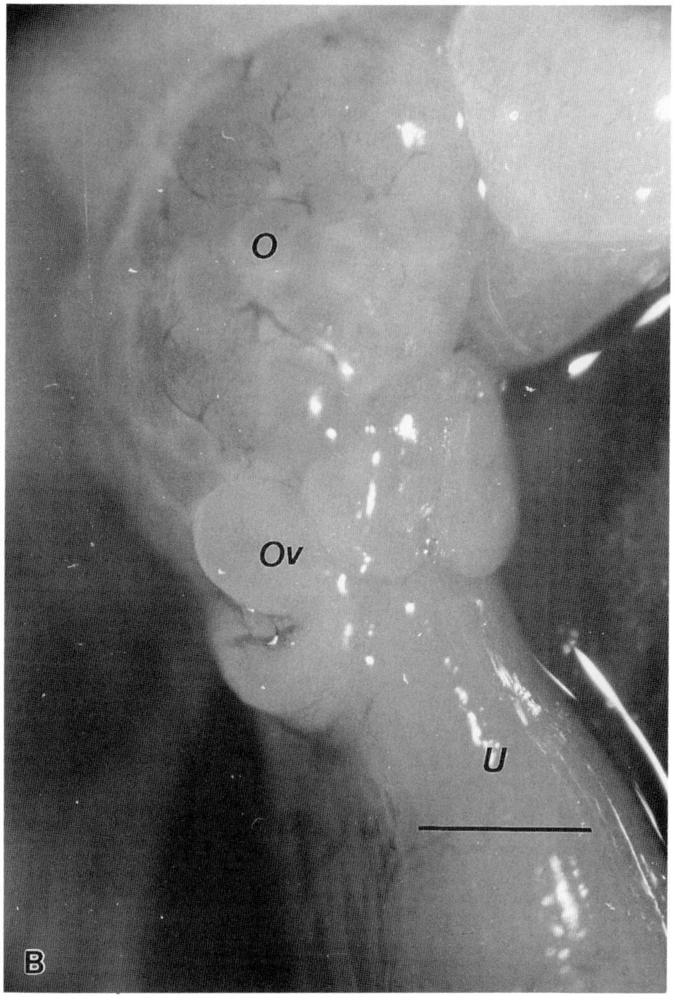

FIG 1 (*continued*)

expected based on the plug date and should be symmetrical. Unfertilized eggs have only one polar body, and one-cell embryos have two, but it is still sometimes difficult to tell them apart. Unfertilized eggs (or nonviable embryos) degenerate over a period of several days, sometimes vaguely resembling multicellular embryos, except that the "cells" are of various sizes and the "embryo" looks abnormal. The cells of normal two-, four-, and eight-cell embryos are equal in size and the em-

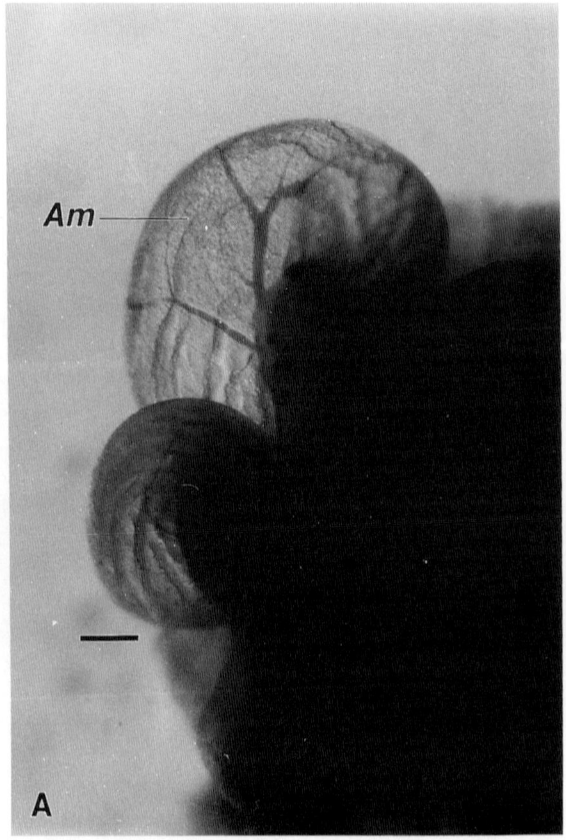

FIG 2 (A) The oviduct from a day-0 pregnant female. Am, Swollen ampulla, containing cumulus mass.

bryos are symmetrical. Unfertilized eggs may also be found in a clutch of multicellular embryos. However, because normal one-cell embryos are never found mixed in with a cohort of four- or eight-cell embryos, the identity of the abnormal ones is never in question. A reasonable rule to apply when evaluating embryos is the following: *When in doubt, throw them out.*

Once healthy embryos are chosen for collection, it is necessary to wash them free of contaminating maternal cells. A pipette with a small (150 μm), cleanly broken-off tip makes embryo handling easier because it allows fine control and minimal carryover of medium. The embryos are manipulated into a small pile and are col-

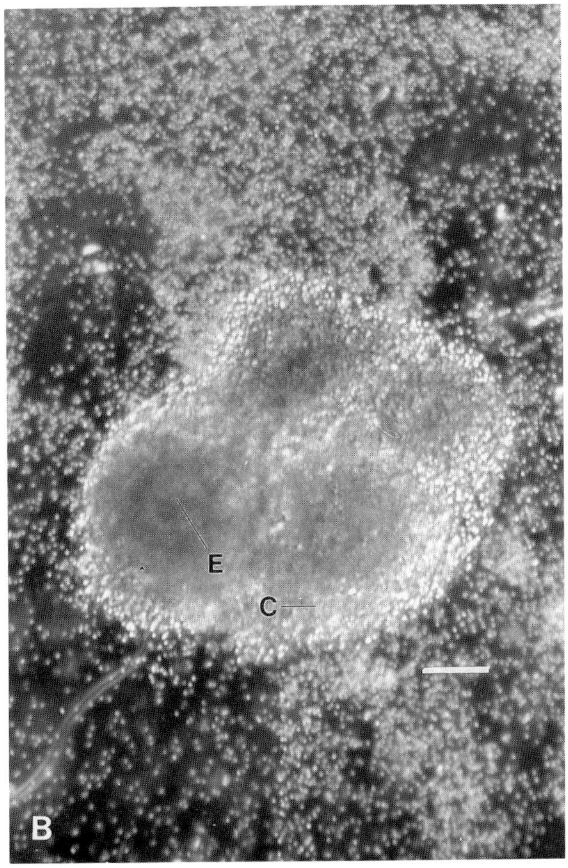

FIG 2 (*continued*) (B) Cumulus mass isolated from day-0 oviduct. E, One-cell embryo; C, cumulus cells. Bar: 100 μm.

lected together in only a few microliters. Then, after positioning the microdrop dish under the microscope, the tip of the pipette is placed in a drop of fresh medium and the embryos are *gently* transferred into the drop. It is crucial to avoid blowing air bubbles into the drop; the embryos may float away in the oil. The pipette tip should be rinsed with clean medium, and the washing procedure repeated until there are no contaminating cells remaining. Thorough washing is most necessary for one-cell embryos after hyaluronidase treatment because of the numerous cumulus cells, and may require passage through many drops of medium.

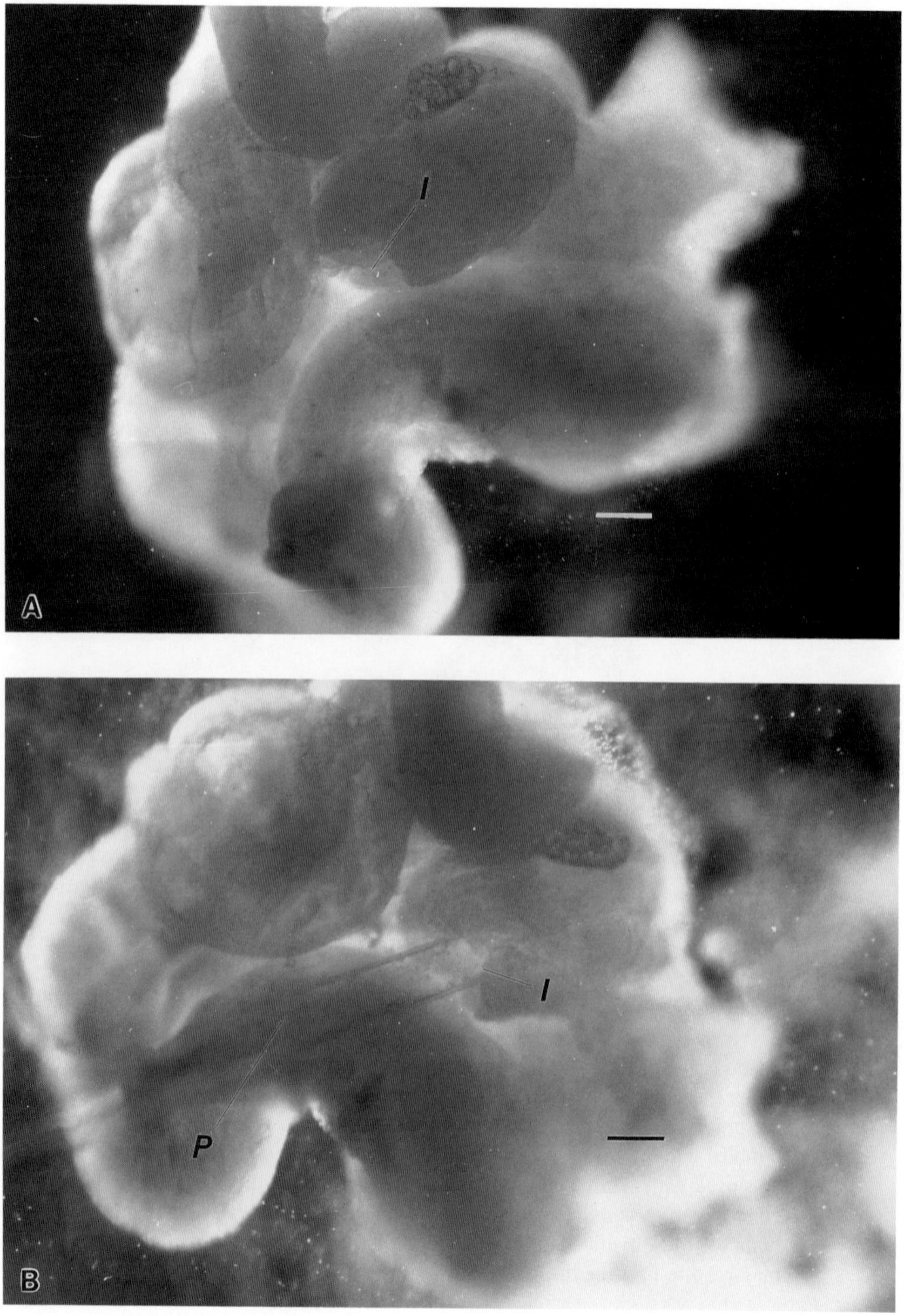

I
A
I
P
B

Embryo Extraction

Individual embryos in as small a volume as possible (1 μl) are transferred into 0.5-ml snap-cap microcentrifuge tubes (PCR compatible) containing 5 μl extraction buffer and then frozen on crushed dry ice. After all embryos are collected, each is put through two more freeze/thaw cycles to ensure lysis. These freeze/thaw cycles free both the RNA and the genomic DNA for PCR amplification. Our limited experience in storing these frozen extracts for more than a few hours suggests that they should be stored at −80°C with RNasin in the extraction buffer.

Reverse Transcription

The reverse transcription reaction is performed by the addition of 15 μl of RT cocktail to the embryo extract and incubation at 42°C for 1 hr. The reaction is terminated by heating at 65°C for 5 min. The cDNA synthesized in the reverse transcription reaction can be divided into aliquots and used as template in multiple PCR reactions.

Polymerase Chain Reaction and Primer Design

A universal feature of the PCR is that optimal reaction conditions must be determined experimentally for each primer–template combination. Because considerable effort is necessary to isolate large numbers of experimental embryos, pilot experiments should be performed on readily obtainable templates such as genomic DNA or RNA from adult tissues. This will allow the efficiency and specificity of the reaction to be optimized, without having to waste valuable biological material. Ideally, PCR primer pairs should be at least 18 nucleotides in length, 50% G·C, and have no homology to each other. The primer pair should not be grossly dissimilar in their G·C content. They should be designed to amplify a fragment large enough to be visualized on a standard agarose gel, but small enough not to be a challenge to efficient amplification by *Taq* DNA polymerase (i.e., 250–500 bp). Unless specifically required, primers should not be designed to amplify sequences at the extreme 5′ end of the cDNA, because 5′ ends are typically underrepresented in an oligo[d(T)$_{12-18}$]-primed RT reaction. If the intron–exon structure of a gene is known, primers can be designed that span an intron, making it theoretically simple to determine whether a product on a gel comes from amplification of the genomic

---

FIG 3 (A) The oviduct from a day-2 pregnant female. (B) Oviduct with pipette inserted into infundibulum. I, Infundibulum; P, pipette. Bar: 100 μm.

locus or from cDNA. However, several groups have reported products of the size expected for rtPCR in addition to the legitimate product from genomic DNA template (5, 6). The subsequent cloning and sequencing of these fragments revealed that they were amplification products derived from processed pseudogenes located elsewhere in the genome. Detection of PCR product from a genomic locus of an early embryo will generally require more extensive amplification than that required to quantitate a moderately abundant cDNA. However, if it is essential that genomic DNA be removed prior to PCR, samples can be treated with RNase-free DNase just prior to the reverse transcription reaction (7). Additionally, omitting the reverse transcriptase and obtaining no product is considered proof that the genomic locus is not the template in the rtPCR reaction. Finally, caution should be used when designing primers to amplify transcripts specifically from one member of a family of related genes, or when amplifying a region of one transcript containing a conserved motif common to many genes. Primers should be designed to anneal to unique DNA sequences in these instances or the wrong product may be amplified. Occasionally, a pair of primers will not work for unknown reasons: redesigning the primers so they anneal to a slightly different region of the target usually solves this problem.

Quantitative Polymerase Chain Reaction

Achieving quantitative PCR relies on avoiding plateaus in synthesis due to a shortage of reagents, saturation by product, or cycling past the useful life of the *Taq* polymerase. Because each transcript–primer combination is unique, due to differences in template abundance and efficiency of amplification, optimal conditions need to be determined for each combination. Two reports, by Gilliland *et al.* (8) and Gaudette and Crain (9), serve as good examples for a number of published reports describing the quantitation of individual transcripts by rtPCR. Two common features of the protocols used are limiting the number of cycles performed during the PCR reaction and adding an internal control for quantitation. To achieve good quantitation, the total yield in a PCR reaction should be at the low end of detectability on ethidium bromide-stained (0.5 μg/ml) agarose gels, that is, 10 ng. For a rare message, quantitation may require two limited rounds of PCR to generate detectable product. For example, a 25-cycle, 50-μl reaction is performed, then 1 μl of this reaction is added to a fresh 50-μl reaction, and 25 more cycles are performed. This ensures that the PCR does not reach a plateau, and that the amount of product reflects the quantity of input template. It stands to reason, however, that it may not be possible to quantitate extremely rare messages reliably because too many cycles would be required for the reaction to remain quantitative. Figure 4 is a photograph of an ethidium bromide-stained agarose gel in which the products of a series of PCR reactions have been electrophoresed. The five lanes in each panel represent PCR reactions with doubling amounts of mouse oocyte cDNA template. The samples in Fig. 4A–D were cycled for 25, 30, 35, and 40 cycles, respectively.

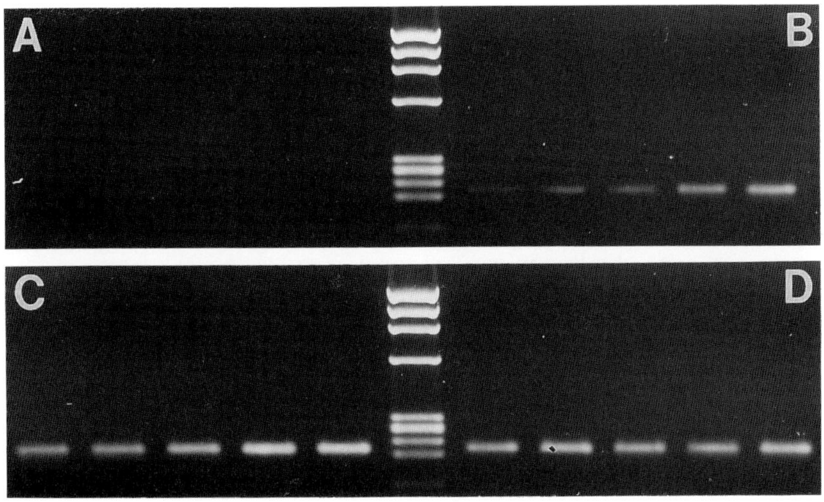

FIG 4 *Oct-4* rtPCR products from mouse oocytes, resolved by agarose gel electrophoresis. (A) Lanes 1–5: one half-, one-, two-, four-, and eight-oocyte equivalents of oligo(dT)-primed template cDNA, PCR amplified for 25 cycles. Lane 6 is ϕX*Hae*III molecular weight standard. (B) As (A), but PCR amplified for 30 cycles. (C) As (A), but PCR amplified for 35 cycles. (D) As (A), but PCR amplified for 40 cycles.

For this template–primer concentration, 25 cycles produces enough product to be visualized and quantitated by Southern transfer and autoradiography. Amplifying past 30 cycles results in a nonquantitative yield.

As mentioned above, quantitative PCR often includes internal controls by simultaneously amplifying more than one related or unrelated product in the same reaction tube. However, some combinations of primer pairs in the same PCR tube interact and yield unexpected results such as one product missing or the yield being far lower than expected. For example, in our hands, if a pair of *Hsp70* primers (10) is included in our *Oct-4* reactions as an internal control, the yield of *Oct-4* product is greatly reduced or absent. However, when a stromelysin fragment (11) is coamplified with *Oct-4,* the yield of each is unaffected by the other. We have no simple explanation for this phenomenon, but this unexpected interaction between template–primer pairs suggests that combinations of primer pairs be thoroughly tested to avoid artifactual results.

Polymerase chain reaction products can be recovered by a variety of standard techniques, such as chloroform extraction and ethanol precipitation. Products are usually run in agarose gels and transferred to membranes for hybridization and autoradiography. For quantitation, an aliquot of each reaction can be loaded onto a membrane by vacuum dot/slot transfer and hybridized with a radiolabeled probe. The resulting autoradiograph can be scanned on a densitometer for accurate quantitation.

Concerns

Although many of the potential pitfalls of rtPCR analysis are shared by all rtPCR experiments, two are unique to the analysis of gene expression during early mammalian development. A mouse egg on average accumulates 83 pg of poly(A)$^+$ RNA during oogenesis. Historically, it was believed that this RNA is not stored for use by the embryo during development, because much of it is degraded soon after ovulation (58%) or shortly following fertilization (>92% degraded) (12). However, these studies also illustrate that 8% of the maternal mRNA persists for an indeterminate time in the early embryo. There is no evidence to suggest whether this residual population of RNAs is simply the vestige of the initial population of the egg or whether the remaining mRNA is a specific subpopulation of transcripts essential for early embryological development. Obviously, rtPCR amplification will permit these maternal transcripts to be amplified from early embryos for detailed analysis. However, although amplifying specific maternal transcripts by rt PCR may provide useful information, if they are unknowingly amplified it may be misleading when studying zygotic transcription.

Latham and colleagues (13) have published a study of preimplantation mouse embryos using high-resolution two-dimensional protein gel analysis. Not surprisingly, they found an extensive remodeling of the protein complement of the embryo at the two-cell stage, following the activation of the zygotic genome. At this stage almost every detectable protein underwent either qualitative or quantitative changes or both. These observations should be kept in mind when designing experiments in which two-cell embryos will be analyzed by rtPCR, because early two-cell and late two-cell embryos are transcriptionally quite different. If circumstances require the accurate quantification of a particular mRNA in two-cell embryos, *in vitro* fertilization and culture may be required for precise timing of embryo stages.

Genotyping Chromosomal Loci by Polymerase Chain Reaction

A modification of the basic rtPCR technique allows the allelic state of a genomic locus to be determined from the same sample that is being examined by rtPCR. This permits one to address how a heterozygous or homozygous mutation at one locus affects expression from the same or another locus. Depending on the nature of the differences between alleles at the locus in question, one of several PCR methods that discriminate between alleles can be used. Which method is chosen depends on the exact nature of the difference between alleles. Below, we present brief descriptions of several methods currently in use.

Before one can accurately determine the genotype of an embryo by PCR, it is essential to remove the polar bodies prior to embryo lysis. Because the polar bodies carry the genetic by-products of meiosis, the DNA that they contain will serve as templates in addition to that of the embryo proper. Removal of polar bodies requires a brief incubation in a drop of M199 with 0.5% (w/v) pronase to remove the zona pellucidae from the embryos. As soon as the zonae appear to be dissolving, the embryos are transferred to a fresh drop of medium. The embryos lacking zonae will remain intact, but their polar bodies will fall off, either spontaneously or after brief pipetting. The embryos should not remain in the pronase solution for too long, because overdigestion with pronase will cause cell lysis. After digestion, the embryos are washed through several drops of medium to remove all traces of pronase.

Simple Sequence Length Repeat Polymorphism Mapping

Simple sequence length repeat polymorphisms (SSLPs) are derived from dinucleotide repeats, $(C{\cdot}A)_n$ or $(G{\cdot}T)_n$, found in most eukaryotic genomes. They are small regions flanked by unique DNA and are highly variable in length. As originally described by Weber and May (14), and as reviewed by Aitman *et al.* (15), SSLPs provide a potential source of polymorphic markers for almost any locus. In brief, to generate an SSLP for a chosen locus, a cosmid clone containing the gene of interest is digested with restriction enzymes (usually those with 4-base recognition sites) to yield small fragments, which are then probed with a radiolabeled $(C{\cdot}A)_n$ or $(G{\cdot}T)_n$ oligonucleotide. Fragments that hybridize to the probe are subcloned and sequenced. Oligonucleotides that hybridize to unique sequences that flank the dinucleotide repeat are synthesized and used as PCR primers. The size of the PCR product, which includes the polymorphic repeat, is highly variable between inbred strains and can be used as a genetic marker for alleles at a locus. Dietrich *et al.* (16) have published a list of over 300 SSLP mouse PCR primer pairs, half of which are polymorphic between the inbred strains C57BL/6J and DBA/2J. On average, this represents one SSLP per 4.3 centimorgans. In the future, as the number of mapped SSLPs grow, ones closely linked to any given loci should become available.

Allele-Specific Hybridization

Commonly used for PCR discrimination of human HLA types, allele-specific hybridization may be used when the difference between alleles is as small as a single base change (17). Primers are designed to amplify a fragment containing the allele-specific polymorphism, then the product is dot blotted onto a membrane and hybridized with radiolabeled oligonucleotides specific to each allele. Careful control of wash stringency (temperature and salt concentration) will melt mismatched probe off the DNA sample, while allowing perfectly matched probe to remain bound. Comparing the intensity of the signal obtained with the two probes will reveal the genotype of the embryo.

[20] Gene Transfer into Adult Rat Heart *in Vivo*

Richard N. Kitsis, Peter M. Buttrick, Alyson A. Kass, Matthew L. Kaplan, and Leslie A. Leinwand

Introduction

Direct gene transfer into rodent skeletal and cardiac muscle *in vivo* has been demonstrated (1–4). The procedure, which is marked by its simplicity, involves injection of naked plasmid DNA with a hypodermic needle into muscle tissue of the intact animal. Thus far, only skeletal and cardiac myocytes have demonstrated the ability to take up and express DNA presented extracellularly (1, 3). Although the mechanism whereby DNA enters the cell remains obscure, the following general observations have been made: injection of closed circular DNA results in significantly more efficient gene transfer than does linear DNA (5). The spatial extent of gene transfer resulting from a single injection is localized to the area surrounding the injection site (5). The DNA appears to persist as an episome (1) and gene expression can be detected for months following a single injection of DNA (1, 2, 5, 5a).

The potential applications of direct gene transfer include both phenotypic modification of the target organ and the study of muscle gene regulation *in vivo*. The former application has not yet been reported. However, the expression of dystrophin in a small number of skeletal muscle fibers of an *mdx* mouse has been demonstrated by direct DNA injection (6). The study of cardiac gene regulation in the intact animal has been accomplished by introducing reporter genes driven by a cellular promoter and potential regulatory sequences into hearts of animals in various physiological or pathological states (4). The purpose of this article is to provide detailed methods for achieving gene transfer in the adult rat heart and to discuss approaches for mapping the regulatory sequences of cardiac genes *in vivo*.

Methodological Approaches to Cardiac Gene Transfer

The basic protocol of the injection experiments involves exteriorization of the heart through a thoracotomy, a single injection of closed circular DNA into the cardiac apex, replacement of the heart into the chest and repair of the thoracotomy, and assay of reporter gene activities in a homogenate or tissue section of the heart 2 hr to 60 days later. The reporter constructs used for quantitative analysis contain the coding regions of the firefly luciferase (*luc*) or bacterial chloramphenicol acetyltransferase (*CAT*) genes linked to cellular or viral promoters (see below). These reporters have the advantage of being quantitative, sensitive, and absent from wild-

Methods in Molecular Genetics, Volume 1

type mammalian cells. The bacterial β-galactosidase gene linked to a viral promoter was used for histochemical studies.

Surgical Procedures and Injection Methods

The surgical procedures are relatively simple. Approximately five to eight animals can be operated on per hour. Mortality due to all causes is 10–15%. Greater than 95% of the injections are successful. The surgical procedure is presented in detail to be instructive to individuals with minimal animal surgery experience.

Most of our experiments have been performed in nonintubated rats. The procedure has been modified slightly for mice (P. M. Buttrick and M. Kaplan, unpublished observations, 1992). Others have employed a subdiaphragmatic approach in intubated rats (3), resulting in expression of reporter genes, although at levels that appear to be less than that resulting from a left thoracotomy approach. This may be due to the fact that the heart is not directly visualized using the subdiaphragmatic approach. Genes have also been expressed in the hearts of dogs (P. M. Buttrick, M. L. Kaplan, and R. Welikson, unpublished observations, 1992), cats (D. Menick and G. Cooper, personal communication, 1991), and rabbits (S. Fisher and M. Periasamy, personal communication, 1991).

The protocol that follows was developed for ~4-month-old female Wistar rats weighing between 180 and 250 g. The first critical parameter is the level of anesthesia. Animals are anesthetized with 0.75 ml/100 g body weight of 4% chloral hydrate in water administered intraperitoneally. Anesthesia takes effect within ~10 min. The absence of corneal reflexes and withdrawal to painful stimulation are indicative of adequate anesthesia. This anesthetic results in ~30 min of deep, followed by 1–2 hr of partial, anesthesia. Occasionally, the appropriate dose of chloral hydrate will be underestimated, requiring administration of a booster of 0.1–0.2 ml. However, whenever possible, a single bolus should be administered because additional doses may result in additive effects at later times.

The animal is placed in a supine position (Fig. 1). Both upper limbs are gently extended laterally and restrained with tape. A left intercostal thoracotomy is performed as follows. A vertical incision centered at the level of the axilla is made slightly to the left of midline. Using blunt dissection, the layers of chest musculature are dissected to the level of the ribs. An efficient way of accomplishing this is to dissect along a raphe, which is found over the left midchest running inferomedially to superolaterally. The extent of the dissection should be large enough to facilitate the externalization of the heart but not so large as to interfere with evacuation of the pneumothorax at the time of chest closure (see below). A general rule of thumb is that as minimal a dissection as possible should be performed, even if this requires foregoing anatomical precision.

Having exposed the ribs, the apical impulse is located visually. Generally, this is

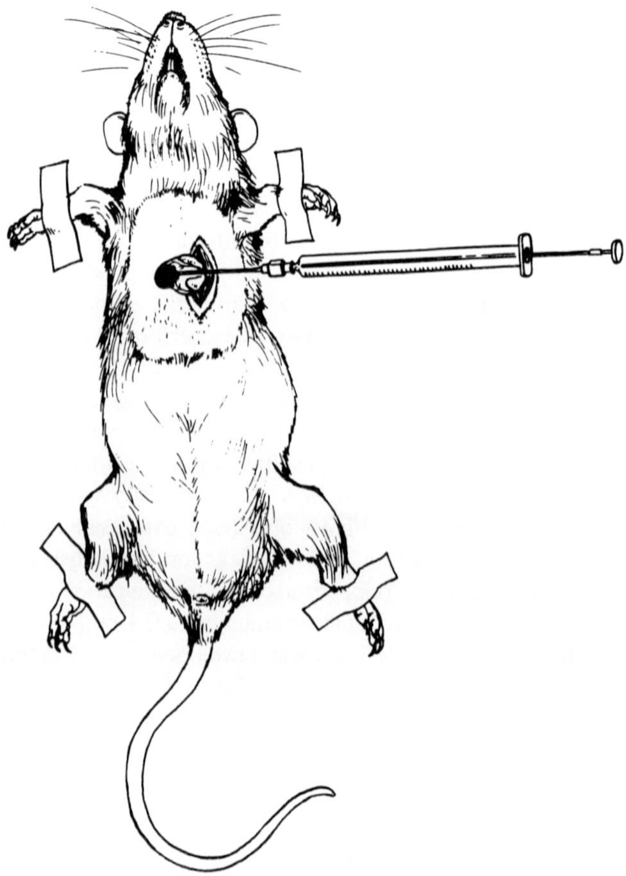

FIG 1 Direct injection of genes into rat heart *in vivo*. Following exteriorization through a thoracotomy, DNA is injected into the apical–lateral aspect of the heart. Evan's blue dye demarcates the spread of the injectate.

the most pronounced pulsation seen; sometimes, however, the great vessels produce a significant impulse in the adjacent superior interspace, which should be avoided. The rib just below the apical impulse is cut approximately 5 mm to the left of the sternum, using a small pair of scissors oriented perpendicular to the chest wall. This snip should be deep enough to result in the complete incision of the rib. Care should be taken not to damage the heart or lungs. Additionally, the rib should not be cut too medially so as to avoid cutting the internal mammary artery, which runs on the internal aspect of the chest wall near the midline.

A Crile forceps, which is a large, blunt, curved forceps with internal serrated edges, is inserted into the chest cavity at the rib incision perpendicular to the chest wall. Although there should be no question that the chest has been entered, care should be taken to avoid contusing the heart or lungs. The gap is widened by gently opening the forceps. Due to the equalization of intrapleural and external pressures, the lungs deflate, preventing respirations.

From this point, the procedure should be performed as quickly as possible. Gentle pressure placed on the right side of the chest helps to move the heart into the field of view. The heart is then gently exteriorized by grasping its apex with the Crile forceps. We have found this instrument to be especially useful in this step. It is important to avoid grasping the heart at the level of the thin-walled atria or at the great vessels. The heart is immobilized by holding it between the operator's thumb and left index finger and laid out across the midline.

Using a 100-μl Hamilton syringe and a 27-gauge needle, the apical–lateral aspect of the left ventricular wall is injected by a second person with a solution containing the DNA of interest in a volume of 50 μl. (Doses of DNA and choice of vehicle will be discussed below.) The injection is delivered over approximately 3 sec. Wolff *et al.* (7) found that varying the speed of delivery of injected DNA to skeletal muscle did not significantly influence reporter gene expression. We have made up the injectate to include Evan's blue dye at a final concentration of 3% (v/v) to provide transient visualization of the area injected. However, we have not observed a clear correlation between the intensity of blue and the adequacy of the injection as gauged by subsequent reporter gene activity.

After the injection is complete, the heart is replaced in the chest. The chest cavity is then squeezed to expel any trapped air and the lungs are immediately reinflated briefly using a Harvard small-animal respirator adapted with a nose cone to fit over the snout of the rat. (The nose cone is made by cutting off the bottom 4 cm of a 30-ml syringe and stretching and securing with tape a latex glove across the cut side. A small incision is then made in the glove large enough to accommodate the nose of the rat. The Luer-lock end of the syringe is attached to tubing, which in turn is attached to the small animal respirator.) The nose cone provides positive-pressure ventilation and obviates the need for intubating the trachea, thus decreasing the length of time for the surgical procedure and eliminating the morbidity associated with intubation. Resuscitation, however, is one of the trickiest aspects of the procedure. Two important points require special emphasis. First, the pneumothorax should be evacuated as completely as possible by squeezing the chest. Second, the tendency to overventilate should be avoided. Two to four respirations with the assistance of the respirator will usually induce spontaneous respiration. The procedure takes approximately 25 sec from the cutting of the rib to the expulsion of air from the chest.

Once spontaneous respirations begin, the muscle layers of the chest are closed as a group using approximately three sutures of 3-O silk. Care is taken to prevent the formation of a tension pneumothorax. The skin is closed with staples.

Harvesting and Homogenization of Tissue

Prior to sacrifice, rats are anesthetized with chloral hydrate as described above. The superior abdomen is entered through a midline incision. While the xiphoid process of the sternum is held with forceps, the right and left anterior ribs are cut, resulting in bilateral pneumothoraces and the cessation of respirations. The ribs are severed lateral to the midclavicular line to avoid damaging the heart. The cardiac apex is lifted out of the opened chest and the great vessels at the base are severed. The heart is removed and washed in 30 ml of iced phosphate-buffered saline (PBS), pH 7.4, and the atria and great vessels trimmed off.

In many experiments, it is adequate to analyze the left and right ventricles together. In some experiments in which a hemodynamic stimulus has been imposed primarily on a specific chamber (e.g., left ventricle in aortic constriction), it may be useful to separate the right ventricular free wall from the left ventricular free wall and interventricular septum to facilitate determination of the weight of and promoter activity in only the chamber subjected to the stimulus.

In addition to measuring the activities of reporter genes, it is often desirable to be able to measure changes in the steady state mRNA levels of the endogenous gene whose promoter is under study. It is technically difficult to isolate RNA and perform reporter gene assays on the same tissue specimen. Therefore, tissue for RNA isolation is obtained from an area of the ventricle separate from that injected with DNA. This can be accomplished by excising 100–200 mg of tissue from the basal aspect of the ventricle (corresponding to approximately one-third of the specimen) and freezing it immediately in liquid nitrogen. The inadvertant inclusion of a small portion of the injected area in the sample for RNA will not affect the analysis of promoter strength, as reporter gene activities directed from the construct containing the cellular promoter and from the internal control plasmid (see below) will be lost in parallel. However, care must be taken to avoid losing too much of the injected area so that reasonable amounts of reporter gene activity remain for analysis.

We homogenize the tissue in a buffer composed of 25 m*M* glycylglycine (pH 7.8), 15 m*M* $MgSO_4$, 4 m*M* ethylene glycol bis(β-aminoethyl ether)-*N,N,N',N'*-tetraacetic acid (EGTA) (pH 7.8), 1 m*M* dithiothreitol (DTT) (added at time of use) at 4°C in a volume proportionate to the mass of the tissue [e.g., volume (μl) = 1.5 × mass (mg)]. This has the theoretical advantage of creating a similar protein (as opposed to DNA) concentration in each tube (see below). Homogenization is performed with a Tissue Tek (Cincinnati, OH) homogenizer at the maximal setting for 30 sec. During homogenization, the sample tube is kept cool by embedding it in a container of ice. Each sample is homogenized as it is harvested. Homogenates are kept on ice until all samples are obtained, following which they are centrifuged at 5000 *g* for 1 hr at 4°C. Supernatants are recovered and the volume of each is measured at 4°C (in the cold room). Samples are then placed at −70°C until assayed.

Reporter Gene Assays

We have found that the reproducibility of the *in vitro* luciferase assay improves after the homogenates have been subjected to a freeze/thaw cycle. When assayed prior to this, a time-dependent increase in luciferase activity is observed for several hours. Although some luciferase activity may be lost following freeze/thawing of homogenates, this loss tends to occur to the same proportion in different homogenates and to be of relatively small magnitude. Therefore, after 12–15 hr at −70°C, homogenates are allowed to thaw on ice in the cold room over 3–4 hr. If maintained at 4°C, luciferase activity remains reasonably constant over the following 8 hr.

Time-dependent variability in CAT activities has not been observed, possibly due to the fact that we routinely heat the aliquot of the homogenate undergoing CAT determination at 65°C for 10 min prior to starting the reaction. This step is included in our protocol because it helps inactivate endogenous tissue deacetylases, which can skew the results of the assay, but does not affect the CAT protein itself. The portion of the homogenate to be assayed for luciferase activity cannot be heated because the luciferase protein is quite heat labile.

The quantity of homogenate to be assayed in any reporter assay depends on several factors. Foremost is that the amount of reporter activity be at least severalfold higher than background (see below). At the other end of the spectrum, assay of too large a proportion of the sample may take the *in vitro* assay out of the range in which activity increases linearly as a function of the amount of reporter gene product being assayed.

Several other, somewhat more subtle factors also influence the quantity of homogenate to be assayed. The relative transcriptional activities of different cellular promoter sequences or of the same sequence in different physiological states should be independent of the actual amount of each sample assayed. In practice, however, there exists some nonlinearity in luciferase activity (and, to a lesser extent, CAT activity) in tissue homogenates as a function of the amount of sample assayed. We hypothesize that this nonlinearity may be the result of nonspecific inhibition of the reporter assay by impurities in the crude homogenate, which may occur to different extents when different amounts are assayed. Regardless of the explanation of this phenomenon, it seems prudent to use a uniform amount of each sample for the reporter gene assays.

We routinely assay a fixed percentage of the volume of each sample. As many of the stimuli that we are interested in studying elicit cardiac hypertrophy (e.g., thyroid hormone and pressure overload) and as the total number of cardiac myocytes in each homogenate should be constant, assaying a fixed quantity of protein results in including the reporter gene products from fewer myocytes when assaying homogenates from hypertrophic as compared with control hearts. Although this should not influence luciferase:CAT ratios, it will affect the luciferase and CAT values themselves and precludes comparison of these between control and stimulated groups.For this reason, we have chosen to assay a percentage of each sample. Hearts coinjected with 3 μg of a *luc* construct driven by various portions of rat cardiac α-myosin heavy chain (α-MHC) 5′ flanking sequences (8) and with 1.5 μg of pRSV*CAT*, consisting of the *CAT* gene spliced to Rous sarcoma virus (RSV) long terminal repeat (LTR) sequences, demonstrated luciferase and CAT activities severalfold above background when 5% of the sample volume was assayed for each of these reporters. Although the optimal proportion of sample to be assayed will vary as a function of the strength of the cellular promoter, it is reasonable to assay between 5 and 10% as a starting point. The amount of homogenate assayed for luciferase activity, however, need not be the same as that assayed for CAT activity as long as the same percentage is used for each of the samples within an assay.

Several detailed protocols for assaying luciferase in transfected cells have been published previously. We fashioned the tissue luciferase assay after that described by Brasier *et al.* (9), with minor modifications. The aliquot of homogenate to be assayed is placed at 4°C in a cuvette compatible with the particular luminometer being used, and a quantity of homogenization buffer (described above) sufficient to bring the volume to 100 μl is added. Next is added 360 μl of a buffer composed of 25 m*M* glycylglycine (pH 7.8), 15 m*M* $MgSO_4$, 4 m*M* EGTA (pH 7.8), 1 m*M* DTT, 15 m*M* potassium phosphate (a mixture of mono- and dibasic salts, pH 7.8), 2 m*M* ATP (pH 7.0), and 0.27% (v/v) Triton X-100 at 4°C (with the DTT and ATP included immediately before use). Many luminometers, including the one we use (Monolight 2010; Analytical Luminescence Laboratory, San Diego, CA), have the capability of automatically delivering the solution containing the luciferin substrate into the cuvette at the time of assay. We use the sodium salt of luciferin, which is reconstituted in distilled deionized water to a convenient 20 m*M* stock solution, aliquoted, and stored at −70°C. A solution consisting of 0.2 m*M* luciferin (sodium salt), 25 m*M* glycylglycine (pH 7.8), 15 m*M* $MgSO_4$, 4 m*M* EGTA (pH 7.8), and 2 m*M* DTT (with the DTT and luciferin added immediately before use) is loaded into the appropriate reservoir in the luminometer. The luciferin solution is stable at room temperature for at least several hours. However, as luciferin is somewhat light sensitive, the solution is protected with foil. Each assay tube is placed in turn into the luminometer, following which 100 μl of the luciferin buffer is injected automatically into the cuvette. Light production is measured by the luminometer. As performed in this way, the reaction produces an early burst of light followed by a relatively low-level trailoff. We generally have the luminometer integrate the total light production over the 20 sec following addition of the luciferin solution, which it reports as raw luminometer units (RLU).

Several technical points deserve emphasis to ensure maximal reproducibility. First, except for the solution containing luciferin, the samples to be assayed and all buffers are maintained at 4°C until the moment of assay. (The luminometer itself is kept at room temperature.) Second, given the inherent time delay in assaying a large number of samples, it has been our practice to add the ATP-containing buffer to only 10 assay tubes at a time. Third, as a check, we always assay each sample in duplicate. Generally, the variation between duplicates is $<10\%$.

The luciferase activity resulting from the assay of a homogenate prepared from a sham-injected or noninjected heart or even from the homogenization buffer alone varies from 200 to 600 RLU over a 20-sec integration. We measure this background several times during the course of performing multiple assays. The average of these background measurements is subtracted from the activity assayed in each tube, although for many samples this correction is relatively unimportant.

In studies of gene regulation, it is usually sufficient to compare the *relative* strengths of various promoter constructs in various physiological contexts. In these types of experiments, therefore, it is not necessary to convert the number of raw luminometer units (corrected for background) to the quantity of luciferase protein corresponding to this amount of light output. In other types of experiments, it may be necessary to quantitate the actual amount of foreign gene product expressed in an injected heart. If so, a standard curve relating light output to amount of luciferase protein should be generated with each set of assays.

Many protocols for CAT assays have been previously published (10, 11). We will outline the protocol used here. The aliquot to be assayed is heated at 65°C for 10 min, spun at 10,000 *g* for 5 min at room temperature, and the supernatant transferred to a new microcentrifuge tube. To this is added a sufficient quantity of homogenization buffer to bring the total volume to 103 μl. Next, 47 μl of a cocktail containing (on a per sample basis) 8 μl of [^{14}C]chloramphenicol (55 mCi/mmol, 25 μCi/ml, Amersham, Arlington Heights, IL), 20 μl of 4 m*M* acetyl-coenzyme A, and 19 μl of 2 *M* Tris, (pH 7.4) is added. This mixture is incubated for 2 hr (can be varied depending on promoter driving the *CAT* construct), extracted with 1 ml of ethyl acetate, and the organic phase (top) transferred to a fresh tube. Samples are then dried down in a vacuum desiccator at room temperature, which takes approximately 45 min. One by one, samples are resuspended in 15 μl of ethyl acetate, spotted on thin-layer chromatography (TLC) plates, and chromatographed in 95% chloroform/5% methanol (v/v) for 2 hr. Thin-layer chromatography plates are air dried for 5 min and then placed under film at room temperature overnight. The radioactivity contained in the acetylated and unacetylated moieties is determined by scintillation counting of relevant areas of the TLC plate or with a phosphorimager, and the percentage of total counts resulting from acetylated species determined. It is important to include one or more negative controls (a homogenate from a sham or noninjected heart or homogenization buffer alone) in each series of CAT assays because thin-layer chromatography, although accurate, is not perfect in resolving the acetylated and unacetylated forms. This negative control averages $\sim 0.1\%$ conversion to acetylated forms. Therefore we routinely subtract 0.1% from the percentage conversion measured for each sample to yield a measure of the quantity of CAT activity contained in each sample.

Variability and Normalization

In determining the strength of the cellular promoter in the *luc* construct, an internal standard is coinjected to control for the 5- to 10-fold variability among animals in the efficiency of cardiac injections and for differences in uptake and expression that may exist among physiological states. The internal standard that we have used is pRSVCAT, as the RSV-LTR is constitutively active in many mammalian cell types. Luciferase activity driven by the cellular promoter under study is normalized to CAT activity directed from the RSV-LTR. The mathematical expression of this concept is the ratio of luciferase activity to CAT activity (luciferase:CAT), which is calculated after having subtracted the various backgrounds described above for each reporter assay. After calculating an individual luciferase:CAT ratio for each homogenate assayed, the luciferase:CAT ratios are grouped according to construct and physiological state. Within a group, the ratios often vary two- to threefold, which is significantly less than the variance in luciferase and CAT activities. This suggests that the CAT internal control has succeeded in reducing a significant portion of the variability observed in the luciferase activities attributed to the factors noted above. The explanation for the remaining variability is not clear. The mean and other descriptive statistics are calculated from the luciferase:CAT ratios within each group and these are compared using the appropriate statistical test. In some situations, obvious differences among groups are apparent with no overlap in luciferase:CAT ratios. In other situations, although mean luciferase:CAT ratios differ among groups, the presence of overlap in these ratios among members of different groups mandates statistical analysis to determine whether there is a high probability (e.g., >95%) that these differences are real. Although somewhat dependent on the differences in promoter strength among constructs and the fold induction resulting from the stimulus under study, generally we include 5–10 animals per construct in each physiological state in order to be able to make this determination with a high degree of certainty. The mean luciferase:CAT ratio in each group reflects the relative transcriptional activities of the cellular promoter sequence under consideration in the context of the physiological state in which it was studied.

As is the case for transfections in tissue culture, a small percentage of cardiac injections are of such poor quality that their luciferase:CAT ratios differ markedly from the other homogenates in that group. This situation arises from the fact that a poor injection is reflected by low CAT activity, one that often cannot be distinguished from the background "CAT activity" measured in homogenization buffer alone. Because CAT activity appears in the denominator of the luciferase:CAT ratio, small inaccuracies in the estimation of the CAT activity of the sample change this ratio markedly when the absolute CAT activity is low. Therefore, inclusion of failed or poor injections can introduce significant distortions in the analysis of promoter strengths. For this reason, it is advisable to include only those samples with CAT activities that are clearly distinguishable from background. As noted above, the "activity" of an uninjected cardiac homogenate averages ~0.1% in our hands. Therefore, we analyze only those injections with CAT activities >0.3% (prior to subtracting the 0.1% background). This criterion is applied uniformly and without exception to each of the samples.

Physical State of DNA and Injection Vehicle

In a variety of experimental situations, the physical state of the transfected DNA affects its subsequent uptake and/or expression. For example, in cultured mammalian cells, transfection of linear plasmid DNA results in only 1% of the transcriptional activity obtained from transfection of an equal quantity of the same DNA sequence in a circular configuration (12). We compared the expression of linear and circular plasmid DNA following intracardiac injection (5). Reporter gene expression was 50- to 100-fold lower in the animals injected with a linearized *luc* construct as compared with those receiving the same dose of the same plasmid in a circular state. The uptake, stability, and/or expression of the linearized DNA is, therefore, significantly lower than that of the circular DNA. Another interesting observation was that CAT activity resulting from a circular *CAT* construct coinjected with a linear *luc* construct was lower than that from a circular *CAT* construct coinjected with a circular *luc* construct. This suggests that the presence of a linear molecule could interfere with circular DNA uptake and/or expression. Based on these data, we have injected circular DNA in our experiments.

Wolff *et al.* (7) have reported that a vehicle consisting of 0.9% (w/v) sodium chloride in water (normal saline) resulted in approximately twofold greater expression of reporter genes injected into mouse skeletal muscle than did a solvent composed of 20% (v/v) sucrose in water. Although we have not investigated this variable in rodent cardiac muscle, our injectates are also based in normal saline.

Quantities of DNA Injected

We injected rat left ventricles with quantities of DNA ranging from 1.5 to 150 μg (0.5 to 50 pmol) in a volume of 50 μl. Although reporter gene expression is easily detectable over this range, it varies linearly with dose only at the low end of the spectrum. For example, luciferase expression resulting from injection of pα-MHC*luc*, a construct consisting of 613 bp of rat α-MHC 5′ flanking sequence spliced to the luciferase gene, is linear over the range of 0 to 5 μg (1.5 pmol). In addition, expression is linear over an even lower dose range with a stronger promoter, such as that derived from the RSV-LTR (A. A. Kass, R. N. Kitsis, and L. A. Leinwand, unpublished observations, 1991). A likely explanation for this nonlinearity in expression from injection of a single construct is that the cellular transcription machinery becomes limiting, resulting in competition between active promoters. This phenomenon is important when designing and interpreting experiments that ask quantitative questions concerning gene regulation. For this reason, it will be discussed in detail in Competition between Promoters (below).

Temporal and Spatial Extent of Gene Expression

To investigate the time course of expression of injected genes, the hearts of animals were injected with 100 μg each of pα-MHC*luc* and pRSV*CAT* and reporter gene activities assayed in cardiac homogenates at various time points (5). Although no luc or CAT activity was detectable 2 hr following injection, both reporter gene products were present at low levels at 1 day. Maximal activities were measured 5–10 days postinjection in the hearts of 100% of animals. Six of 10 hearts assayed between days 17 and 23 and 3 of 10 hearts assayed between days 38 and 60 expressed significant levels of reporter genes. Given that the half-life of luc following the injection of polyadenylated *luc* RNA into skeletal muscle was well under 1 day (1), these data suggest that *luc* transcription and translation was ongoing at each time point. The persistence of expression reported above was similar to that observed by Lin *et al.* (2) but differs from that noted by Acsadi *et al.* (3), who reported less than 2 weeks of expression in the hearts of most animals. Significant levels of injected gene expression have also been detected in mouse skeletal muscle 19 months following injection (1, 5a).

To determine the geographic extent of expression following a single injection into the heart apex, the following experiments were done (5; A. A. Kass, R. N. Kitsis, and L. A. Leinwand, unpublished observations, 1991). Two animals were each injected with 100 μg pRSV*luc,* consisting of the *luc* gene spliced to RSV-LTR sequences. Five days postinjection, the hearts were excised, washed in phosphate-buffered saline, trimmed of the atria and great vessels, and sliced perpendicular to the long axis into ten 1-mm thick sections. Each section was then homogenized and assayed for luciferase activity, which was found to be limited to two 1-mm sections surrounding the injection site, although low levels of activity were seen in two other sections. To further delineate the location of luciferase activity, three additional animals were injected with 100 μg pRSV*luc,* sacrificed 5 days postinjection, and their hearts cut into four 2.5-mm sections perpendicular to the long axis. Each of these four sections was then cut into quadrants perpendicular to the long axis and assayed individually for luciferase activity. Approximately 60% of *luc* expression was localized to one quadrant of one section surrounding the injection site (Fig. 2). The remaining activity was found in two to three surrounding quadrants. This suggests that most of the expression of injected DNA was restricted to an area in close proximity to the injection site.

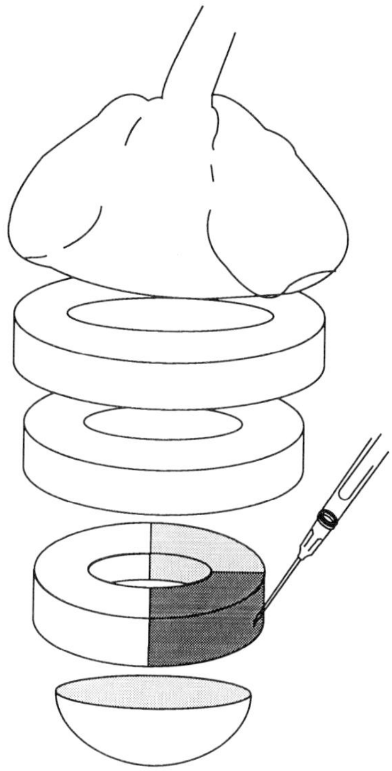

FIG 2 Spatial distribution of luciferase activity following a single injection of pRSV*luc* into the apical–lateral aspect of the heart 5 days earlier. Sixty percent of the total luciferase activity is found in one quadrant of one 2.5-mm section at the injection site (dark shading) and most of the remaining activity is in an adjacent quadrant of the same section or in an adjacent section (light shading).

To provide histological verification (Fig. 3) of these data, 100 μg of pRSVβ-*gal*, a construct consisting of the *Escherichia coli* β-galactosidase gene spliced to RSV-LTR sequences, was injected into the apex of one heart. Five days postinjection this heart was frozen in isopentane for β-galactosidase staining (1). The heart was sliced into 10-μm sections, which were fixed in 1.5% glutaraldehyde in phosphate-buffered saline (PBS) at room temperature and incubated at 37°C for 6 hr in 400 μg/ml X-Gal (Sigma, St. Louis, MO), 5 m*M* potassium ferriferrocyanide, and 1 m*M* $MgCl_2$ in PBS. The staining, which could be observed with the naked eye, was localized to approximately 3 mm of the tissue. Microscopically, the positively staining cells appeared centered around a needle track. These histochemical data yielded results similar to those obtained from luciferase assays performed on quadrants of

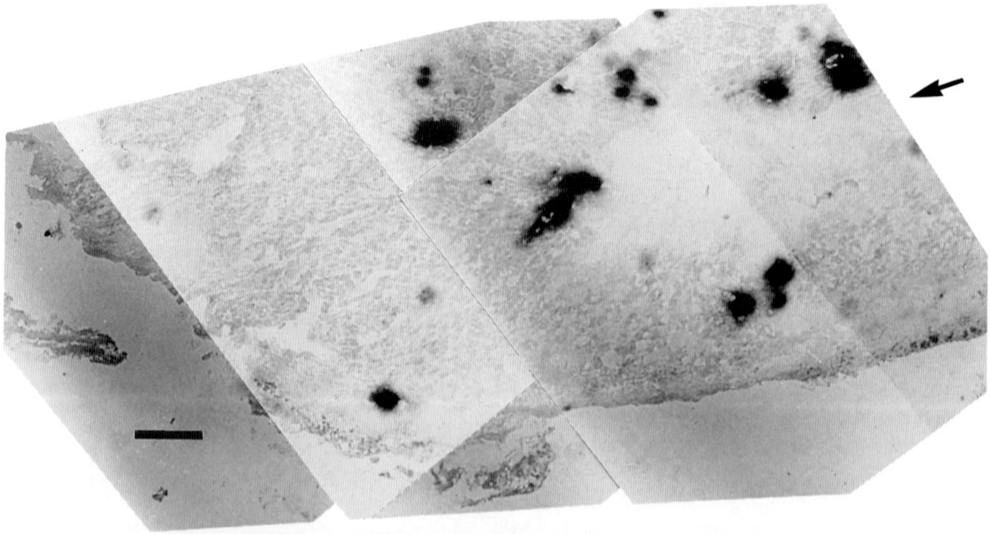

FIG 3 Differential interference contrast image showing β-galactosidase activity in frozen sections of rat heart injected 5 days earlier with pRSVβ-*gal*. Arrow indicates probable needle tract. Bar: 0.2 mm.

tissue described above. Approximately 6000 cells were positively stained for β-galactosidase following a single cardiac injection. This corresponds to approximately 0.03% of the myocytes in an entire heart (13), but a much larger proportion of those cells immediately around the injection site.

When considering a delivery system for somatic cell gene therapy, two parameters must be considered: the number and spatial extent of the cells expressing the gene, and the stability of expression over time. Thus, although the temporal and spatial expression of injected genes might be compatible with some gene therapy applications (see below), it would appear that this method in its present form is inadequate to alter the phenotype of the entire organ.

Use of Gene Injection to Study Cardiac Gene Regulation

The adult heart alters its patterns of gene expression rapidly and dramatically in response to a variety of stimuli (14). This has been demonstrated by quantitative and qualitative shifts at both mRNA and protein levels. Given the changes in steady state mRNA levels, one question is whether the level of regulation is transcriptional, posttranscriptional, or a combination of both. Nuclear run-on assays have been successfully performed in cultured cardiac myocytes and have demonstrated that expression of some cardiac genes is regulated, at least in part, at the transcriptional level (15). However, it has been difficult to measure transcription rates of cardiac myocyte genes *in vivo* due to obstacles in obtaining transcriptionally active myocyte nuclei from cardiac tissue. Although a recent report has demonstrated methodology that may overcome these problems (15a), conditions for performing nuclear run-on assays with cardiac tissue have not been optimized. Cis-acting regulatory sequences for several cardiac genes have been identified by transfecting reporter constructs into cultured cells (16–18). These studies, however, have been limited to phenomena that can be reproduced *in vitro,* precluding the use of this approach to identify regulatory elements that mediate the responses of various cardiac genes to complex stimuli, such as hemodynamic pressure overload. In addition, because it has not been possible to establish reproducible transfection protocols in cultures of adult cardiac myocytes, these *in vitro* studies have been performed in fetal and neonatal cardiocytes, which lack the adult phenotype. Thus, until recently, the dissection of the molecular mechanisms underlying the responses of genes in adult heart cells to complex stimuli necessitated the creation of transgenic animals. Although theoretically possible, this approach is costly and time consuming.

Gene injection provides a means of determining both the relative transcription rates of genes in adult heart cells *in vivo* and identifying the elements of those genes that mediate their responses to complex stimuli. The premise on which this is based is that an injected construct containing cellular promoter and regulatory elements will be expressed in the same manner as the endogenous cellular gene from which the control elements were derived. We have validated this hypothesis with promoter and regulatory sequences from the rat cardiac α-*MHC* gene (4). An injected construct containing these sequences directs reporter gene expression in the muscle-specific and thyroid hormone-responsive pattern exhibited by the endogenous rat cardiac α-*MHC* gene. The general approach to measuring the relative transcriptional activity of reporter constructs injected into hearts in various physiological states has been discussed above. In addition, by injecting wild-type and mutated regulatory elements, those sequences that confer positive or negative responsiveness to a stimulus can be identified. In the next two sections, we discuss issues that impact on gene injection experiments designed to study these quantitative aspects of gene regulation.

Competition between Promoters

As discussed above, quantitative comparisons of the transcriptional activity of various regulatory sequences require that the quantity of DNA injected be in the linear range of expression. The existence of competition between different promoters is supported by an experiment in which rat hearts were coinjected with 15 μg of pRSV*CAT* and 100 μg of one of several constructs, each containing a different promoter linked to the *luc* gene (5). These included the RSV-LTR (pRSV*luc*); 613 bp of rat α-*MHC* upstream sequence as defined above (pα-MHC*luc*); and 5′ flanking sequence (-522 to $+20$) from the mouse α_1-antitrypsin gene (19), which is expressed in liver but not in heart (pα_1-AT*luc*). As expected, no luciferase activity was detected in rat hearts injected with pα_1-AT*luc* ($n=6$) whereas the mean luciferase activities were higher in those injected with pRSV*luc* (297,659 RLU; $n=5$) than in those injected with pα-MHC*luc* (73,668 RLU; $n=6$). If there were no competition, similar CAT activites would be expected with all three pairs of plasmids. Of note, however, is that the mean CAT activities directed from 15 μg of pRSV*CAT* were inversely proportional to the amount of luciferase activity: 3.4, 15.0, and 27.2% in those injected with pRSV*luc*, pα-MHC*luc*, and pα_1-AT*luc*, respectively. This suggests the existence of competition for transcriptional machinery between the pairs of promoters.

To define a dose range at which (1) competition effects could be minimized and (2) expression of *luc* and *CAT* constructs could still be easily detected, we injected rat hearts with 1 μg of pRSV*CAT and either 1.0, 2.5, 5.0, or 10.0* μg of pα-MHC*luc* (A. A. Kass, R. N. Kitsis, and L. A. Leinwand, unpublished observations, 1991). Luciferase activity increased in a linear fashion between the 1.0- and 5.0-μg doses of pα-MHC*luc* but leveled off between the 5.0- and 10.0-μg doses. This suggests the absence of competition among pα-MHC*luc* molecules when <5.0 μg is injected. As only a single dose of pRSV*CAT* was tested, no conclusions can be drawn as to whether pRSV*CAT* molecules were still competing with pα-MHC*luc* molecules for generalized transcriptional machinery needed by the α-*MHC* promoter and upstream sequences. At a minimum, however, it can be stated that, if it exists, the level of this competition is constant over pα-MHC*luc* doses <5.0 μg. A second result of this experiment is that the amount of CAT activity resulting from injection of 1 μg of pRSV*CAT* did not decrease as the dose of pα-MHC*luc* was increased between 1.0 and 10.0 μg. This suggests that, over this dose range, pα-MHC*luc* molecules were not competing with pRSV*CAT* molecules for transcriptional machinery needed by the latter. *CAT* doses <1.0 μg do not reliably result in detectable CAT activity. Therefore, we considered it moot to investigate lower doses.

In summary, it is likely that competition among the same and different promoters for commonly needed transcriptional machinery occurs when quantities of injected DNA exceed a certain dose. This is not particularly surprising considering the limited number of cells transfected with this technique. In fact, this phenomenon could be exploited to identify functionally important regulatory sequences by coinjecting plasmids containing these sequences (without a reporter gene) along with plasmids containing the intact promoter and regulatory sequences linked to a reporter gene. A similar strategy has been employed with transfections in cultured cells (20).

The existence of competition influences the design and interpretation of gene regulation experiments. As the promoter strengths of different genes vary, when beginning injection experiments with the regulatory sequences of a new gene, it is probably wise to perform a dose–response curve with 1.0–15.0 μg of the *luc* construct containing the promoter under consideration coinjected with 1 μg of pRSV*CAT*. The goal is to choose a dose for the *luc* construct that results in expression that is easily detectable but well within the linear range. This will increase the likelihood that the expression of *luc* constructs containing mutations of the regulatory sequence will also be detectable and in the linear range.

Timing of Gene Injection, Stimulus, and Reporter Gene Assay

The temporal relationship of gene transfer, transcriptional stimulus, and reporter gene assay is an important aspect in the design of any gene regulation experiment. Timing may be a particularly important variable in experiments involving nonhumoral stimuli in intact animals, in which the usual lack of knowledge regarding the timing of molecular signals transduced by the stimulus is compounded by uncertainty as to the precise time of onset of the stimulus itself. For example, if one wished to identify those DNA sequences that mediate increases in transcription of the β-*MHC* gene in response to pressure overload, one would have to consider the relationship of gene injection, imposition of pressure overload, and reporter gene assays. What follows is a largely theoretical discussion of factors that merit consideration in designing these types of experiments.

As the identities of the signals that mediate most stimulus-induced changes in cardiac gene expression are unknown, indirect clues are often the only information available as to the timing of these signals. Signaling could begin as early as stimulus onset or at some later time. The *latest* possible moment of signal onset, however, must be prior to the earliest detectable changes in steady state mRNA levels of the endogenous cardiac gene whose promoter is under study.

If reporter constructs are injected after imposition of the stimulus, there is the possibility that the DNA will not be present to receive all of the signals resulting from the stimulus. For this reason, we have opted to have gene injection precede imposition of the stimulus. Part of this decision is based on our observations that reporter gene expression could not be detected 2 hr following injection and was present at low levels even 24 hr postinjection. It should be emphasized, however, that this decision is also based on theoretical considerations. A corollary of adopting this approach is that gene injection is performed at a time when the system has not yet reached a new steady state (see below).

The stimulated state may be a transitory state or a steady state, depending on the amount of time between stimulus and assay of reporter gene expression. Having the stimulated state be a steady state has the theoretical advantage that expression measurements are performed at times of minimum flux, thereby reducing experimental error. However, assaying reporter gene expression prior to the system reaching steady state may be advantageous for other reasons. Little is known about the time course of the intensity of the transducing signals; it is plausible that these signals reach some maximum after stimulus onset but before a new physiological steady state is achieved. Thus, it may be optimal to test the stimulus responsiveness of a given construct before a new steady state is reached. One logical time point would be when the earliest changes occur in the expression of the endogenous gene whose promoter sequences are under study. We are beginning to test these assumptions experimentally.

Summary

With the current technology, it is possible to characterize the sequences of cardiac genes that regulate basal and inducible expression. Because no established ventricular cardiocyte cell lines are currently available and because certain complex states, such as pressure overload and myocardial ischemia, cannot be modeled easily, if at all, *in vitro,* gene injections are a valuable approach with which to investigate the regulation of cardiac growth and development *in vivo*.

To carry out quantitative studies of cardiac gene regulation by direct DNA injection, the following general guidelines should be considered.

1. Of the conditions investigated thus far, closed circular DNA in saline results in the most efficient gene transfer.
2. Two reporter gene constructs should be coinjected. The first reporter (preferably *luc*) should be linked to the promoter sequence under study. The second reporter should be linked to a constitutive promoter to control for variations in transfection efficiency among animals and between physiological states.
3. A dose–response curve should be generated for each promoter under study. The internal control plasmid should be included in these injections at the dose (1–3 μg) to be used in the promoter mapping studies. The amount of experimental plasmid used should be in the linear range of its dose–response curve and not suppress the expression of the internal control plasmid. In general, the stronger the promoter, the less DNA should be injected to avoid competition for the transcriptional machinery.

Future Prospects for Cardiac Gene Transfer

An obvious future goal for this technique is the modification of cardiac and skeletal muscle function. The current technology appears to result in gene transfer to a relatively small number of cells in a restricted area of tissue. The mechanism whereby the DNA is taken up by muscle cells remains obscure. Regardless, direct DNA injection may offer several potential advantages over other approaches to somatic cell gene therapy, First, the method obviates the need for potentially hazardous viral vectors. Second, the DNA appears to remain episomal, thus reducing the risk of mutagenesis to the host genome. Direct injection of a construct containing dystrophin cDNA has been shown to result in the presence of dystrophin in skeletal muscle of *mdx* mice (6). Approaches that result in more widespread gene transfer should be actively pursued. Even with its limited spatial expression, the current approach may suffice to deliver to skeletal muscle genes encoding substances intended for secretion such as serum proteins, hormones, and antigens for vaccination.

Even if multiple injections are undertaken, it appears unlikely that the current direct injection methodology will achieve widespread gene transfer in the heart. Therefore, phenotypic alteration of cardiac muscle seems to demand that the genes being transferred be able to function in a paracrine fashion or that the delivery system be significantly improved. With regard to the former, several genes have been described, including angiotensinogen (21) and IGF-1 (22) among others, whose expression within heart muscle in a spatially restricted fashion could conceivably alter cardiac function. With respect to the methodology of delivery, we are currently exploring alternate systems and vectors to achieve more widespread gene transfer. These include intracoronary administration, catheters with multiple conduits, and so on. Among the most exciting approaches is the use of modified DNA viruses to infect target cells. These vectors may prove to be particularly useful in myocytes, which are not optimal targets for retroviruses, because of their postmitotic state. In these cells DNA viruses may function long-term as episomes, which would also reduce the potential for insertional mutagenesis. Adenoviral vectors have been used to express β-galactosidase in mouse skeletal muscle (23), and α_1-antitrypsin (24) and the cystic fibrosis transmembrane regulator (25) in rat bronchial epithelium. It is certainly possible that various tissues could be infected *in vivo* following intravenous injection. Specificity of expression could be provided by designing genes that carry tissue-specific regulatory elements. If a more widespread delivery system could be developed, the therapeutic implications of gene transfer into the heart would be significant.

Acknowledgments

This work was supported by grants from the National Institutes of Health [5F32HL-07913 (R.N.K.), HL-46034 (P.M.B.), and HL-37412 (L.A.L.)], American Heart Association, New York City Affiliate (R.N.K.), Aaron Diamond Foundation (P.M.B.), and Irma T. Hirshl Trust (L.A.L.). We thank Liang Liguo for preparation of artwork.

References

1. J. A. Wolff, R. W. Malone, P. Williams, W. Chong, G. Acsadi, A. Jani, and P. L. Felgner, *Science* **247,** 1465 (1990).
2. H. Lin, M. S. Parmacek, G. Morle, S. Bolling, and J. M. Leiden, *Circulation* **82,** 2217 (1990).

3. G. Acsadi, S. Jiao, A. Jani, D. Duke, P. Williams, W. Chong, and J. A. Wolff, *New Biol.* **3,** 71 (1991).
4. R. N. Kitsis, P. M. Buttrick, E. M. McNally, M. L. Kaplan, and L. A. Leinwand, *Proc. Natl. Acad. Sci. U.S.A.* **88,** 4138 (1991).
5. P. M. Buttrick, A. Kass, R. N. Kitsis, M. Kaplan, and L. A. Leinwand, *Circ. Res.* **70,** 193 (1992).

5a. J. A. Wolff, J. J. Ludtke, G. Aksadi, P. Williams, and A. Jani, *Hum. Mol. Genet.* **1,** 363 (1992).

6. G. Acsadi, G. Dickson, D. R. Love, A. Jani, F. S. Walsh, A. Gurusinghe, J. A. Wolff, and K. E. Davies, *Nature (London)* **352,** 815 (1991).
7. J. A. Wolff, P. Williams, G. Acsadi, S. Jiao, A. Jani, and W. Chong, *BioTechniques* **11,** 474 (1991).
8. V. Mahdavi, A. P. Chambers, and B. Nadal-Ginard, *Proc. Natl. Acad. Sci. U.S.A.* **81,** 2626 (1984).
9. A. R. Brasier, J. E. Tate, and J. F. Habener, *BioTechniques* **7,** 1116 (1989).
10. C. M. Gorman, G. T. Merlino, M. C. Willingham, I. Pastan, and B. H. Howard, *Proc. Natl. Acad. Sci. U.S.A.* **79,** 6777 (1982).
11. F. M. Ausubel, R. Brent, R. E. Kingston, D. D. Moore, J. G. Seidman, J. A. Smith, and K. Struhl, "Current Protocols in Molecular Biology." Wiley, New York, 1989.
12. C. Chen and H. Okayama, *Mol. Cell. Biol.* **7,** 2745 (1987).
13. F. J. Clubb, P. D. Bell, J. D. Kriseman, and S. P. Bishop, *Lab. Invest.* **56,** 189 (1987).
14. B. Nadal-Ginard and V. Mahdavi, *J. Clin. Invest.* **84,** 1693 (1989).
15. C. S. Long, C. P. Ordahl, and P. C. Simpson, *J. Clin. Invest.* **83,** 1078 (1989).

15a. K. R. Boheler, C. Chassagne, X. Martin, C. Wisnewsky, and K. Schwartz, *J. Biol. Chem.* **267,** 12979 (1992).

16. T. A. Gustafson, B. E. Markham, J. J. Bahl, and E. Morkin, *Proc. Natl. Acad. Sci. U.S.A.* **84,** 3122 (1987).
17. B. E. Markham, J. J. Bahl, T. A. Gustafson, and E. Morkin, *J. Biol. Chem.* **262,** 12856 (1987).
18. R. W. Tsika, J. J. Bahl, L. A. Leinwand, and E. Morkin, *Proc. Natl. Acad. Sci. U.S.A.* **87,** 379 (1990).
19. K. J. Montgomery, J. Tardiff, L. M. Reid, and K. S. Krauter, *Mol. Cell. Biol.* **10,** 2625 (1990).
20. T. Miwa and L. Kedes, *Mol. Cell. Biol.* **7,** 2803 (1987).
21. K. Lindpainter and D. Ganten, *Circ. Res.* **68,** 905 (1991).
22. J. D. Turner, P. Rotwein, J. Novakofski, and P. J. Bechtel, *Am. J. Physiol.* **255,** 513 (1988).
23. B. Quantin, L. D. Perricaudet, S. Tajbakhsh, and J.-L. Mandel, *Proc. Natl. Acad. Sci. U.S.A.* **89,** 2581 (1992).
24. M. A. Rosenfield, W. Siegfried, K. Yoshimura, K. Yoneyama, M. Fukayama, L. Stier, P. Paakko, P. Gilardi, L. D. Stratford-Perricaudet, M. Perricaudet, S. Jallat, A. Pavirani, J.-P. Lecocq, and R. G. Crystal, *Science* **252,** 431 (1991).
25. M. A. Rosenfield, K. Yoshimura, B. C. Trapnell, K. Yoneyama, E. R. Rosenthal, W. Dalemans, M. Fukayama, J. Bargon, L. Stier, L. Stratford-Perricaudet, M. Perricaudet, W. B. Guggino, A. Pavirani, J.-P. Lecocq, and R. G. Crystal, *Cell* **68,** 143 (1992).

Subject Index

E

N

O

P

V

W

X

Y